DIET AND CANCER

Markers, Prevention, and Treatment

ADVANCES IN EXPERIMENTAL MEDICINE AND BIOLOGY

Recent Volumes in this Series

Volume 346
INTERACTIVE PHENOMENA IN THE CARDIAC SYSTEM
Edited by Samuel Sideman and Rafael Beyar

Volume 347
IMMUNOBIOLOGY OF PROTEINS AND PEPTIDES VII: Unwanted Immune Responses
Edited by M. Zouhair Atassi

Volume 348
ADVANCES IN NUTRITION AND CANCER
Edited by Vincenzo Zappia, Marco Salvatore, and Fulvio Della Ragione

Volume 349
ANTIMICROBIAL SUSCEPTIBILITY TESTING: Critical Issues for the 90s
Edited by James A. Poupard, Lori R. Walsh, and Bruce Kleger

Volume 350
LACRIMAL GLAND, TEAR FILM, AND DRY EYE SYNDROMES: Basic Science and Clinical Relevance
Edited by David A. Sullivan

Volume 351
THE CHEMOKINES: Biology of the Inflammatory Peptide Supergene Family II
Edited by I. J. D. Lindley, J. Westwick, and S. Kunkel

Volume 352
NUTRIENT REGULATION DURING PREGNANCY, LACTATION, AND INFANT GROWTH
Edited by Lindsay Allen, Janet King, and Bo Lönnerdal

Volume 353
ANTIGEN AND ANTIBODY MOLECULAR ENGINEERING IN BREAST CANCER DIAGNOSIS AND TREATMENT
Edited by Roberto L. Ceriani

Volume 354
DIET AND CANCER: Markers, Prevention, and Treatment
Edited by Maryce M. Jacobs

DIET AND CANCER

Markers, Prevention, and Treatment

Edited by

Maryce M. Jacobs
American Institute for Cancer Research
Washington, D.C.

Springer Science+Business Media, LLC

Library of Congress Cataloging-in-Publication Data

Diet and cancer : markers, prevention, and treatment / edited by
Maryce M. Jacobs.
p. cm. -- (Advances in experimental medicine and biology ; v.
354)
Proceedings of the American Institute for Cancer Research's
Conference on Diet and Cancer: Markers, Prevention, and Treatment,
held October 29-30, 1992, in McLean, Va.
Includes bibliographical references and index.

1. Cancer--Nutritional aspects--Congresses. 2. Cancer--Diet
therapy--Congresses. 3. Cancer--Pathophysiology--Congresses.
I. Jacobs, Maryce M. II. American Institute for Cancer Research.
III. Conference on Diet and Cancer: Markers, Prevention, and
Treatment (1992 : McLean, Va.) IV. Series.
[DNLM: 1. Nutrition--congresses. 2. Biological Markers-
-congresses. 3. Diet--adverse effects--congresses. 4. Neoplasms-
-etiology--congresses. 5. Neoplasms--prevention & control-
-congresses. 6. Neoplasms--diet therapy--congresses. W1 AD559 v.
354 1994 / QZ 200 D565 1994]
RC267.45.D54 1994
616.99'4--dc20
DNLM/DLC
for Library of Congress 94-9683
CIP

American Institute for Cancer Research

Proceedings of the American Institute for Cancer Research's (AICR) Third Annual Conference on Diet and Cancer: Markers, Prevention, and Treatment, held October 29–30, 1992, in McLean, Virginia

ISBN 978-1-4899-0941-1 ISBN 978-1-4899-0939-8 (eBook)
DOI 10.1007/978-1-4899-0939-8

Originally published by Plenum Press, New York in 1994
Softcover reprint of the hardcover 1st edition 1994

Preface

The American Institute for Cancer Research (AICR) sponsored its third annual conference on nutrition and cancer. The theme was "Diet and Cancer: Markers, Prevention, and Treatment." The conference was held October 29-30, 1992 at the Ritz Carlton Hotel in McLean, Virginia. This proceedings contains chapters from the platform presentations and abstracts from the poster presentations. Several chapters address each of four session topics: Retinoids as Differentiation Agents in Cancer Therapy, Biological Markers of Cancer Risk, Chemoprevention of Cancer by Non-Nutrients in Foods, and Nutritional Problems and Support in the Treatment of Cancer.

The first three chapters discuss in detail different mechanisms by which retinoids influence differentiation and provide evidence to support their use in cancer therapy. *In vivo* and *in vitro* studies show the effects of retinoic acid (RA) on tumorigenicity and cellular/molecular events. A synopsis of data showing the involvement of the retinoblastoma (RB) gene in HL-60 cell differentiation induced by RA and 1,25-dihydroxy vitamin D_3 (VD_3) is presented. In SCID (severe combined immunodeficiency) mice injected with HL-60 human leukemia cells and gavaged daily with RA, the number of tumor sites and number of mice with tumors are reduced. All trans-RA induces myeloid differentiation in HL-60 cells. Similarly, VD_3 induces HL-60 monocytic differentiation. In both cases an early down regulation of retinoblastoma (RB) gene expression precedes the differentiation. The data show the distribution of the phosphorylated RB states with cell cycle and variations in the quotient of RB protein and DNA with cell cycle phases.

Data are discussed that show human keratinocytes (HKc) transfected with human papilloma virus 16 (HPV16) DNA are 100-fold more sensitive to growth inhibition by RA than normal HKcs. Associated with this inhibition is the observation that RA substantially reduces the expression of two HPV16 oncogenes, E6 and E7, as well as of two early messages, E2 and E5. Further studies indicate that HPV immortalization of HKc is inhibited by RA. These and other studies elucidating mechanisms by which RA regulates cell proliferation provide biochemical evidence in support of using retinoids in the chemoprevention of HPV-induced cancers, such as cervical cancer.

One chapter reports that RA can induce protein kinase C (PKC) mRNA and protein, that PKC-α is the isotype induced, that overexpression of PKC-α can mimic the action of RA, and that retinoic acid receptor beta (RAR-β) may mediate the induction of PKC-α by RA. Studies illustrating these findings are carried out in B16 mouse melanoma cells. The increase in PKC-α appears to be an early biochemical change that occurs in the RA induction of differentiation. Low levels of RAR-β are normally seen in B16 cells. Transfection of these cells to overexpress RAR-β is accompanied by an increase in PKC mRNA and protein. A new and rapid method for RNA extraction is described in Chapter 3. This chapter includes extensive methodology and data for evaluation of RA-induced changes in phenotypic expression as well as effects on PKC activity, PKC mRNA, and protein levels in B16 and B16-RAR transfectants. It is noted that VD_3 induces similar increases in PKC in HL-60 and human monoblastoid cell line U937. Whether increased PKC is a result of differentiation or the inhibition of cell proliferation that accompanies this process is unclear.

Chapters 4 through 6 address the identification, development and potential uses of biological markers for cancer risk. Epidemiological data are reported that show patterns of food intake and cancer incidences in different racial/ethnic groups (e.g., especially Black and White populations). These data are used to probe for racial/ethnic differences in susceptibilities to diet-related cancers that may lead to the identification of biological markers for cancer risk. The practical difficulties in distinguishing racial/ethnic differences based on genetically determined biological differences versus environmentally determined differences are raised. Dietary factors that might be more or less prevalent in specific racial/ethnic groups and that appear to be associated with low or high rates of specific cancers observed in these groups are delineated. Confounding factors that influence the collection and interpretation of epidemiological data are elucidated. Provocative issues are raised, such as the associations between high vitamin A consumption and increased risk for prostate cancer and decreased risk for cancer at other sites.

Possible associations among fat and calorie consumption, oxidative stress and DNA damage, and breast cancer risk are examined. Increased oxidative stress and DNA damage have been observed in leukocytes from breast cancer patients and from high risk individuals. Caloric restriction induces enzymes that repair DNA. The levels of oxidative stress and DNA damage are assessed in women consuming the conventional western diet of 40% fat diet, and an intervention group consuming a low, 20%, fat diet, with no statistically significant difference in total calories between groups. These data are compared with similar measurements for oxidative stress and DNA damage in rats fed low fat and high fat diets as well as diets restricted by 40% in calories. Five-hydroxymethyluracil (HMU) was measured in DNA from peripheral nucleated blood cells in humans and in mammary gland and liver from rats as a marker for oxidative stress and DNA damage. Significantly less HMU was observed in women on low fat diets and in rats fed restricted calorie diets. Additional studies are necessary to distinguish the influence of fat and calories on HMU and the low versus high fat effects in rats on HMU. These data support the potential use of HMU as a marker for possibly identifying individuals at increased risk for breast cancer.

Large bowel cancer is a major cancer in the United States that, if detected early, can be surgically removed. Individuals with precursor lesions, or with cancer in the large intestine, secrete mucus that is characterized by a D-Galactose-β[1→3]-N-Acetyl-D-Galactosamine (Gal-GalNAc) component. A new diagnostic test is proposed. It uses the galactose oxidase-Schiff assay to identify individuals with, or at risk for, cancer of the large intestine. Data illustrating the sensitivity and specificity of this test are presented and are discussed in relation to data from other diagnostic tests currently in use.

In Chapters 7 through 11 chemoprevention of cancer by non-nutrients in foods is discussed. In addition, approaches to detecting contaminants in foods and preventing their carcinogenic activities are indicated. In Chapter 7 the chemical and mutagenic specificities of the reactive metabolites from activation of four polycyclic aromatic hydrocarbons (PAHs), contaminants of vegetables and chemicals formed on cooked meats, are described in depth. Data on the differences in the stereochemistry of several reactive epoxides and differences in their preference to form adducts with deoxyguanosine and deoxyadenosine are compared and correlated with mutagenicity. The data on the chemical selectivity for guanine and adenine residues in DNA and the mutagenic selectivity for A-T or G-C base pairs show reasonable correspondence. Understanding the different mechanisms of action of the potent PAH carcinogens might facilitate the development of chemopreventive strategies.

The activation of organosulfur compounds in tea and garlic, and mechanisms by which they might inhibit nitrosamine-induced tumorigenesis are discussed in two chapters. Diallyl sulfide (DAS), in garlic, is the most extensively studied organosulfur compound. For many, but not all, nitrosamines the data indicate that inhibition of tumorigenicity is attributable to the selective inhibition by DAS of specific P_{450} enzyme (e.g., P_{450} 2E1-dependent N-nitrodimethylamine demethylase) activities necessary for carcinogen activation. Evidence is provided illustrating that (-)epicatechin gallate and related compounds found in green and black teas are inhibitors of nitrosamine-induced tumorigenesis in lung, forestomach, and esophagus. Other inhibitors isolated from black tea are the isoflavins. Data on the inhibition of tumor incidence and tumor multiplicity with tea extracts are presented and mechanisms for the inhibition caused by the various non-nutrient extracts are postulated.

Treatment of cancer is often ineffective once a tumor is diagnosed; hence, research on chemoprevention, especially with natural foodstuffs, is emphasized. Data are reported demonstrating that both dietary and topical application of polyphenols from green tea, (e.g., water extracts) and from epigallocatechin-3-gallate, a non-nutrient component in green tea, can inhibit chemical carcinogen- and/or ultraviolet-induced tumorigenesis in mice. Evidence is provided that the inhibition occurs at initiation and promotion stages. Administration of the polyphenols as the water extract of green tea and sole source of drinking water, protects against lung and forestomach tumorigenicity induced by several different chemical species of carcinogens. The proposed mechanisms of inhibition include inhibition of specific P_{450} enzymes, prevention of DNA adduct formation, and effects on ornithine decarboxylase, cyclooxygenase, and phase II enzymes.

Epidemiological data have associated consumption of soybean based foods with lower risk for breast and prostate cancer. *In vivo* and *in vitro* studies on soybean extracts and their isoflavone components have aided in the identification of the active inhibitors and have suggested possible mechanisms for their action. In rats, soybean extracts provided in the diet are reported to inhibit 7,12-dimethylbenz[a]anthracene (DMBA)-induced mammary tumorigenesis. Inhibition is attributed to the isoflavone components of the soybean, genistein and daidzein. These isoflavones inhibit the growth of both estrogen receptor-positive and estrogen receptor-negative human breast cancer cell lines and of human prostate cancer cell lines in culture. Based on data from the *in vitro* studies, hormone-dependent (e.g., antiestrogenic activity) and other unrelated mechanisms are proposed for the inhibition by isoflavones.

Epidemiological data from different populations worldwide associate dietary aflatoxin exposure with increased risk for liver cancer. Parallel studies in humans and animals support the hypothesis that quantitation of aflatoxin-N^7-guanine (AFB-N^7-gua) in urine is a potential biological marker that is indicative of aflatoxin B_1 (AFB_1) exposure and may be predictive of increased risk for developing primary hepatocellular carcinoma. Provision of oltipraz in the diet reduces the level of AFB-N^7-gua in urine and serum. Human data and clinical trials to assess both the utility of AFB-N^7-gua in identifying AFB_1 exposed individuals, and the efficacy of using oltipraz in aflatoxin exposed individuals to protect against liver cancer, are discussed.

The last four chapters address nutritional problems and support in the treatment of cancer. Studies on iron, amino acid restriction, prevention of anorexia, and insulin resistance are discussed. Epidemiological studies suggest that people with high body stores of iron may have an increased risk for cancer. Animal studies indicate that tumor cells require iron for growth and in some reports increases in dietary iron promote tumor growth. These data suggest tumor cells might more effectively use available iron than normal cells, and suggest that iron depletion might be useful in cancer therapy. Although clinical reports differ, cancer patients frequently have depressed levels of serum iron and increased serum ferritin. Changes in serum ferritin may be useful: as a marker of disease, for monitoring advancement or remission, or for prognosis. Increased serum ferritin is associated with a poor prognosis in Hodgkin's and a fall in serum ferritin has been indicative of a positive response to anticancer therapy in some cancers. How iron and ferritin might be involved in modulation of the immune system and the host's defense against malignancy is discussed.

Dietary restriction of tyrosine and phenylalanine is reported to suppress tumor growth and metastasis and to increase survival in tumor bearing mice. The antimetastatic effect of restricting these amino acids has been observed in melanoma, Lewis lung carcinoma, hepatocarcinoma, and L1210 leukemia tumor cell systems. Detailed studies in B16 murine melanoma cells *in vitro* and in mice *in vivo* are described that suggest that antimetastasis results from the direct modulation of the tumor cell phenotype by these amino acids. Modulation of host immune responses, which in turn suppress metastasis, does not appear to contribute significantly to the altered phenotype caused by tyrosine and phenylalanine restriction in these systems. Given the current data available, a variety of speculative mechanisms by which amino acid restriction might modulate host and tumor cell growth is evaluated.

Development of anorexia and loss of lean body mass are primary obstacles to the aggressive treatment of cancer patients. An interesting new approach to reversing anorexia and the resultant cachexia common to cancer patients is the study of neuropeptide Y (NPY). Both plasma and hypothalamic levels of NPY are depressed in anorectic tumor bearing animals. Intrahypothalamic injection of this peptide stimulates appetite in rats, significantly increasing food intake and body weight gain. Based on preliminary data, the number of NPY receptors is unchanged in anorexia, however, the binding affinity of these receptors is decreased ten-fold in anorectic tumor bearing animals compared with normal controls. Additional studies reported in rats suggest that hyperammonemia may be involved in the etiology of experimental cancer anorexia. In support of this hypothesis are data from studies in rats infused with ammonia, showing that this results in depressed food intake, anorexia, and neurochemical alterations similar to those seen in tumor bearing rats.

The final chapter reviews the various causes of cancer cachexia and focusses on the concurrent observations of elevated lactate, elevated insulin, elevated glucose intolerance, and insulin resistance in lung cancer patients. Among possible explanations offered for this paradox are that in tumor tissue there is an excessive conversion of glucose to lactate, and in the host liver tissue a diminished restraining effect of insulin on glycogenolysis that leads to increased hepatic conversion of glucose to lactate. Insulin resistance is defined as a reduced ability to respond to normal amounts of insulin. As a consequence, there is an impaired ability of insulin to stimulate glucose uptake and metabolism, lipid synthesis, and most other anabolic processes. The wide-ranging impact of the phenomenon of insulin resistance on the nutritional status of cancer patients is elucidated.

Important contributions were made in both platform and poster presentations at the American Institute for Cancer Research Conference on "Diet and Cancer: Markers, Prevention, and Treatment." In this volume, much can be learned about the mechanisms by which retinoids influence differentiation and about their possible use in cancer therapy. Novel biological markers are proposed for identification of individuals with, or at risk for, several cancers. Data are presented on carcinogenic contaminants in foods and their activation, and about the identification and chemopreventive potential of a number of non-nutrient components in foods. Insight into the etiology of serious nutritional problems in cancer patients is presented, and innovative therapeutic approaches to their reversal are proposed. Although much has been learned about the potential roles of nutrients and non-nutrients in the prevention and adjuvant therapy of cancer, the need for continued research on this important subject cannot be overemphasized.

The Editor

Maryce M. Jacobs, Ph.D. received her B.S. in Chemistry from New Mexico State University in 1966 and her Ph.D. in Biological Chemistry from the University of California at Los Angeles in 1970. She has had an active career in research, publishing more than 50 scientific articles, five chapters, and fourteen books. Dr. Jacobs has served as a consultant to government and the private sector, and has worked in research management for twelve years. In 1991 Dr. Jacobs earned the M.S. degree in Business Administration from Strayer College.

From 1988 to 1993 Dr. Jacobs served as Vice President for Research at the American Institute for Cancer Research (AICR) in Washington, D.C. During this period, she initiated the Small Grant program, developed a Cooperative Research and Development Agreement (CRADA) between AICR and the Food and Drug Administration's National Center for Toxicological Research, initiated a Matching Grant program between AICR and industry, and initiated the Annual AICR Conference series. This proceedings volume represents AICR's third conference.

Prior to joining AICR, Dr. Jacobs was employed for five years at the MITRE Corporation in McLean, VA as a Biochemical Toxicologist, for six years at the Eppley Institute for Cancer Research at the University of Nebraska in Omaha as Associate Professor and Director of the Industrial Contract Testing laboratories, and for six years at M.D. Anderson Hospital and Tumor Institute in Houston where she also served as Cochairman of the Biochemistry Area of the University of Texas Graduate School of Biomedical Sciences for two years.

Dr. Jacobs' primary area of research interest is inhibition of chemical carcinogenesis with dietary factors, particularly selenium. She is a leader in this research area, having published some of the earliest studies showing selenium inhibition of colon carcinogenesis, as well as of liver and lung carcinogenesis. Dr. Jacobs has published on the antimutagenic, anticlastogenic, and antiangiogenic properties of selenium, and has described acute, subchronic, and chronic toxicity parameters of selenium in rodents. In recent years, Dr. Jacobs has lectured and published reviews on the role of diet/nutrition in the prevention and adjuvant therapy of cancer.

Dr. Jacobs is a member of several professional organizations including the American Association for Cancer Research, the Society of Toxicology, the American Academy of Clinical Toxicology, the American Chemical Society, American Men and Women in Science, American Association for the Advancement of Science, and Sigma Xi. She received the Distinguished Alumna Award from the College of Arts and Sciences at New Mexico State University and the Dr. Anna-Lisa Barofsky Lectureship Award at Howard University.

Contents

Chapter 1

Vitamin-Regulated Retinoblastoma Tumor Suppressor Gene Expression in Leukemic Cells

ANDREW YEN, TRACY FRENCH, KAREN RUSSELL, SUSI VARVAYANIS and MARY FORBES

I. Introduction

Vitamins A and D are well known to perform important functions in growth and development. The mechanisms by which they act at the cellular level to control proliferation and differentiation thus become of importance in understanding how these dietary factors work. Using tissue culture, it has been found that a variety of cultured cells respond in particular to the metabolites, retinoic acid, and 1,25-dihydroxy vitamin D_3. These cells have formed convenient experimental systems in which to study the mechanism of action of these metabolites at the molecular level. At this level, it is known that vitamins A and D have receptors that are related as members of the family of steroid-thyroid hormone receptors.[1,2] Ligand receptor binding and receptor complex translocation to the nucleus thus regulate gene expression. The ultimate result of a lengthy metabolic cascade initiated this way can be control of cell proliferation or differentiation. The identity and roles of the genes that mediate this metabolic cascade are of obvious interest in understanding the mechanism by which these vitamins or their metabolites act in growth and development.

Various *in vitro* cell lines have been developed that can be exploited to study the mechanism of action of vitamin A or D metabolites in cell proliferation or differentiation. One that has been extensively used and studied is the human promyelocytic leukemia cell, HL-60.[3,4] Like normal promyelocytes, they retain the capability *in vitro* to undergo differentiation along the myeloid or monocytic pathways. Retinoic acid elicits a metabolic cascade culminating in terminal myeloid differentiation. 1,25-dihydroxy vitamin D_3 in contrast causes monocytic differentiation. Certain broad features of these metabolic cascades have been determined. The duration of the cascade extends over a period corresponding to two division cycles before onset of G1/0 specific growth arrest and phenotypic differentiation.[5,6]

Andrew Yen, Tracy French, Karen Russell, Susi Varvayanis, and Mary Forbes • Cancer Biology Laboratories, Department of Pathology, College of Veterinary Medicine, Cornell University, Ithaca, NY 14853

Diet and Cancer: Markers, Prevention, and Treatment
Edited by M.M Jacobs, Plenum Press, New York, 1994

In the HL-60 subline, which we routinely use, the division cycle is approximately 24 hr. The metabolic cascade segregates into two steps. Initiation of the cascade involves an S phase-specific event. Early events prime the cell to differentiate without regard to lineage specificity, and extend over a period equal to one division cycle. The early events culminate in a metastable "precommitment" state,[5,7] where cells can differentiate after only an abbreviated subsequent exposure to the inducer. Late events, occurring subsequently and prior to differentiation, specify the actual differentiation lineage. These late events occur over a period that is one division cycle in duration. The regulatory involvement of various genes has been found to segregate with early or late events.[4-7]

Given the presumed need during cell differentiation for nuclear reprogramming, certain classes of genes become implicated with significant regulatory roles. Tumor cell biology indicates that dysfunction in certain genes—oncogenes—leads to aberrant regulation of cell proliferation or differentiation, indicating a critical regulatory function for these genes in normal cells. Among these genes, the subset of nuclear oncogenes have nuclear translocation sequences enabling nuclear localization. In the instances where some known function can be ascribed to the gene, these genes can act as transcriptional regulators. For example, c-*fos* or c-*jun* (aka AP-1) can form a hetero-duplex and regulate transcription through the AP-1-binding DNA consensus sequence.[8-11] A gene related to these in certain ways is the retinoblastoma (RB) tumor suppressor gene.[12-16] The RB protein also has a nuclear translocation sequence and bears a leucine zipper motif. The latter can be a characteristic of nuclear proteins indicating ability to complex with other such proteins and so bind to DNA regulatory regions. It is a phosphoprotein whose phosphorylation is cell-cycle dependent. Significantly, the hypophosphorylated RB protein can negatively regulate the expression of c-*fos*, c-*myc*, and RB itself through E2 elements in the promoters of the latter two.[17-20] Thus, it has the potential for regulatory involvement with nuclear oncogenes.

RB was originally found as the recessive gene, encoding a 105-kD protein with multiple phosphorylation sites, whose biallelic loss of function conferred susceptibility to the occurrence of retinoblastoma.[12-14] Because the tumor occurs almost exclusively in early childhood and because retinas are still developmentally evolving at this time, it is possible that RB may perform a normal developmental regulatory function. Loss of this function may thus result in the aberrant developmental pattern expressed as the retinoblastoma. If this gene has any significant function in regulating cell differentiation and proliferation, then one may anticipate that its expression should be regulated during an elicited process of cell differentiation *in vitro*. One particular *in vitro* model, the HL-60 cell line, provides a context in which to test this hypothesis because cellular and other molecular features of its differentiation are so well characterized. Mitogenic stimulation and viral transformation have been associated with altered RB expression in other leukocytes.[21] Given that retinoic acid and 1,25-dihydroxy vitamin D_3 cause HL-60 myeloid or monocytic differentiation, respectively, an obvious possibility is that a downstream regulator of their action may be the RB tumor suppressor gene product. In the following report a synopsis of new and old findings of the involvement of RB in HL-60 cell differentiation induced by retinoic acid and 1,25-dihydroxy vitamin D_3 is presented, as well as their potential implications on the function of the RB protein.

II. Materials and Methods

A. Cells and Culture Conditions

HL-60 promyelocytic leukemia cells were maintained as stock cultures between initial and final cell densities of 0.2×10^6 cells/ml to 2×10^6 cells/ml in RPMI 1640 medium supplemented with 10% heat-inactivated fetal calf serum as previously described.[5-7,21] Cells were treated with 10^{-6} M β-all *trans*-retinoic acid (Sigma Chemical Co.) or 0.5×10^{-6} M 1,25-dihydroxy vitamin D_3 (Hoffmann-LaRoche Co.) as previously described.[5-7]

For cultures treated with antibody (anti-CSF-1R) against the CSF-1 receptor, at the indicated times, 70 µl of a previously described rat monoclonal antibody (Oncogene Science, Inc., Uniondale, NY) was added to various experimental cultures, using the manufacturer's stock solution of 100 µg/ml IgG_{2b} in 1.0 ml of 0.05 M sodium phosphate buffer containing 0.1% sodium azide and 0.2% gelatin. This concentration of antibody failed to impair cell viability exceeding 95%, but was enough to have a biological effect as evidenced by a slight increase in cell doubling time not evident at lower concentrations. A nonspecific affinity-purified murine IgG_{2b} used at the same concentration to treat control cells had no effect on rate of cell growth or differentiation, indicating the specificity of the effects of the anti-CSF-1R.

B. Assay for c-*fms*/CSF-1 Receptor Expression

For the flow cytometric assay for cell surface expression of the c-*fms* gene product, 5×10^5 cells were harvested from the experimental culture. The cells were washed once in 150 µl phosphate-buffered saline (PBS) (pH 7.2), the supernatant removed and gently resuspended in 85 µl NGS/PBS (50% normal goat serum (heat inactivated 60°C for 30 min), 0.002% Triton X-100 (in PBS). A monoclonal antibody against the c-*fms* protein was added, using 15 µl of the manufacturer's (Ab-2 Oncogene Science) stock of 100 µg of IgG_{2b} per ml. The suspension was gently mixed and incubated on ice for 20 min in the dark. The cells were then washed three times with 150 µl PBS, with gentle mixing and resuspension between each wash. For secondary immunofluorescent staining, the cells were then resuspended in 95 µl NGS/PBS and 5 µl FITC-GAR (fluorescein-conjugated goat anti-rat) IgG antibody, whole molecule, reconstituted per the manufacturer's directions to 2 ml in sterile distilled water and then diluted 1:10 in NGS/PBS (Cappel Division Organon Teknika Corp., West Chester, PA). The cells were incubated on ice for 20 min and protected from light. They were then washed three times in 150 µl PBS, with gentle resuspension after each wash, and finally resuspended in 500 µl PBS for flow-cytometric analysis. An EPICS 752 (Coulter Electronics, Hialeah, FL) dual-laser multiparameter flow cytometer was used. Excitation was provided by an argon ion laser operating at 488 nm, 100-mW output. Emitted fluorescence was collected through a 457-502-nm band laser-blocking filter and a 550-nm long pass dichroic mirror. The green fluorescence (fluorescein) was detected by a photomultiplier masked with a 525-nm band pass filter. The photomultiplier had a high voltage setting of 1700 V and an amplifier gain of 10 or 20, depending on which cell line was being analyzed. The forward and wide-angle light scatter of the cells were detected by solid-state detectors. Forward-angle light scatter (a measure of size) was used as the trigger signal and as a discriminator to eliminate signal from debris. Five to ten thousand events were collected for each sample. The data were stored in list mode files for further analysis. A bit map defining the cell population in dual parameter plots of forward- versus wide-angle light scatter was used as a logic gate to define cells free of debris.

C. Cell Cycle and Differentiation

The distribution of cells in the cell cycle was measured using flow cytometry of cells stained with hypotonic propidium iodide as previously described.[5-7] Cell differentiation was assayed by the inducible oxidative metabolism characteristic of mature myelomonocytic cells using nitroblue tetrazolium as a detecting agent as previously described.[5-7]

D. RB Detection Assays

RB protein was detected in Western blots using a modification of the method of Zhang et al.[22] The cells were lysed in a buffer of 6% SDS, 4M Urea, 125mM Tris, 4mM EDTA, pH 6.9 with β-mercaptoethanol and bromophenol blue and run on a 6% SDS polyacrylamide gel using a 37.5:1 ratio of acrylamide:bis. After transfer to nitrocellulose or PVDF membranes, the RB protein was detected using a monclonal antibody (Triton Diagnostics, Alameda, CA) and a chemiluminescence detection kit (Amersham, Arlington Heights, IL).

RB was detected in bivariate flow-cytometric assays in conjunction with DNA, using immunofluorescence and a fluoresceinated secondary antibody to detect RB and propidium iodide to detect DNA. The staining and flow-cytometric assay are exactly as described previously.[21,23,24] The same samples were analyzed by laser confocal microscopy using an ACAS 570 laser cytometer (Meridian Instruments, Okemos MI). Excitation was with an argon ion laser tuned to 488 nm. Green and red fluorescence were segregated using a dichroic mirror and band-pass filters similar to those described previously.[21,23,24]

E. Xenografts

Male mice with severe combined immunodeficiency (SCID) were obtained from Taconic Laboratory Animals (Germantown, NY) and were randomly assigned to various treatment groups. At 11 weeks of age, 10 mice were given 7×10^6 untreated proliferating HL-60 cells suspended in 200 μl PBS by intravenous injection into the tail vein. Half of these mice were subsequently treated, starting at 2 weeks postinjection, with β-all *trans*-retinoic acid. Retinoic acid was dissolved in olive oil and delivered by gavage at dose of 0.26 mg/mouse per day. An additional group of five mice was injected in the same manner with 7×10^6 HL-60 cells that had been pretreated with retinoic acid in culture for 24 hr to induce the previously characterized "precommitment" state. These mice received no further treatment after inoculation. Mice were examined daily and were sacrificed when clinical signs of distress due to tumor growth were observed. Complete necropsies were performed and cell suspensions were prepared from tumor masses. These cells were examined by flow cytometry using a phycoerythrin-conjugated monoclonal antibody to CD 33 (My-9 RD-1, Coulter Immunology, Hialeah, FL) to confirm their human origin, and following treatment with hypotonic propidium iodide for DNA/cell-cycle analysis.

III. Results

A. RB Expression during Exponential Cell Growth

The RB protein has characteristics that indicate that it is regulated with progression through the cell division cycle as cells mature from G1 through S and G2 to M (mitosis). The amount of the protein per cell increases as cells progress through the cell cycle.[23,24] If HL-60 cells are sustained in maximal exponential growth and then harvested, then they can be analyzed by multiparameter flow cytometry for their DNA and RB protein content simultaneously on a per-cell basis. The harvested cells are stained with propidium iodide to detect DNA and a fluorescein-conjugated antibody to detect RB protein. Figure 1 shows a bivariate dot-density correlation plot between RB protein and DNA content per cell. It can be seen that maturation through the cell cycle is generally associated with increasing RB protein content.

The protein is distributed between the hypophosphorylated and a continuum of phosphorylated states that give rise to the migratory heterogeneity seen as a diffuse band by Western analysis as shown in Figure 2, consistent with previous reports.[15,16,22] The relative numbers of G1 cells in this proliferating cell population determined by flow cytometry of propidium

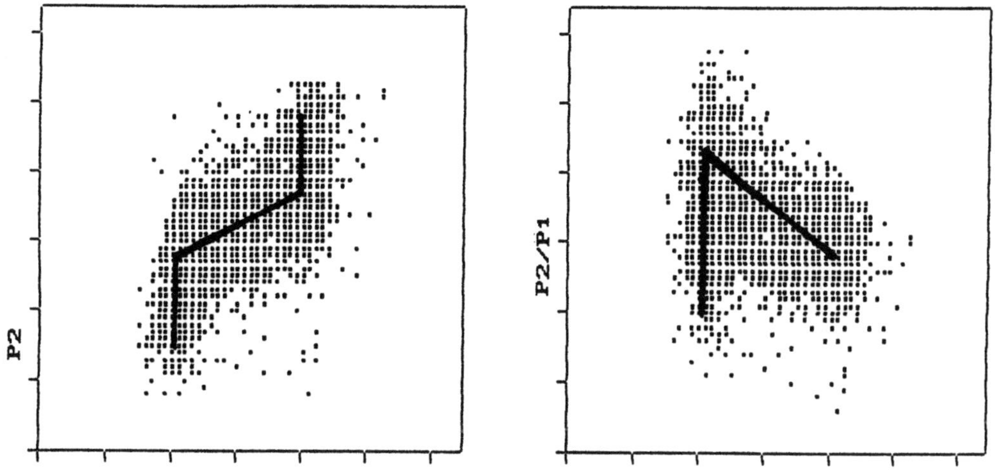

Figure 1. Bivariate dot-density correlation plots of RB protein per cell (vertical) versus genomic DNA content (horizontal) (left panel), and normalized ratio of RB protein to DNA per cell (vertical) versus DNA content (horizontal) (right panel).

iodide-stained cells is approximately 45%. A densitometric scan of the Western blot (Figure 2) shows that approximately 7% of the RB seen in exponential cells may be hypophosphorylated. The hypophosphorylated protein is seen as the leading band on a broad continuum of hyperphosphorylated protein in the densitometric scans.[25,26] The relative amount of hypophosphorylated RB protein is thus much less than the relative number of G1 cells in exponential populations. Although there are previous results that indicate that the RB protein is hypophosphorylated in G1 and hyperphosphorylated thereafter in certain cells, it appears that in this case not all G1 cells could have just hypophosphorylated RB protein. Regardless of how the unphosphorylated form of RB is distributed with respect to cell-cycle phases, there is a minimal amount of it in cycling cells. Most of the RB molecules are phosphorylated albeit to varying degrees.

The amount of cellular RB protein per amount of genomic DNA increases to a maximum in G1 and then progressively decreases. Bivariate flow-cytometric analysis of the amount of RB protein and DNA per cell was used to analyze exponentially growing HL-60 cells. The dot-density correlation plot of the normalized quotient of the RB protein and DNA content versus the DNA content is shown in Figure 1. The RB protein per DNA reaches a maximum in G1, and then progressively decreases with entry and progression through S and the remainder of the cell cycle. Thus, although the amount of RB protein increases as cells mature through the cell cycle, the increase is slower than the rate of doubling of DNA once S phase begins. This behavior is not a general feature of all potential nuclear localizing regulatory genes, such as c-*myc*. Figure 3 shows the corresponding plots for the c-*myc* nuclear oncogene. The flow-cytometric analysis was performed as previously described.[24] Although the amount of c-*myc* protein also increases with progression through the cell cycle, the quotient of c-*myc* protein and DNA is relatively stable throughout the cell cycle. Analysis of the expression of two other nuclear oncogenes, c-*fos* and c-*myb*, in the same way resulted in bivariate plots indistinguishable from those of c-*myc* (data not shown).

The RB protein contains a nuclear translocation sequence and is found in the nucleus as well as the cytoplasm. The localization of the RB protein can be characterized using laser confocal microscopy with image analysis. Figure 4 shows the confocal image of an

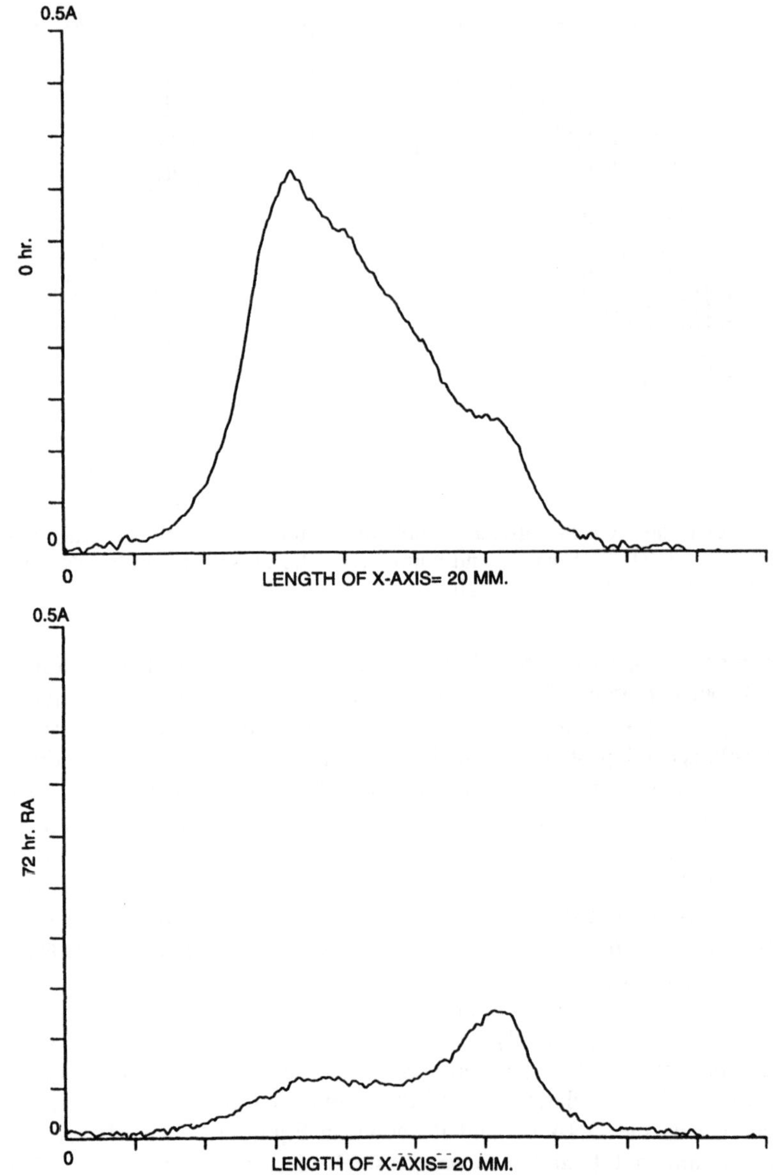

Figure 2. Western analysis of RB protein. Densitometric scans of films exposed to a membrane treated with a chemiluminescence developing agent show the top of the gel on the left. The plots show the amount of RB protein (vertical) versus apparent molecular weight (horizontal).

equatorial section of an HL-60 cell fluorescently stained for DNA with propidium iodide, delineating the nucleus, and for RB protein with fluorescein. A plot of RB protein signal intensity and of DNA signal intensity along a cross section of this optical slice shows certain features of the RB protein distribution. The protein is present in both the nucleus and the cytoplasm. The density of the protein is several-fold greater in the cytoplasm compared to the nucleus. There are occasional discrete concentrations of the protein within the nucleus. Overlay of the fluorescence and the phase-contrast images indicates that these concentrations of RB protein are within nucleoli.

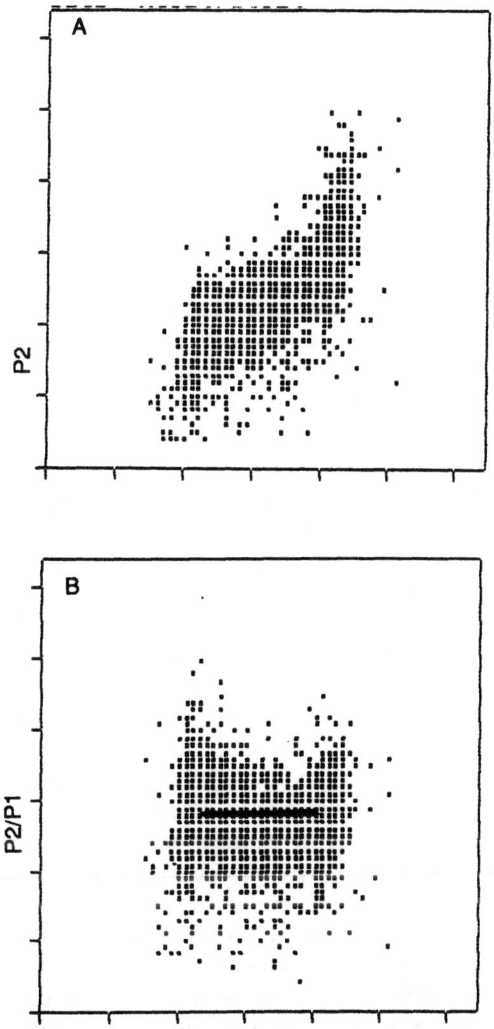

Figure 3. Bivariate dot-density correlation plots of *c-myc* protein per cell (vertical) versus genomic DNA content (horizontal) (panel A), and normalized ratio of *c-myc* protein to DNA per cell (vertical) versus DNA content (horizontal) (panel B).

B. Retinoic Acid-Induced Effects

If HL-60 cells are cultured in the presence of 10^{-6} M β-all *trans*-retinoic acid, they undergo myeloid differentiation. Onset of G0-specific growth arrest and differentiation occurs after a period corresponding to two division cycles (Table 1). The same is the case for monocytic differentiation induced by 1,25-dihydroxy vitamin D_3. It has been found previously that in both cases, an early downregulation of RB occurred before any overt modulation of cell cycling or differentiation.[24] The reduction in expression occurred similarly for cells in all cell-cycle phases and showed no differential regulation in the cell cycle. At the reduced levels, the relative RB expression per cell with progression through the cell cycle is the same as shown in Figure 1, but at a lower overall RB level. That is, it is tantamount to a change in scale for RB. The same is the case for the quotient of RB protein and DNA also shown in Figure 1. Again the relative relationship during the cell cycle is preserved, although there is an overall downregulation of RB expression.

The accumulation of cells in G1/0 due to treatment with retinoic acid is associated with differential enrichment in the hypophosphorylated form of the RB protein. A Western analysis of the RB protein expressed by HL-60 cells treated for 72 hr with retinoic acid is shown in Figure 2. There is an enrichment in the relative amount of the faster migrating hypophosphorylated form of the molecules. A crude approximation from integration of the densitometric scan of the Western analysis shows approximately 50% of the RB protein is in the hypophosphorylated form. In contrast, approximately 90% of the cells are in G1/0, as determined by flow cytometry of propidium iodide-stained cells. As for exponentially growing cells, it is unlikely that G1/0 cells express only the unphosphorylated form; ergo, it appears G1/0 cells have the hypo- and hyperhosphorylated RB protein. Although some of the RB in G1/0 arrested cells must be hypophosphorylated, the present data do not distinguish between heterogeneity of RB molecules within the same cell or heterogeneity among G0/1 cells. It cannot be distinguished if all G0 cells have some hyperhosphorylated RB or if only some G0 cells have hyperphosphorylated RB with present common techniques. Nevertheless, the present results clearly indicate that the potential relationship between quantitation of hypophosphorylated RB and G1/0 cells in exponential populations is altered in retinoic acid-induced G0/1 arrested populations. It is likely that this reflects the differentiation of the cells induced by retinoic acid. Thus, it appears that retinoic acid may induce a cellular state where the role of RB hypophosphorylation is distinguishable from that occurring in G1 proliferating cells.

Retinoic acid also induces the differential relocalization of the amount of RB protein in the cytoplasmic and nuclear compartments with reduced expression. Retinoic acid-treated cells were fluorescently stained for RB protein with fluorescein and DNA with propidium iodide and the image of an equatorial plane analyzed by laser confocal microscopy as shown in Figure 3. Measuring RB protein density as well as DNA along a diagonal line through the cell in this plane shows a relative reduction in the RB protein density in the cytoplasm and an increase in the nucleus compared to untreated cells. In other cell types it has been observed that the RB protein is not redistributed during the cell cycle.[27] This suggests that the presently observed redistribution may reflect cell differentiation more than any effect of cell-cycle arrest.

C. Dependence on c-*fms*

The retinoic acid elicited regulation of RB depends on the function of other genes. One

CONTROL

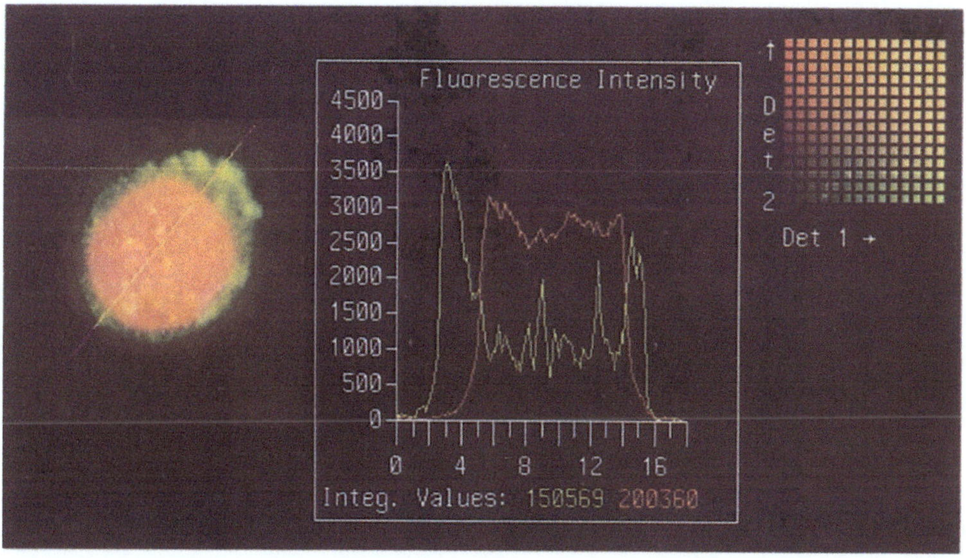

RA,72 hr

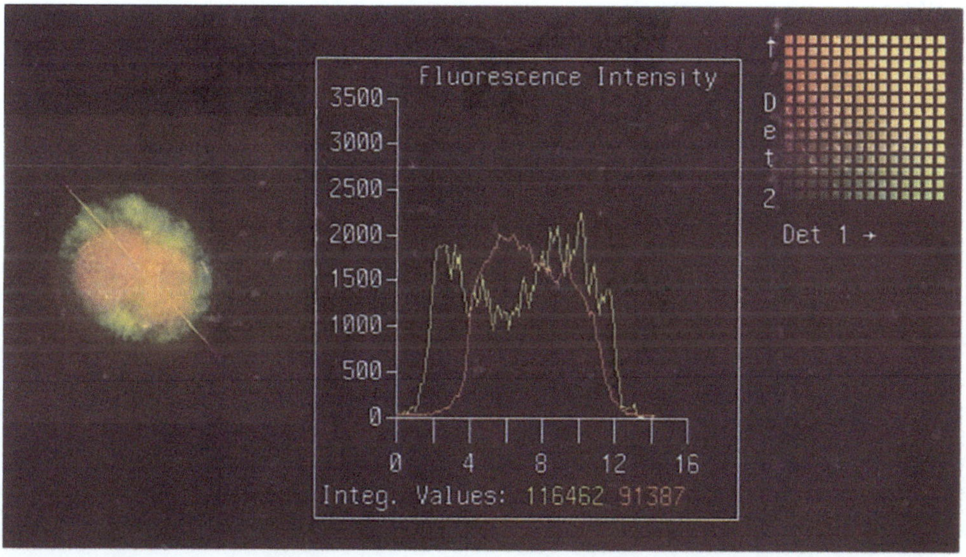

Figure 4. Image analysis of HL-60 cells stained for RB protein with fluorscein and for genomic DNA with propidium iodide, which fluoresce green and red, respectively. The image shows an equatorial plane derived by laser confocal microscopy. The intensity of the RB protein or DNA signal along the indicated diagonal in that plane is plotted, showing the density of the RB protein in the cytoplasm and the nucleus. The cell treated with retinoic acid has a greater amount of RB protein in the nucleus than the untreated control cell.

Table 1. Response to Retinoic Acid and 1,25-Dihydroxy Vitamin D_3

	Time	%G1	%NBT
Control	0	46	3.3
	24	43	1.9
	48	44	1.8
	72	47	7.2
	96	71	3.4
RA	24	47	6.9
	48	59	24.2
	72	89	89.1
	96	96	97.5
VD_3	24	46	5.1
	48	54	52.2
	72	87	95.4
	96	95	99.0

of these is c-*fms*. Retinoic acid increases the expression of c-*fms*, the gene encoding the receptor for CSF-1, detected at the plasma membrane. CSF-1 is a lymphokine that is well known to positively regulate the myeloid and monocytic lineages of hematopoiesis[28-30] and has been found to be involved in early developmental regulation during embryogenesis as well.[31] The amount of c-*fms* at the plasma membrane was measured by immunofluorescence and flow cytometry using a fluorescently labeled antibody recognizing the c-*fms* protein. The amount of c-*fms* protein expressed at the membrane of HL-60 cells was weak, but increased several-fold after addition of retinoic acid. Figure 5 shows the mean relative expression level per cell as a function of time. The response was faster than that elicited by 1,25-dihydroxy vitamin D_3, which has previously been reported to also cause the upregulation of c-*fms* expression in HL-60 cells.[32] As seen in Figure 5, the slower upregulation of c-*fms* expression at the plasma membrane by 1,25-dihydroxy vitamin D_3 lagged by approximately 48 hr, a period equal to two division cycles in this subline of HL-60 cells.

The retinoic acid induced early down regulation of RB expression and the subsequent differentiation of HL-60 cells are both dependent on a function of the c-*fms* gene product. HL-60 cells were exposed to retinoic acid in the presence or absence of an antibody[33] directed against the CSF-1 receptor. The antibody was thus used to disrupt the function of the receptor during the retinoic acid-induced metabolic cascade. Figure 6 shows the percent of cells differentiated at 48 hr when onset of differentiation typically occurs as indicated in Table 1. The fraction of cells differentiating due to retinoic acid was greatly reduced by the presence of antibody against the CSF-1 receptor. The 1,25-dihydroxy vitamin D_3-induced moncytic differentiation was also inhibited by the presence of the antibody. In control experiments, the antibody by itself caused no apparent change in the proliferation of HL-60 cells. Nor did it cause any apparent differentiation. The concentration of antibody was chosen using dose-response trials to determine the maximum allowable concentration without

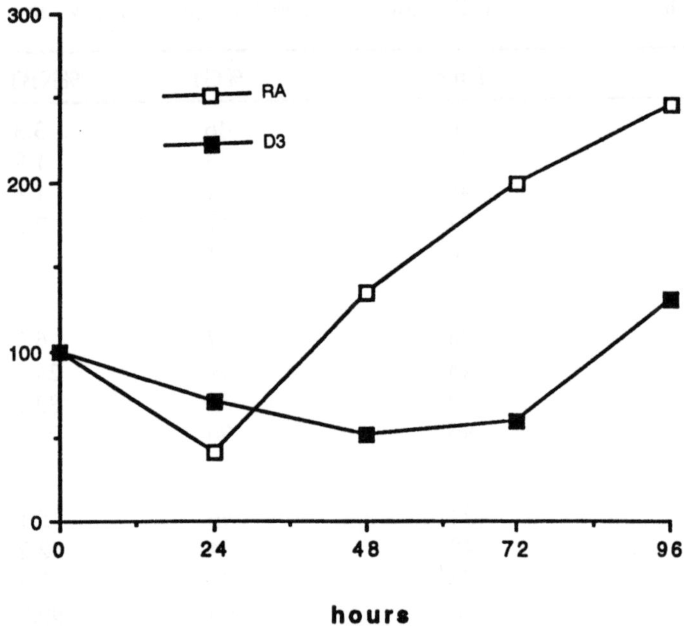

Figure 5. Expression of the CSF-1 receptor per cell (vertical, relative units of immunofluorescence) at the plasma membrane in HL-60 cells treated with retinoic acid (open symbol) or 1,25-dihydroxy vitamin D_3 (closed symbol) for the indicated (horizontal) times. The vertical axis is in arbitrary units of fluorescence. The c-*fms* expression measured is normalized to that of untreated control cells at t equals 0.

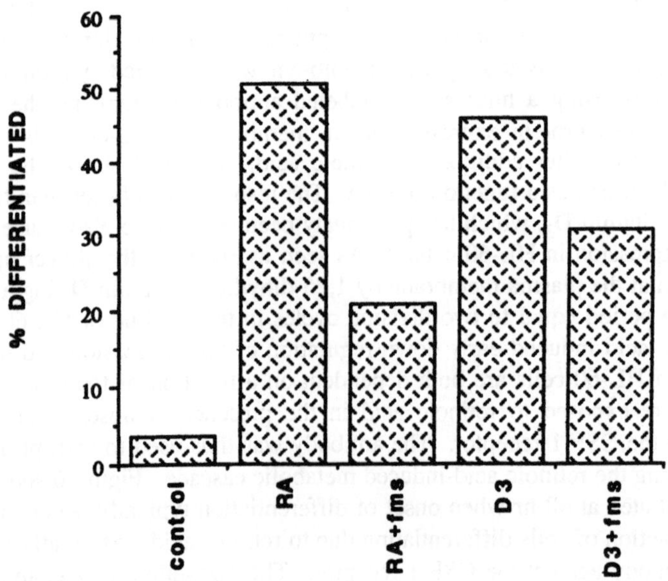

Figure 6. Percent of cells that are functionally differentiated for untreated (control), retinoic acid-treated (RA), retinoic acid treated in the presence of antibody against the CSF-1 receptor (RA+*fms*), 1,25-dihydroxy vitamin D_3-treated in the presence of antibody against the CSF-1 receptor (D3+*fms*). Cells were treated or not for 48 hr.

any overt effect on cell proliferation, differentiation, or viability. A nonspecific immunoglobulin used at the same concentration had no effect (data not shown).

Inhibition of differentiation correlated with inhibition of downregulation of RB expression. The downregulation of RB expression typically is evident before any overt change in cell proliferation or differentiation due to retinoic acid and continues progressively thereafter.[24] The expression of RB per cell for retinoic acid-treated cells in the presence or absence of the antibody against the CSF-1 receptor is shown in Figure 7. The cells were harvested at the same time (48 hr) at onset of differentiation (Figure 6) and immunofluorescently stained with an antibody to measure RB expression per cell and with propidium iodide to measure DNA content of the same cell. As can be seen, the presence of antibody against the CSF-1 receptor disrupted the downregulation of RB expression normally induced by retinoic acid at this time. The same was true for the downregulation of RB expression typically induced by 1,25-dihydroxy vitamin D_3. The observed modulation of RB expression was not a manifestation of changes in the cell cycle and can be isolated from cell-cycle effects by considering the RB expression of cells restricted to just one cell-cycle phase, for example G1/0. Figure 8 shows that the same inhibition of RB downregulation due to retinoic acid or 1,25-dihydroxy vitamin D_3 occurs when the analysis is restricted to G1/0 cells. Identical relationships persist regardless of which cell-cycle phase is chosen for analysis (data not shown). Thus both the retinoic acid-induced downregulation of RB expression and the differentiation of the cells were inhibited by disrupting the function of the CSF-1 receptor. A function of this receptor and its ligand may thus be an intermediate in effecting the action of retinoic acid on altering RB gene expression and cell differentiation.

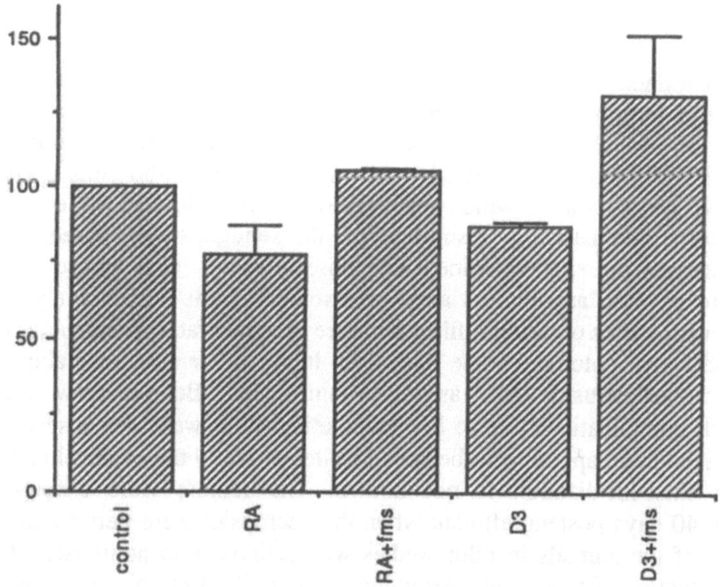

Figure 7. Average RB protein expression level per cell for untreated (control), retinoic acid-treated (RA), retinoic acid-treated in the presence of antibody against the CSF-1 receptor (RA+*fms*), 1,25-dihydroxy vitamin D_3-treated (D3), 1,25-dihydroxy vitamin D_3-treated in the presence of antibody against the CSF-1 receptor (D3+*fms*). Cells were treated or not for 48 hr.

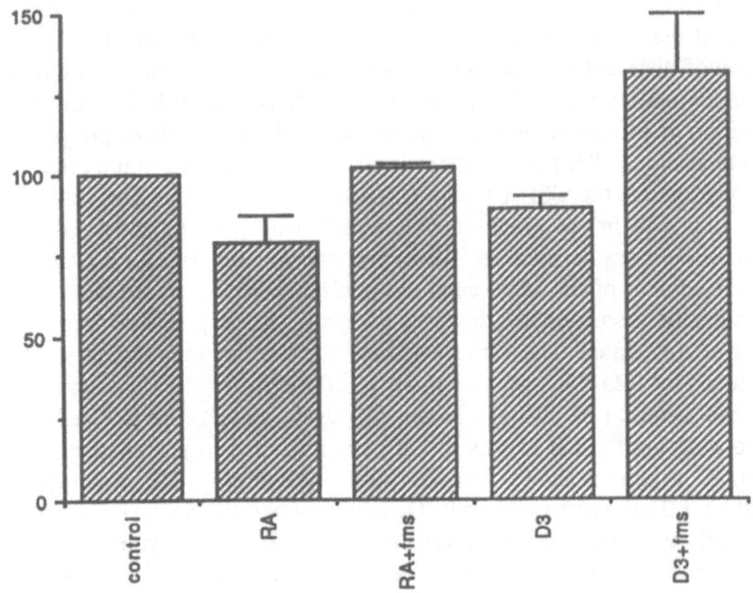

Figure 8. Average RB protein expression level per cell for G1 cells in cell populations that were untreated (control), retinoic acid-treated (RA), retinoic acid-treated in the presence of antibody against the CSF-1 receptor (RA+*fms*), 1,25-dihydroxy vitamin D_3-treated (D3), 1,25-dihydroxy vitamin D_3-treated in the presence of antibody against the CSF-1 receptor (D3+*fms*). Cells were treated or not for 48 hr.

D. Effect on Tumorigenicity

Given that retinoic acid quickly reduces expression of RB and primes these leukemic cells to undergo terminal myeloid differentiation, an obvious question is whether the tumorigenicity of the leukemic cells in mice is affected by exposure to retinoic acid. In pilot experiments, SCID mice were injected intravenously with HL-60 cells. The mice were divided into two groups, one of which was treated with retinoic acid while the other was an untreated control. The pathology resulting from the xenograft was subsequently observed. Table 2 summarizes the sites examined and those affected. One mouse in the untreated group failed to develop any tumors and was excluded from further analysis. The most frequently occurring sites of tumor infiltration were perirenal fat, suprascapular lymph nodes, mesentery, and spinal column. Some mice with large and/or numerous abdominal tumors also had a peritoneal effusion that was rich in tumor cells. Bone marrow and spleen were rarely affected; complications due to hematopoietic failure were not observed. Based on these initial trials, there appeared to be specific sites prone to tumor development, although the cause of this localization is not known. The tumors were strongly evident at approximately 40 days postengraftment when the necropsies were performed.

Treatment of the animals in pilot studies with retinoic acid administered as a gavage resulted in a reduction of tumor sites per animal (Figure 9). Retinoic acid was administered daily unless the animal exhibited clinical signs of toxicity, most typically hyperkeratosis and lethargy in which case treatment was suspended until the animal's condition improved. The frequency of tumor occurrence at the most frequent sites for retinoic acid-untreated and -treated animals is shown in Figure 10. There is an apparent reduction in the frequency of tumors at all these sites except the perirenal fat. Thus, it appears that retinoic acid may have

Table 2. Distribution of HL-60 Engraftment Sites in Various Experimental Groups

Untreated HL-60 Injected, No Subsequent Treatment of Mice															
Mouse No.	Liver	Spleen	Bone Marrow	Peri-renal	Lung	Brain	Peri-ocular	Scapular L.N.	Axilliary L.N.	Submand. L.N.	Caudal Mesentery	Root of Mesentery	Spinal Column	Ascites	Misc.
27	–	–	–	–	–	–	–	–	–	–	+	–	–	+	rear leg
28	–	–	–	+	–	–	–	+	+	–	+	+	+	+	–
29	–	–	–	+	–	–	–	+	–	–	+	+	+	+	–
30	–	–	–	+	–	–	–	–	–	+	+	+	+	–	rear leg, mandible

"Precommitted" HL-60 Injected, No Subsequent Treatment of Mice															
Mouse No.	Liver	Spleen	Bone Marrow	Peri-renal	Lung	Brain	Peri-ocular	Scapular L.N.	Axilliary L.N.	Submand. L.N.	Caudal Mesentery	Root of Mesentery	Spinal Column	Ascites	Misc.
37	–	–	–	+	–	+	–	–	–	–	–	+	–	+	peri-anal
39	+	–	–	+	–	–	+	–	–	–	+	–	–	+	–
40	–	+	+	+	+	–	–	–	–	–	+	+	+	+	–

Untreated HL-60 Injected, Mice Treated with Retinoic Acid															
Mouse No.	Liver	Spleen	Bone Marrow	Peri-renal	Lung	Brain	Peri-ocular	Scapular L.N.	Axilliary L.N.	Submand. L.N.	Caudal Mesentery	Root of Mesentery	Spinal Column	Ascites	Misc.
31	–	–	–	+	–	–	–	–	–	–	+	–	–	–	rear leg
32	–	–	–	+	–	–	–	+	–	–	+	–	–	–	inguinal L.N.
33	–	–	–	+	–	–	–	–	–	–	–	–	–	–	–
34	–	–	–	+	–	–	–	–	–	–	+	–	–	–	–
35	–	–	–	–	–	+	–	–	–	–	+	–	+	–	–

an effect on the tumorigenicity of the cells as well as the cellular-molecular aspects described above.

It was previously shown that HL-60 cells briefly exposed to retinoic acid entered a "precommitment state" in which they were essentially primed to differentiate and exhibited a variety of the molecular features presaging differentiation, including downregulation of RB and c-*myc* expression.[5,6,24] To test if these cellular-molecular changes by themselves were

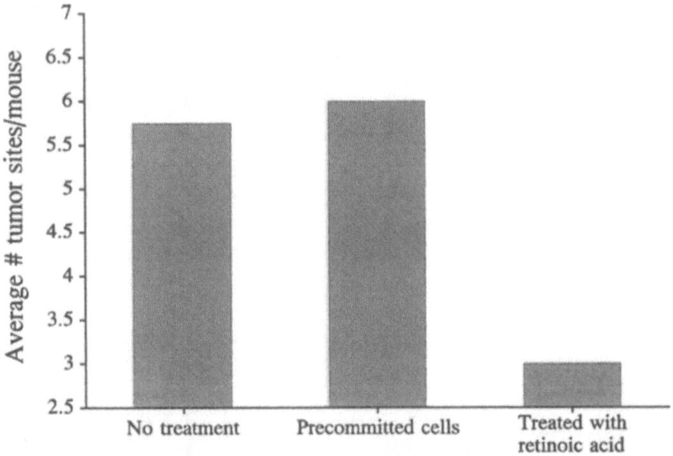

Figure 9. Mean number of tumor sites per mouse for untreated mice engrafted with untreated HL-60 cells, untreated mice engrafted with HL-60 cells previously treated with retinoic acid for 24 hr (precommitment), retinoic acid-treated mice engrafted with untreated HL-60 cells.

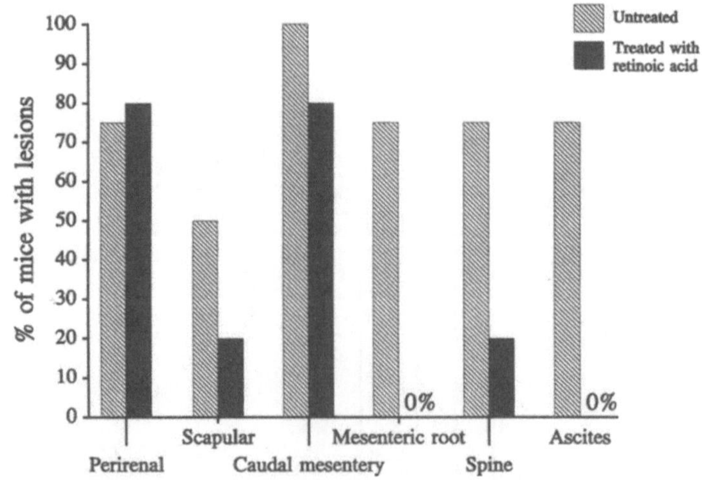

Figure 10. Effect of retinoic acid on percent of mice with tumors at the most frequently occurring sites. Percent of mice with lesions (vertical) for untreated (hatched) and retinoic acid treated (shaded) mice occurring as perirenal, scapular, caudal mesentery, mesenteric root, spine, and ascites.

sufficient to cause any reduction in tumorigenicity, HL-60 cells were treated for 24 hr with retinoic acid prior to their injection into mice. Of the five mice in this group, two developed murine tumors (one mediastinal lymphoma, one acute leukemia) but no HL-60 masses as determined by staining for MY-9 (CD 33). These mice were excluded from further analysis. Of the three mice successfully engrafted with HL-60 cells in this group, the pretreatment of the cells with retinoic acid did not significantly affect tumorigenicity (Figure 9). The mice injected with these cells showed no difference in numbers of tumors per animal compared to mice injected with untreated HL-60 cells. This is in contrast to the reduction in the numbers of tumors per animal observed in mice that were treated continuously with retinoic acid after untreated HL-60 cells were injected.

IV. Discussion

These results show that retinoic acid as well as 1,25-dihydroxy vitamin D_3, which are related in that both receptor ligand complexes bind related DNA consensus sequences and both are members of the steroid-thyroid hormone family, regulate the expression of the RB gene in the metabolic cascade leading to terminal cell differentiation in HL-60 human promyelocytic leukemia cells. The regulation is superimposed on the relative relationships existing for differential RB expression during the proliferative cell cycle; that is, the regular division cycle. During the progression of cells from G1 to M, the amount of RB protein progressively increases, roughly maintaining homeostasis with the total cell mass. The protein is present in a continuum of hypophosphorylated to hyperphosphorylated forms. The relative amount of the hypophosphorylated form is low (approximately 7% or less) and is less than the fraction of cells in G1. The ratio of the amount of RB protein relative to DNA increases during G1, reaching a maximum, and then decreases as cells enter and progress through S, and the rest of the cell cycle. If as will be discussed below, the RB protein serves as an alternative binding site for transcriptional regulators, then this ratio specifies the stoichiometry that determines if such regulatory factors bind their DNA consensus sequences or are held in abeyance from them. The protein is located in both the cytoplasm and the nucleus. The density of protein, which is of course distinguished from the amount, is greater in the cytoplasm than the nucleus. It is yet undetermined how the localization of RB occurs with respect to cell-cycle phase or phosphorylation state, although it is of obvious interest given the above considerations. Since the hypophosphorylated forms of the RB protein bind more tightly to the nuclear compartment[34] than the hyperphosphorylated protein, the localization of the protein may in part be driven by its phosphorylation state, as is suggested by the effects of retinoic acid discussed below.

Retinoic acid and likewise 1,25-dihydroxy vitamin D_3 cause certain changes in the above regulation of RB protein in HL-60 cells. The total amount of the protein per cell is downregulated for cells in all cell-cycle phases before any overt modulation of cell cycle or differentiation state of the cell. The downregulation preserves the relative relationships of RB protein during the cell cycle; that is, increasing with progression and having a maximum in the ratio of RB protein to DNA at the G1/S boundary. However, in the later resulting G1/0 cell populations, the fraction of RB protein that is hypophosphorylated is much greater than in proliferating G1 cell populations. Thus, in retinoic acid-induced differentiated HL-60 cell populations, the relationship of phosphorylation to the cell cycle is apparently altered from that in actively proliferating cells. This may reflect a different function of the RB protein in quiescent, differentiated cells compared to proliferating, immature cells. With downregulation in the total amount of protein, the remaining protein is differentially relocated to the nucleus. Thus the RB protein density increased in the nucleus and decreased in the cytoplasm. This is associated with an increase in the relative amount of the hypophosphorylated protein consistent with the above suggestion that phosphorylation may in part determine its relative nuclear-cytoplasmic localization during differentiation. The early reduction in RB protein induced by retinoic acid and the ultimate occurrence of cellular differentiation both depend on a function of the c-*fms* protein, the receptor for CSF-1. This is consistent with a regulatory coupling between RB downregulation and subsequent cell differentiation. The retinoic acid induced effect on RB expression thus appears to involve an intermediate step dependent on a c-*fms* protein function. This may implicate the involvement of portions of growth factor signal transduction pathways in the retinoic acid-induced metabolic cascade. Finally the cumulative effect of exposure to retinoic acid resulting in these molecular alterations causes a macroscopic alteration apparent as the reduced tumorigenicity of the HL-60 cells as a xenograft in SCID mice.

The explicit function of RB in the metabolic cascade leading to cell differentiation and the significance of the downregulation in response to retinoic acid is not known. Given the present results and that it is known that the hypophosphorylated form of the RB protein binds to known transcriptional regulatory factors and also that the transforming agents of various viruses bind the hypophosphorylated RB protein, then an obvious speculation can be synthesized. It may be that retinoic acid exerts its effect on gene regulation at least in part by reducing the amount of RB protein per cell thereby altering stoichiometry to favor the binding of transcriptional regulatory factors needed in differentiation to their DNA consensus sequences. If the avidity of RB for binding such regulatory factors depends on the extent of phosphorylation, then there is a continuum of varying binding affinity dependent on not just if the RB protein is phosphorylated, but the extent of phosphorylation. Thus there are two modes of regulation—reduction in amount of total protein or phosphorylation—to control the relative availability of transcriptional regulators. The latter may be favored for cell-cycle control in proliferating cells because it can be effected faster both positively and negatively, whereas the former appears to be favored for differentiation by parsimony. The observed changes in the ratio of the amount RB protein to DNA during the cell cycle with a maximum at the G1/S interface and subsequent decline are consistent with a potential regulatory role for the relative stoichiometry of the two as suggested here. Other molecules binding to the RB protein may thus alter its capacity to function in this respect and so deregulate proliferation or differentiation.

In this capacity, RB is functioning as previously described[21,24] as a "status quo" gene or a brake against change of a given cellular state. One can hypothesize that its downregulation is thus permissive of change such as in cell differentiation. Likewise phosphorylating the RB protein also reduces the amount of hypophosphorylated RB protein thereby possibly promoting conversion from a 2n DNA (diploid) to a 4n DNA cell through DNA synthesis. If its function is to prevent change, then its downregulation is needed to effect changes in cell proliferation and differentiation by retinoic acid. This relatively basic perception of its potential function is consistent with its observed early downregulation by both retinoic acid and 1,25-dihydroxy vitamin D_3 that affect different pathways of differentiation in the HL-60 cell line.

Acknowledgments

We are grateful for support to do these studies to the American Institute for Cancer Research and the National Institutes of Health. We thank Amy Pellegrino for skillful secretarial assistance in preparing this manuscript.

This research is supported in part by grants from the USPHS (NIH), American Institute for Cancer Research, and the Cornell Consolidated and Biotechnology Programs. In particular, this work was supported (in part) by a grant from the Cornell Biotechnology Program, which is sponsored by the New York State Science and Technology Foundation, a consortium of industries, the U.S. Army Research Office, and the National Science Foundation. We thank Edwin DeFeijter and Dr. Peggy Wade of Meridian Instruments, Okemos, MI for performing the image analysis and for having us as guests in their laboratory.

Address correspondence to A. Yen.

References

1. R.M. Evans, The steroid and thyroid hormone receptor superfamily, *Science* 240:889-895 (1988).
2. R. Schule, K. Umesono, D.J. Mangelsdorf, J. Bolado, J.W. Pike, R.M. Evans, *Jun-fos* and receptors for vitamins A and D recognize a common response element in the human osteocalcin gene, *Cell* 61:497-504 (1990).
3. S.J. Collins, R.C. Gallo, R.E. Gallagher, Continuous growth and differentiation of human myloid leukemia cells in suspension culture, *Nature* 270: 347-349 (1977).
4. A. Yen, HL-60 cells as a model of growth control and differentiation: The significance of variant cells, *Hemat Rev* 4:5-46 (1990).
5. A. Yen, S.L. Reece, K.L. Albright, Dependence of HL-60 myeloid cell differentiation on continuous and split retinoic acid exposures: Pre-commitment memory associated with altered nuclear structure, *J Cell Physiol* 118:277-286 (1984).
6. A. Yen, M.E. Forbes, c-*myc* Down regulation and precommitment in HL-60 cells due to bromodeoxyuridine, *Cancer Res* 50:1411-1420 (1990).
7. A. Yen, M. Forbes, G. deGala, J. Fishbaugh, Control of HL-60 cell differentiation lineage specificity: A late event occurring after precommitment, *Cancer Res* 47:129-134 (1987).
8. R. Chiu, W.J. Boyle, J. Meek, T. Smeal, T. Hunter, M. Karin, The c-*fos* protein interacts with c-*jun*/AP-1 to stimulate transcription of AP-1 responsive genes, *Cell* 54:541-552 (1988).
9. T. Curran, B.R. Franza, Jr., *fos* and *jun*: The AP-1 connection, *Cell* 55:395-397 (1988).
10. R. Turner, R. Tjian, Leucine repeats and an adjacent DNA binding domain mediate the formation of functional *cfos-cjun* heterodimers, *Science* 243:1689-1694 (1989).
11. R. Gentz, F.J. Rauscher, III, C. Abate, T. Curran, Parallel association of *fos* and *jun* leucine zippers juxtaposes DNA binding domains, *Science* 243:1695-1699 (1989).
12. T.-K.T. Fung, A.L. Murphree, A. T'Ang, J. Qian, S.H. Hinrichs, W.F. Benedict, Structural evidence for the authenticity of the human retinoblastoma gene, *Science* 236:1657-1661 (1987).
13. W.-H. Lee, R. Bookstein, F. Hong, L.-J. Young, J.-Y. Shew, E.Y.-H.P. Lee, Human retinoblastoma susceptibility gene: Cloning, identification, and sequence, *Science* 235:1394-1399 (1987).
14. S.H. Friend, R. Bernards, S. Rogelj, R.A. Weinberg, J.M. Rapaport, D.M. Albert, T.P. Dryja, A human DNA segment with properties of the gene that predisposes to retinoblastoma and osteosarcoma, *Nature* 323:643-646 (1986).
15. K. Mihara, X.-R. Cao, A. Yen, S. Chandler, B. Driscoll, A.L. Murphree, A. T'Ang, Y.K.T. Fung, Cell-cycle dependent regulation of phosphorylation of the human retinoblastoma gene product, *Science* 246:1300-1303 (1989).
16. P.-L. Chen, P. Scully, J.-Y. Shew, J.Y.J. Wang, W.-H. Lee, Phosphorylation of the retinoblastoma gene product is modulated during the cell cycle and cellular differentiation, *Cell* 58:1193-1198 (1989).
17. P.D. Robbins, J.M. Horowitz, R.C. Mulligan, Negative regulation of human c-*fos* expression by the retinoblastoma gene product, *Nature* 346:668-671 (1990) and *ibid. Nature* 351:419 (1991).
18. M.M. Ouellette, J. Chen, W.E. Wright, J.W. Shay, Complexes containing the retinoblastoma gene product recognize different DNA motifs related to the E2F binding site, *Oncogene* 7:1075-1081 (1992).

19. P.A. Hamel, R.M. Gill, R.A. Phillips, B.L. Gallie, Transcriptional repression of the E2-containing promoters EIIaE, c-*myc*, and RB1 by the product of the RB1 gene, *Molec Cell Biol* 12:3431-3438 (1992).
20. S.J. Weintraub, C.A. Prater, D.C. Dean, Retinoblastoma protein switches the E2F site from positive to negative element, *Nature* 358:259-261 (1992).
21. A. Yen, S. Chandler, S. Sturzenegger-Varvayanis, Regulated expression of the RB "tumor suppressor gene" in normal lymphocyte mitogenesis: Elevated expression in transformed leukocytes and role as a "status quo" gene, *Experimental Cell Research* 192:289-297 (1991).
22. W. Zhang, W. Hittelman, N. Van, M. Andreeff, A. Deisseroth, The phosphorylation of retinoblastoma gene product in human myeloid leukemia cells during the cell cycle, *Biochem Biophys Res Comm* 184:212-216 (1992).
23. A. Yen, S. Chandler, Inducers of leukemic cell differentiation cause down-regulation of RB gene expression, *Proc Soc Exper Bio Med* 199:291-297 (1992).
24. A. Yen, S. Chandler, M.E. Forbes, Y.-K. Fung, A. T'Ang, R. Pearson, Coupled down-regulation of the RB retinoblastoma and c-*myc* genes antecedes cell differentiation: Possible role of RB as a "status quo" gene, *Eur J Cell Biol* 57:210-221 (1992).
25. J.W. Ludlow, J.A. DeCaprio, C.-M. Huang, W.-H. Lee, E. Paucha, D.M. Livingston, SV40 large T antigen binds preferentially to an underphosphorylated member of the retinoblastoma susceptibility gene product family, *Cell* 56:57-65 (1989).
26. J.W. Ludlow, J. Shon, J.M. Pipas, D.M. Livingston, J.A. DeCaprio, The retinoblastoma susceptibility gene product undergoes cell cycle-dependent dephosphorylation and binding to and release from SV40 large T, *Cell* 60:387-396 (1990).
27. L. Szekely, E. Uzvolgyi, W.-Q. Jiang, M. Durko, K.G. Wiman, G. Klein, J. Sumegi, Subcellular localization of the retinoblastoma protein, *Cell Growth and Differentiation* 2:287-295 (1991).
28. B. Alberts, D. Bray, J. Lewis, M. Raff, K. Roberts, J.D. Watson, Renewal by pluripotent stem cells: Blood cell formation, *in*: *Molecular Biology of the Cell*, 2nd Ed. p.973, Garland Publishing, New York (1989).
29. P.J. Quesenberry, Hemopoietic stem cells, progenitor cells, and growth factors, *in*: *Hematology*, 4th Ed. W. J. Williams, E. Beutler, A.J. Erslev, M.A. Lichtmen, eds. p. 129, McGraw-Hill, New York (1990).
30. H.E. Broxmeyer, D.E. Williams, G. Hangoc, S. Cooper, S. Gillis, R.K. Shadduck, D.C. Bicknell, Synergistic myelopoietic actions *in vivo* after administration to mice of combinations of purified natural murine colony-stimulating factor 1, recombinant murine interleukin 3, and recombinant murine granulocyte/macrophage colony-stimulating factor, *Proc Natl Acad Sci* 84:3871-3875 (1987).
31. R. Muller, D.J. Slamon, E.D. Adamson, J.M. Tremblay, D. Muller, M.J. Cline, I.M. Verma, Transcription of c-*onc* genes c-*ras*^{Ki} and c-*fms* during mouse development, *Molec Cell Biol* 3:1062-1069 (1983).
32. P.T. Rowley, B. Farley, R. Giuliano, S. LaBella, J.F. Leary, Induction of the FMS proto-oncogene product in HL-60 cells by vitamin D: A flow cytometric analysis, *Leuk Res* 16:403-410 (1992).
33. C.J. Sherr, R.A. Ahmun, J.R. Downing, M. Ohtsuka, S.G. Quan, D.W. Golde, M.F. Roussel, Inhibition of colony stimulating factor-1 activity by monoclonal antibodies to the human CSF-1 receptor, *Blood* 73:1786-1793 (1989).
34. S. Mittnacht, R.A. Weinberg, G1/S phosphorylation of the retinoblastoma protein is associated with an altered affinity for the nuclear compartment, *Cell* 65:381-393 (1991).

Chapter 2

Retinoic Acid Suppresses Human Papillomavirus Type 16 (HPV16)-Mediated Transformation of Human Keratinocytes and Inhibits the Expression of the HPV16 Oncogenes

KIM E. CREEK, G. RONALD JENKINS, MOHAMMAD A. KHAN, AYSE BATOVA, JOHN R. HODAM, WILLIAM H. TOLLESON and LUCIA PIRISI

I. Introduction

It has been estimated that approximately 35% of cancer deaths occurring now may be related to diet, and that diet may constitute the greatest lifestyle contribution to cancer mortality.[1] Of the numerous dietary constituents related to cancer risk, β-carotene and/or vitamin A have consistently been implicated as potential agents for the chemoprevention of cancer in humans.[2-7] Vitamin A and its natural and synthetic derivatives (collectively known as retinoids) have been shown to inhibit both chemical[8] and viral carcinogenesis.[9,10] For example, retinoids inhibit skin carcinogenesis in the classic two-stage system[11] and exert a marked inhibitory effect on the induction and development of virus-induced papilloma (Shope) of rabbit skin.[9,10] In addition, numerous epidemiological studies have demonstrated an inverse correlation between dietary intake or blood levels of vitamin A and/or its precursors, the carotenoids, and cancer risk at several epithelial sites,[12-17] including the cervix.[18-21] Retinoids also may be effective in treating premalignant lesions such as cervical intraepithelial neoplasia (CIN).[22-27] In fact, phase I and phase II clinical trials of all-*trans*-retinoic acid (RA) for CIN have shown that RA is capable of reversing cervical dysplasia in some patients, indicating that retinoids may be useful in the prevention and treatment of cervical malignancies.[22-27]

Kim E. Creek, G. Ronald Jenkins, Ayse Batova, and John R. Hodam • Department of Chemistry and Biochemistry, University of South Carolina, Columbia, SC 29208; Mohammad A. Khan, William H. Tolleson, and Lucia Pirisi • Department of Pathology, University of South Carolina School of Medicine, Columbia, SC 29208

Diet and Cancer: Markers, Prevention, and Treatment
Edited by M.M Jacobs, Plenum Press, New York, 1994

The finding that greater than 90% of all cervical cancers harbor human papillomavirus (HPV), especially HPV types 16, 18, and 33, suggests an etiologic role of this virus in the genesis of cervical cancer.[28-29] In addition, we,[30-31] as well as others,[32-35] have shown that DNA from these HPV types immortalize human keratinocytes (HKc) and human cervical cells in culture. However, other HPV types such as 1, 6, and 11, which are responsible for benign proliferative lesions but are not associated with malignancies, will not produce immortalized HKc lines.[33,36] Immortalization of normal HKc by HPVs is directly due to the presence of viral sequences, which are maintained as multiple copies per genome, integrated into the cellular DNA, and expressed as specific viral messages.[30,31,33] The major messages produced by both HPV-immortalized human epithelial cells *in vitro* and human cervical cancer cells are encoded by two open reading frames (ORFs) of the early region of the viral genome, E6 and E7.[31,37-39] While the precise function of the products of these genes is not completely understood, the constant expression of these two ORFs is necessary to maintain continuous growth of HPV16/EJ-*ras*-transformed rat fibroblasts.[40] In addition, plasmids containing only E6 and E7 are sufficient to immortalize HKc *in vitro*, and both ORFs seem to be required for efficient immortalization, although in rare cases E7 alone is sufficient.[41-44] E7 shows high homology to the adenovirus E1A gene, in those portions that are important for the transforming ability of this gene.[45] Both HPV16 E7 and the adenovirus E1A proteins are able to bind to the retinoblastoma gene product,[46-48] and HPV16 E6 binds to the product of another tumor suppressor gene, p53.[49] It is therefore the current view that E6 and E7 are the oncogenes of HPV16 and that the constant expression of these proteins is necessary to maintain continuous growth in cells transformed by HPV16.

We have used a model system of normal HKc and several independently derived immortalized cell lines (HKc/HPV16), obtained by transfection of different normal HKc strains with HPV16 DNA,[30,31] to investigate the control by RA of cellular growth and HPV16 gene expression. In this chapter, we demonstrate that HKc/HPV16 lines are more sensitive than normal HKc to growth inhibition by RA. Furthermore, RA treatment of HKc/HPV16 dramatically inhibits the expression of the HPV16 oncogenes E6 and E7 as well as other HPV16 early messages (E2 and E5). In addition, physiological concentrations of RA (1 nM) inhibited by almost 95% HVP16-mediated immortalization of HKc. Overall these results suggest that endogenous RA may serve as a natural inhibitor of proliferation of cells harboring HPV, possibly through an inhibition by RA of the expression of the HPV oncogenes, and provide further biochemical evidence in support of a role of dietary vitamin A or its precursor, β-carotene, in the chemoprevention of HPV-induced cancers.

II. Materials and Methods

A. Materials

RA was from Eastman Kodak Co., and the β-carotene was from Sigma Chemical Co. [11,12(n)-^{3}H]All-*trans*-retinol (49.8 Ci/mmol) and [11,12(n)-^{3}H]All-*trans*-RA (55.7 Ci/mmol) were from New England Nuclear. Cesium trifluoroacetate was from Pharmacia LKB Biotechnology, Inc.

B. Cell Culture and Cell Lines

Normal HKc were isolated from newborn foreskins as described previously,[30,31] except the epidermis was separated from the dermis by incubation overnight at 4°C in 0.25% trypsin (Gibco/BRL) instead of collagenase. Isolation and characterization of the immortalized HKc/HPV16 lines has been described in detail in previous publications.[30,31] These cell lines were obtained by transfecting normal HKc strains, each derived from a different individual, with the plasmid pMHPV16d: this is a head-to-tail dimer of the full length HPV16 DNA cloned into the *Bam*H1 site of the vector pdMMTneo, which carries a gene for resistance to the antibiotic G418.[30] The different immortalized lines were selected with G418 and were designated HKc/HPV16d–1 to -5.

Both normal HKc and HKc/HPV16 were cultured in serum-free MCDB153-LB medium, supplemented with hydrocortisone (0.2 μM), insulin (5 μg/ml), transferrin (10 μg/ml), triiodothyronine (10 nM), $CaCl_2$ (0.1 mM), epidermal growth factor (5 ng/ml), and bovine pituitary extract (35-50 μg protein/ml) with medium changes every 48 hr.[30,31] This medium will be referred to as complete MCDB153-LB medium.

C. Clonal Growth Assay

Normal HKc or HKc/HPV16d–1 were plated at 1000 cells/60-mm culture dish and the following day re-fed with complete MCDB153-LB medium containing various concentrations of RA. Retinoic acid was added to the medium in dimethyl sulfoxide (DMSO) and the controls contained DMSO only. The final DMSO concentration was 0.1%. The cells were re-fed and treated with RA (as above) 6 days after plating and stained with Giemsa 5 days later. In this assay, since the cells are allowed to attach before the initiation of RA treatment, plating efficiency is not affected by RA, and the area of the colonies is a direct measure of the extent of growth. The total area of the colonies relative to the area of the dish was determined using an image analysis system consisting of a SIT video camera and an IBM AT computer equipped with a DataCube IVG-128 frame grabber.

D. Mass Culture Growth

Normal HKc or HKc/HPV16 were plated at 20,000 cells/35-mm culture dish and re-fed 24 hr after plating in complete MCDB153-LB medium containing DMSO only, or retinoids. Triplicate dishes for each experimental condition were trypsinized and the cell numbers were determined by counting in a hemocytometer.

E. Uptake of [^{3}H]Retinol and [^{3}H]Retinoic Acid by Normal HKc and HKc/HPV16

Procedures for the uptake of [^{3}H]retinol and [^{3}H]RA by normal HKc and HKc/HPV16d–1 were similar to those described previously for the uptake of [^{3}H]retinol by mouse keratinocytes.[50] Briefly, normal HKc or HKc/HPV16d–1 were cultured in 35-mm dishes to a density of about 1×10^6 cells/dish in complete MCDB153-LB medium. Complete MCDB153-LB medium containing 1 mg/ml bovine serum albumin (BSA) (1 ml) and either [^{3}H]retinol (0.16 μCi/dish) or [^{3}H]RA (0.16 μCi/dish) was then added to the cells. The cells were incubated at 37°C for various times (1 to 24 hr) and placed on ice. The cells were then washed twice with 2 ml of ice-cold Dulbecco's phosphate-buffered saline (PBS) containing 1 mg/ml BSA followed by two additional washes with 2 ml of ice-cold PBS. Phosphate-buffered saline (1 ml) was then added, and the cells were lysed by freezing and thawing. Radioactivity was measured in an aliquot (0.8-0.9 ml) of the cell lysate in 4 ml of ScintiVerse II (Fisher Scientific).

F. Northern Blot Analysis

HKc/HPV16 were plated in 150-mm dishes in complete MCDB153-LB medium without RA. When cells reached about 40% confluence, groups of 5 dishes/condition were fed complete MCDB153-LB medium containing 10^{-7}M RA (added in DMSO, final DMSO concentration of 0.1%) or with complete MCDB153-LB medium without RA (containing 0.1% DMSO). Media were replaced daily, with the addition of fresh RA. The RNA was extracted after 72 hr of RA treatment according to the guanidine thiocyanate-cesium trifluoroacetate technique, modified as described by Yasumoto *et al.*[51] Gel electrophoresis of total RNA was performed in 1.2% agarose-2.2 M formaldehyde gels. The RNA was transferred to a Gene-Screen filter membrane (New England Nuclear) by capillary transfer and hybridized with polymerase chain reaction (PCR)-generated probes specific for HPV16 E6, E7, E2, and E5 (see below). A human β_2-microglobulin cDNA probe, labeled by random priming (Promega), was used as control for RNA loading. The hybridization buffer consisted of: 5X SSPE (1X SSPE is 0.15 M NaCl-10 mM NaH_2PO_4, pH 7.4-1 mM EDTA)-50% formamide-10% dextran sulfate-5X Denhardt's solution-1% sodium dodecyl sulfate (SDS)-100 µg/ml yeast tRNA; hybridization was conducted at 42°C for 16 hr. The filters were washed twice in 2X standard saline citrate (SSC)-0.5% SDS at room temperature for 30 min, and once with 0.1X SSC-0.1% SDS at 42°C for 30 min. Filters were exposed to Kodak XAR 2 films with an enhancing screen at –70°C.

G. Probes

The ORF-specific probes were generated and labeled by PCR. Primers identical to nucleotides 50-73 and complementary to positions 602-625 of the HPV16 genome were used to amplify a 576 base-pair fragment corresponding to the full-length E6 ORF. Primers identical to nucleotides 653-676 and complementary to nucleotides 823-846 were used to generate a 194 base-pair probe specific for E7. The primers used to produce a probe for E2 were identical to nucleotides 2844-2867 and complementary to nucleotides 3105-3128. The E2-specific probe generated by these primers was 285 nucleotides long. For E5, the sense primer was derived from bases 3849-3872 and the antisense primer was derived from bases 4071-4094 such that a 246 base pair PCR product was formed.

Each PCR reaction mixture (100 µl) contained 0.02 nmol of each primer, 0.25 nmol of dATP, dGTP, and dTTP, 50 µCi of [α–^{32}P]dCTP (3000 Ci/mmol; Amersham), 1 unit of Taq DNA polymerase, and 4.0 ng of *Bam*HI-linearized pMHPV16d template DNA. Amplification was carried out in a Coy TempCycler for 32 cycles, (denaturing at 94°C for 1 + 2 min, annealing at 55°C for 2 + 2 min, elongating at 72°C for 1 + 1 min). The final product was separated from unincorporated nucleotides by chromatography on Sephadex G50, denatured by boiling, and added to the hybridization mix. Typical PCR labeling reactions produced 10-25 ng PCR product at 3.5-4.5 × 10^9 cpm/µg of DNA.

H. Analysis of E6 and E7 Protein Expression

HKc/HPV16 were plated into 35-mm dishes and cultured to 50% confluence. Groups of three dishes per condition were then treated with 10^{-7} M RA for 72 hr. The dishes were washed with PBS and the cells were fixed and permeabilized with methanol at –20°C for 30 min. The dishes were treated with 2% BSA in PBS for 30 min, and then incubated for 30 min with 1 µg/plate of an anti-HPV16 E7 monoclonal antibody (Triton Diagnostics), an anti-HPV16/18 E6 monoclonal antibody (Oncogene Science), or an anti-human ß-actin monoclonal antibody (Amersham) in 1 ml of PBS containing 0.2% BSA. Plates were then washed three times with PBS containing 0.2% BSA and incubated for 30 min with 10 µg/dish of a fluorescein-labeled goat anti-mouse IgG (heavy and light chain) (Research Plus, Inc.). Cells were then washed extensively with PBS containing 0.2% BSA. The fluorescence intensity per cell was quantified with an adherent cell analysis and sorting system (ACAS), consisting of a Model 570 Interactive Laser Cytometer (Meridian Instruments, Inc.). Scanning was performed in triplicate plates, using three randomly selected areas per plate. A total of 300-1000 cells per experimental condition were analyzed using an excitation wavelength of 488 nm.

I. Western Blot Analysis of HPV16 E7

HKc/HPV16 were placed into 100-mm dishes, cultured to about 50% confluence, and then treated for 72 hr with the indicated concentrations of RA. Cells were then lysed in 125 mM Tris, 2% SDS, 10% glycerol, pH 6.8. Protein concentration in the lysates was determined by the BCA protein assay (Pierce) in 96-well microtiter plates using BSA as the protein standard (Pierce). Proteins in the cell lysate (150 µg protein/lane) were separated on a denaturing polyacrylamide gel (5% stacking gel, 15% separating gel) as described by Laemmli[52] and electrophoretically transferred to polyvinyldifluoride membranes in carbonate buffer as described by Dunn.[53] The blots were probed using the Photoblot Chemiluminescence Kit (Gibco/BRL) with an HPV16 E7-specific monoclonal antibody (Triton Diagnostics) at 1:250 dilution in Photoblot blocking buffer.

J. Transformation Assay

Equal numbers of cells (25,000/dish) from second passage cultures of individual normal HKc strains, were plated into 35-mm dishes. Triplicate 35-mm dishes were transfected overnight by lipofectin (Gibco/BRL). Transfection mixtures contained 5 µg/plate of either pMHPV16d or control (calf thymus) DNA, in a total volume of 50 µl of transfection mixture/dish (25 µl of lipofectin reagent/dish). Cultures were split 24 hr after transfection into quadruplicate 100-mm dishes, and then cultured without further passaging, in the presence or in the absence of 1 nM RA, until calf thymus DNA-transfected cells had senesced. Cells immortalized by HPV16 DNA form rapidly growing colonies. When negative control cells (those transfected with calf thymus DNA) had senesced, dishes of HPV16-transfected cells were stained with Giemsa as colonies reached 4-8 mm in diameter (20 to 30 days after plating). The number of colonies per plate was determined and taken as a measure of the efficiency of immortalization, since in our experience each colony at this stage represents an immortalized clone. An outline of the transformation assay is presented in Figure 1.

III. Results

A. Effect of Retinoic Acid on Clonal and Mass Culture Growth of Normal HKc and HKc/HPV16

The influence of increasing concentrations of RA on the growth of normal HKc and HKc/HPV16d–1 was determined using a clonal growth assay. In this procedure, cells are

Flow Chart of Immortalization Assay

Day 1. Plate 25,000 cells/dish into 35 mm dishes (6 well plates).

Day 2. Transfect with:

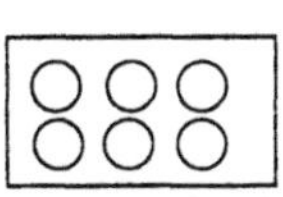

HPV16 DNA

Control (calf thymus) DNA

Day 3. Split cells from each 35 mm well into 4 100 mm dishes. Feed 2 dishes with RA (1 nM) and 2 without RA.

Incubate at 37° C, with media changes every 48h, until control cells senesce and colonies are visible in HPV16-transfected controls.

About day 30. Stain with Giemsa and count colonies in RA+ and RA- dishes.

Figure 1. Schematic of transformation assay.

plated at low density and allowed to attach before the initiation of RA treatment, thus, plating efficiency is not affected by RA, and the area of the colonies is a direct measure of the extent of growth. The clonal growth assay was quantified using an image analysis system that determines the area of the culture dish occupied by colonies, relative to the total area of the culture dish. As shown in Figure 2, HKc/HPV16d–1 showed approximately 25% inhibition of clonal growth at 10^{-9} M RA, and growth was completely inhibited at 10^{-6} M RA. In contrast, inhibition of clonal growth of normal HKc required 10^{-6} M RA, 1000-fold more RA than required for similar inhibition of HKc/HPV16d–1 (Figure 2).

We next compared the influence of RA on the mass culture growth of normal HKc and four HKc/HPV16 lines, independently derived from different individuals.[31] In the mass culture assays the cells were plated at high density and cell numbers were determined five

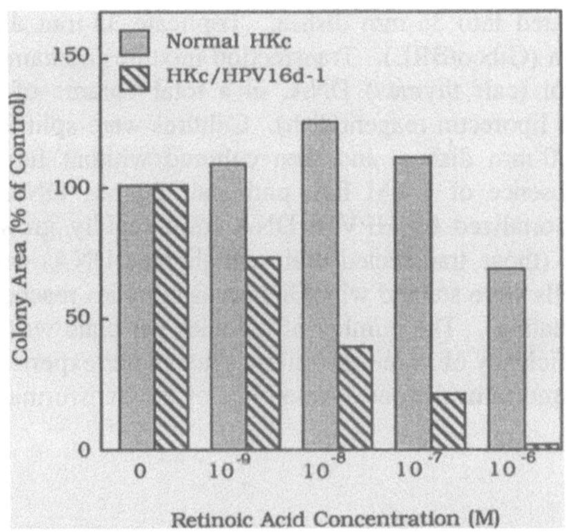

Figure 2. Colony area of normal HKc and HKc/HPV16d-1 treated with RA. Normal HKc and HKc/HPV16d-1 were plated at 1000 cells/60-mm culture dish and re-fed the following day with complete MCDB153-LB medium (8ml) containing the indicated concentrations of RA. The cells were re-fed as above 6 days after plating and stained with Giemsa 5 days later. The total area of the colonies relative to the surface area of the culture dish was determined by an image analysis system. The area occupied by the colonies cultured in the absence of RA is normalized to 100%. Reproduced with permission from Pirisi *et al.*[54]

days after initiation of RA treatment. Similar to the results in the clonal growth assay with the HKc/HPV16d–1 line, all of the HKc/HPV16 lines analyzed were more sensitive than normal HKc to inhibition of mass culture growth by both 10^{-7} and 10^{-6} M RA (Table 1). In fact, mass culture growth of normal HKc was not inhibited by 10^{-7} M RA, while growth inhibition of HKc/HPV16 averaged 40% at this RA concentration (Table 1). The effect of a wider range of concentrations of RA (10^{-9} to 10^{-6} M) on the mass culture growth of normal HKc, HKc/HPV16d–1 and HKc/HPV16d–2 was also investigated. Mass culture growth of these HKc/HPV16 lines was significantly inhibited by 10^{-9} M RA. In contrast, significant growth inhibition of normal HKc required 10^{-6} M RA (Figure 3).

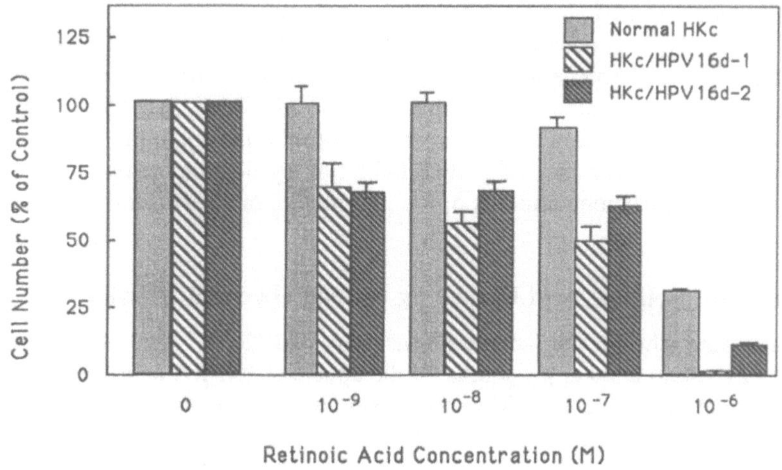

Figure 3. Effect of various RA concentrations on mass culture growth of normal HKc, HKc/HPV16d-1, and HKc/HPV16d-2. Keratinocytes were plated at a density of 20,000 cells/35-mm culture dish. Cells were re-fed 24 hr after plating and every 48 hr thereafter with complete MCDB153-LB medium containing the indicated concentrations of RA. Cells (3 dishes per experimental group) were trypsinized and cell numbers were determined by counting in a hemocytometer 6 days after plating. The cell number in dishes cultured in the absence of RA is normalized to 100%. Bars represent SEM.

Table 1. Increased Sensitivity of HKc/HPV16 Lines to Inhibition of Mass Culture Growth by Retinoic Acid[a]

		Growth inhibition (%)[b,c]	
Cells	RA (µM)	0.1	1.0
Normal HKc		-7.0[c] ± 5.4[d] (15)	57.7 ± 3.7 (10)
HKc/HPV16d-1		51.0 ± 3.5 (6)	92.6 ± 2.2 (8)
HKc/HPV16d-2		33.0 ± 3.5 (6)	88.9 ± 0.2 (3)
HKc/HPV16d-3		50.4 ± 5.4 (3)	84.8 ± 1.8 (3)
HKc/HPV16d-5		24.9 ± 4.3 (8)	86.1 ± 1.0 (5)

[a] Normal HKc and the various HKc/HPV16 lines were plated at a density of 20,000 cells/35-mm dish. Cells were re-fed 24 hr after plating and every 48 hr thereafter with complete MCDB153-LB media containing the indicated concentration of RA. Cells were trypsinized and cell numbers were determined by counting in a hemocytometer 6 days after plating. (Adapted from Pirisi *et al.*[54])

[b] $\% \text{ Growth inihibition} = 100 - \left(\frac{\text{cell number in RA-treated dishes}}{\text{average cell number in DMSO-treated controls}} \times 100 \right)$

[c] Data are means ± SEM. Numbers in parentheses indicate the number of determinations.

[d] A negative sign indicates growth stimulation.

We also investigated the growth inhibitory effect of retinol and ß-carotene in the mass culture growth assay to determine if HKc/HPV16 were more sensitive than normal HKc to these precursors of RA. Like RA, retinol was found to be a more potent inhibitor of growth of HKc/HPV16 than of normal HKc (Table 2). Retinol at 10^{-6} M inhibited the growth of normal HKc by about 15%, but inhibited on the average the growth of HKc/HPV16 by 47% (Table 2). Retinol at a concentration of 10^{-5} M was found to be toxic to both normal HKc and HKc/HPV16. ß-Carotene was found not to be growth inhibitory to either normal HKc or HKc/HPV16 at concentrations as high as 10^{-6} M (Table 2). The testing of higher concentrations of ß-carotene was problematic, due to limited solubility of this compound in the aqueous medium. The lack of growth inhibition by ß-carotene may reflect an inability of HKc to take up sufficient amounts of the provitamin, or to efficiently convert ß-carotene to retinol.

B. Retinol and Retinoic Acid Uptake by Normal HKc and HKc/HPV16

To determine whether the increased sensitivity of HKc/HPV16 to RA might be related to the rate of retinoid uptake, we compared the uptake of [^{3}H]RA and [^{3}H]retinol by normal HKc and HKc/HPV16d–1. No difference was found in the overall rate of uptake, or in the total cell-associated [^{3}H]retinol or [^{3}H]RA between normal HKc and HKc/HPV16d–1 (Figure 4). This observation indicates that the increased sensitivity of HKc/HPV16 to RA and retinol is not due to increased retinoid accumulation in HKc/HPV16.

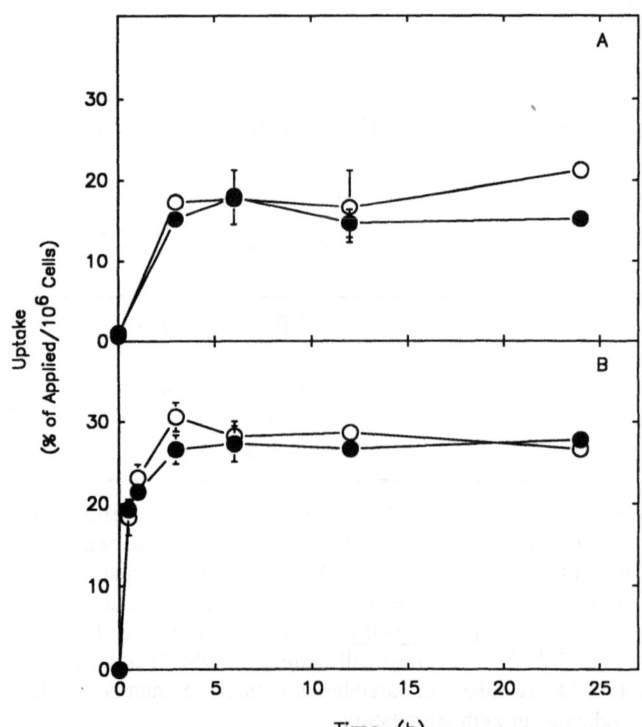

Figure 4. RA and retinol uptake by normal HKc and HKc/HPV16d-1. Confluent cultures of normal HKc (●) or HKc/HPV16d-1 (○) were fed with complete MCDB153-LB medium containing 1 mg/ml BSA and [^{3}H]RA (A) or [^{3}H]retinol (B). The dishes were incubated at 37° C for the indicated times and cell-associated radioactivity determined as described in "Materials and Methods."

Table 2. Effect of Retinol and β-Carotene on Mass Culture Growth of Normal HKc and HKc/HPV16 Lines[a]

Cells	Growth inhibition (%)[b,c]	
	Retinol	β-Carotene[d]
Normal HKc	14.7 ± 4.2	-3.9 ± 5.7[e]
HKc/HPV16d-2	ND[f]	0.4 ± 5.6
HKc/HPV16d-3	53.6 ± 1.9	ND
HKc/HPV16d-5	40.6 ± 4.3	5.7 ± 2.8

[a] Normal HKc and the various HKc/HPV16 lines were plated at a density of 20,000 cells/35-mm dish. Cells were re-fed 24 hr after plating and every 48 hr thereafter with complete MCDB153-LB medium containing 1 μM retinol or 1 μM β-carotene. Cells were trypsinized and cells numbers were determined by counting in a hemocytometer 6 days after plating. (Adapted from Pirisi *et al.*[54])
[b] % Growth inhibition was calculated as described in Table 1.
[c] Data are means ± SEM of at least triplicate determinations.
[d] A stock solution of 1mM β-carotene was prepared in butanol and then diluted 1000-fold in complete MCDB153-LB medium.
[e] A negative sign indicates growth stimulation.
[f] ND = not determined.

C. Inhibition of HPV16 Early Gene Expression by Retinoic Acid

We next sought to determine if RA treatment of HKc/HPV16 inhibited the expression of the HPV16 oncogenes E6 and E7, as well as other early ORFs of HPV16. Using PCR-generated probes specific for the E6, E7, E2, and E5 ORFs of HPV16, we conducted Northern blot analysis on total mRNA isolated from HKc/HPV16 cultured for 72 hr in the absence or the presence of 10^{-7} M RA. As shown in Figure 5, RA treatment dramatically inhibited the expression of the E6 and E7 oncogenes as well as the E2 and E5 ORFs. Retinoic acid inhibited the expression of the two major RNA species (4.2 and 1.8 kb), which we detected with all four early ORF-specific probes, to a similar extent (Figure 5). We have also shown that the inhibition of E6 and E7 expression by RA is both dose- and time-dependent.[54]

D. Retinoic Acid Treatment of HKc/HPV16 Reduces E6 and E7 Protein Levels

We next determined the effect of RA treatment of HKc/HPV16 on the cellular levels of the E6 and E7 proteins. HKc/HPV16 were cultured for 72 hr in the absence or the presence of 10^{-7} M RA, and E6 and E7 protein levels were detected by indirect immunofluorescence using monoclonal antibodies specific for HPV16/HPV18 E6 and for HPV16 E7. Average fluorescence per cell was quantified by ACAS analysis as described in "Materials and Methods." As shown in Figure 6, RA treatment of HKc/HPV16 substantially reduced E6 and E7 protein levels. The antibodies used in these assays did not produce any specific fluorescent signal in normal (HPV-negative) HKc as determined by ACAS analysis (data not shown). In addition, a similar ACAS analysis using an anti-ß-actin monoclonal antibody showed no difference between RA-treated and -untreated cells, indicating that RA treatment of HKc/HPV16 did not produce a general nonspecific depression of protein synthesis (Figure 6).

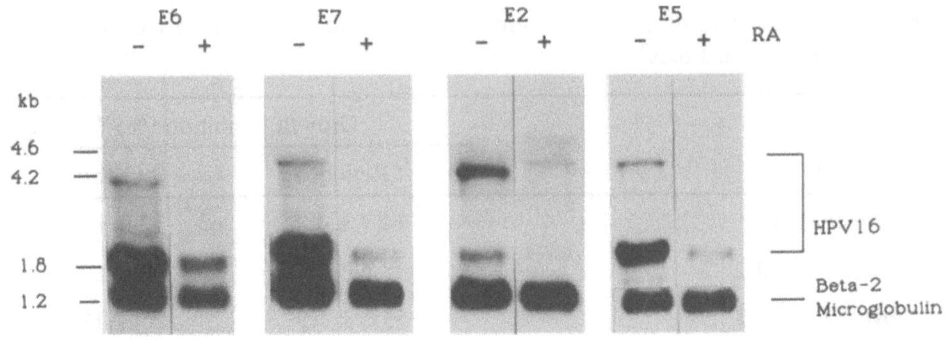

Figure 5. Inhibition by RA of HPV16 early gene expression. Northern blot analysis was conducted on total RNA (30 µg/lane) extracted from HKc/HPV16 cultured for 72 hr in the absence or the presence of 10^{-7} M RA. PCR-generated probes specific for the E6, E7, E5, and E2 ORFs were used. The blotting and hybridization conditions are described in "Materials and Methods." A β2-microglobulin cDNA probe was used as a control for RNA loading. The molecular size of the bands (kb) was determined based on RNA molecular size markers (BRL).

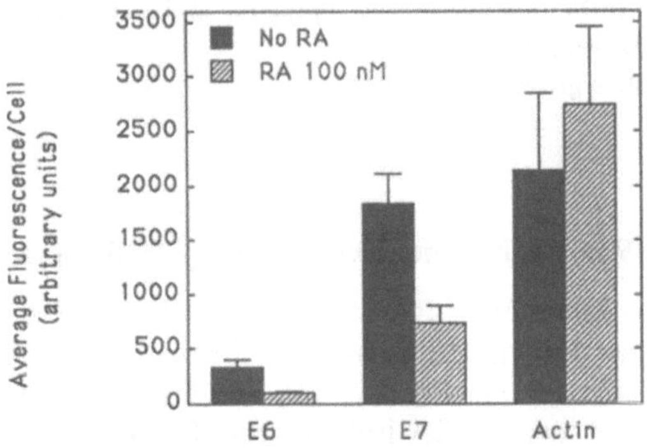

Figure 6. Inhibition by RA of E6 and E7 protein expression. HKc/HPV16 plated in triplicate 35-mm dishes/condition were treated with 10^{-7} M RA for 72 hr, then washed and fixed in methanol. Immunofluorescence and ACAS analysis were conducted as described in "Materials and Methods," using monoclonal anti-HPV16 E7, anti-HPV16/18 E6, and anti-human β-actin antibodies. Bars represent SD.

The ACAS results were confirmed by Western blot analysis with the anti-E7 monoclonal antibody, on total protein extracts from HKc/HPV16 treated with various concentrations of RA for 72 hr. Western blot analysis demonstrated a dose-dependent decrease of E7 protein levels in HKc/HPV16 treated with RA (Figure 7). As expected, the anti-E7 antibody did not detect E7 in protein extracts from normal HKc (Figure 7).

E. Retinoic Acid Inhibits HPV16-Induced Immortalization of Normal HKc

We developed a transformation assay to determine if the reduction by RA of HPV16 early gene expression would result in an inhibition of HPV16-mediated transformation of normal HKc. An outline of the transformation protocol used in these studies is shown in Figure 1. Normal HKc strains, each derived from a single individual, were transfected with HPV16 DNA. Following transfection, the cells were passed and maintained in the absence

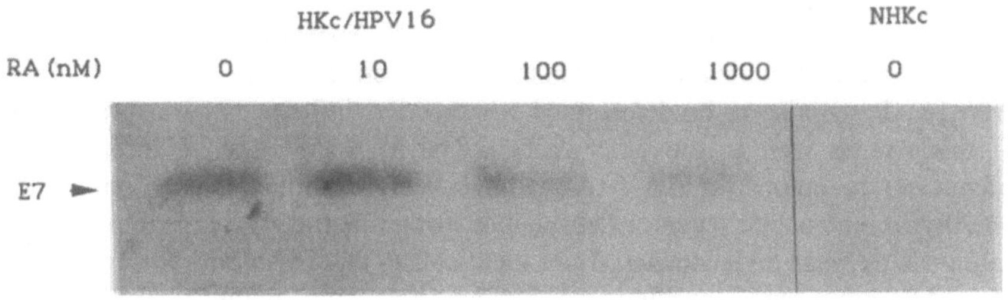

Figure 7. Western blot analysis of E7 in RA-treated HKc/HPV16. Protein extracts were prepared from normal HKc (NHKc) and from HKc/HPV16 treated with the indicated concentrations of RA for 72 hr. Western blot analysis was conducted as described in "Materials and Methods," using an anti-HPV16 E7 antibody.

or in the presence of 10^{-9} M RA. This concentration of RA was selected following preliminary studies that indicated it would not negatively affect the growth rate or lifespan of normal HKc following long-term exposure. In addition, this RA concentration only slightly inhibits clonal growth of established HKc/HPV16 lines. The cells were maintained in the presence or in the absence of RA until negative controls (transfected with calf thymus DNA) had senesced. Cells were then stained with Giemsa and immortalized colonies in RA-treated and control dishes were counted. As shown in Tables 3, RA dramatically inhibited (by almost 95%) HPV16-mediated immortalization of normal HKc.

Table 3. RA Inhibition of HPV16-Mediated Transformation of Normal HKc[a]

RA (nM)	Immortalized colonies/dish	% Inhibition
0	118 ± 14.1	–
1	6.6 ± 1.5	94.4

[a] Normal HKc were plated in 35-mm dishes, transfected with pMHPV16d by lipofection, passaged 24 hr after transfection, and the effect of RA on the development of HPV16-induced immortalized colonies was determined, as described in detail in "Materials and Methods." The data in this table are averages ± SD of triplicate determinations from a single normal HKc strain. The experiment has been repeated using three different normal HKc strains, with similar results.

IV. Discussion

We have used a model system of normal and HPV16-immortalized HKc to investigate the biochemical basis by which retinoids may act as chemopreventive agents of HPV-induced lesions. The results summarized in this chapter demonstrate the following four important points: (1) that HKc/HPV16 are more sensitive than normal HKc to growth control by RA and retinol (vitamin A); (2) that RA treatment of HKc/HPV16 reduces the expression of mRNA for the viral oncogenes E6 and E7, as well as other early ORFs including E2 and E5; (3) that RA treatment of HKc/HPV16 reduces cellular levels of E6 and E7 protein; and (4) that a physiological concentration of RA (1 nM) suppresses (about 95%) HPV-mediated immortalization of normal HKc.

The constant expression of the E6 and E7 ORFs is required for the maintenance of continuous growth of cells transformed by HPV16.[40] Therefore, the modulation of E6/E7 expression by RA provides a mechanism by which RA can control proliferation of HKc/HPV16. It remains to be determined whether RA influences the expression of the E6 and E7 mRNAs as well as the other early ORFs by a direct or an indirect mechanism. Several nuclear receptors for RA (RARs) have been identified that belong to the steroid hormone/thyroxine nuclear receptor family, indicating that the mechanism by which retinoids modulate gene expression is similar, if not identical, to that of steroid hormones.[55–59] Three different RARs have been identified (RAR-alpha, -beta, and -gamma), all of which bind RA with high affinity,[55–64] as well as three retinoid X receptors (RXRs) (RXR-alpha, -beta, -gamma) that bind 9-*cis*-retinoic acid.[58] The RARs and RXRs modulate gene expression by binding to specific hormone-responsive elements (HREs) on target genes. Recently, several genes that are modulated by RA have been identified and the DNA sequences responsible for RA modulation have been determined.[58,59] The upstream regulatory region (URR) of HPV16 carries the promoter and the regulatory elements responsible for controlling the expression of the E6 and E7 oncogenes as well as the other early ORFs. Although the URR contains a steroid HRE and nuclear factor 1 and AP1 sites,[65,66] no obvious RA-responsive elements (RAREs) are located in the URR of HPV16. However, repression of AP1-regulated promoters by RARs has been reported.[67] In addition, a recent report by Bartsch *et al.*,[68] clearly showed that RA can regulate HPV18 E6/E7 expression at the transcriptional level in HeLa cell hybrids, and that RA inhibits chloramphenicol acetyl transferase activity expressed under the control of the HPV16URR. The fact that the expression of all the early ORFs is inhibited by RA to more or less the same extent suggests that modulation of the early ORFs likely occurs through the URR. However, further studies are clearly needed to determine whether RA inhibits HPV16 early gene expression directly, or by modulating the levels of other transcription factors that control the activity of the viral P97 promoter. Clearly though, whatever the mechanism by which RA decreases the levels of E6 and E7 mRNA, these studies have demonstrated that the protein levels of these HPV16 oncogenes are reduced to a similar extent following RA treatment of HKc/HPV16.

In addition to retinoids modulating growth of HKc/HPV16 by directly controlling the expression of HPV16 genes through the URR, more indirect mechanisms may also exist. We have recently found that RA induces the secretion of latent transforming growth factor-beta (TGF-ß) in normal HKc and HKc/HPV16,[69] and Glick *et al.*[70] reported that RA induced the secretion of TGF-ß in mouse keratinocytes. Since TFG-ß is a potent inhibitor of keratinocyte proliferation,[71–73] RA-induced secretion of TGF-ß would result in growth inhibition once TGF-ß has been activated at the cell surface.[74] It recently has been shown that TGF-ß markedly inhibits the expression of HPV16 E6 and E7 in HPV16-immortalized human genital epithelial cells.[75] Therefore, an autocrine loop may exist whereby retinoids induce the secretion of TGF-ß, which in turn inhibits growth, in part through an inhibition of E6 and E7 expression in HKc/HPV16. It is not yet clear whether this mechanism could explain the increased sensitivity of the HKc/HPV16 lines to retinoids. Some HPV16-immortalized HKc lines have been found to be resistant to TGF-ß[76] but the HKc/HPV16 lines used in these studies remain sensitive to TGF-ß.[69] However, we have observed a decreased sensitivity to the antiproliferative effect of TGF-ß at late passages in HKc/HPV16d–1, and we are currently investigating if this resistance to TGF-ß is associated with a resistance to RA.

Perhaps the most dramatic result of these studies was the finding that a physiological concentration of RA nearly suppresses HPV16-mediated immortalization of HKc (Table 3). The results of the immortalization assay could be affected, but not completely explained, by a simple inhibition of growth of the transformed cells by RA. The RA concentration used in the immortalization assay (1 nM) inhibits to a certain extent clonal growth of HKc/HPV16 (Figure 2) but still allows for colony development. Therefore, we expected to obtain smaller colonies in the transformation assay in the RA-treated dishes. However the *number* of immortalized colonies was drastically reduced by RA treatment, suggesting an effect of RA on transformation itself and not only on the proliferation of the transformed cells. In addition to the effects of RA on cellular proliferation, it also profoundly influences cellular differentiation and keratin expression as well as other aspects of cellular gene expression.[54,77] Thus RA may act at various levels in the host cell, to inhibit HPV16-mediated immortalization. Whatever the mechanism, this marked effect of physiological levels of RA on HPV16-mediated transformation shows promise for a future use of RA in the chemoprevention and/or treatment of HPV-induced lesions.

In conclusion, although the exact mechanism(s) by which RA dramatically inhibits the growth of HPV16-immortalized HKc, the expression of the HPV16 oncogenes in HKc/HPV16, and HPV16-mediated immortalization of normal HKc remain to be determined, these findings provide a direct biochemical basis for the epidemiological evidence that links a diet poor in vitamin A or its precursor, ß-carotene, to an increased risk of cervical cancer. Overall, we feel that these results constitute a solid support for a role of dietary retinoids (carotenoids) in the chemoprevention of HPV-induced lesions such as cervical dysplasia and, possibly, cervical cancer, and provide the basis for additional studies on the use of retinoids in the prevention and early treatment of these lesions.

V. Summary

We have used a model system of normal HKc and HKc immortalized by transfection with HPV16 DNA (HKc/HPV16) to investigate the effect of RA on the growth of HKc/HPV16 and the expression of the HPV16 oncogenes E6 and E7. These studies found that HKc/HPV16 are about 100-fold more sensitive than normal HKc to growth inhibition by RA in both clonal and mass culture growth assays. The precursor to RA, retinol, was also found to be a more potent inhibitor of growth of HKc/HPV16 than normal HKc while ß-carotene did not inhibit growth of either normal HKc or HKc/HPV16. No differences were observed in the rate of uptake of $[^3H]$RA or $[^3H]$retinol between normal HKc and HKc/HPV16. Northern blot analysis of mRNA extracted from HKc/IIPV16 cultured in the absence or in the presence of 10^{-7} M RA showed that the expression of the HPV16 oncogenes E6 and E7 as well as the early ORFs E2 and E5 is substantially reduced following RA treatment. In addition, protein levels of E6 and E7, as measured by immunofluorescence (E6 and E7) and Western blot (E7) are also decreased by RA treatment of HKc/HPV16. Since E6 and E7 are considered the oncogenes of HPV16, we explored the possibility that RA may interfere with HPV16-mediated immortalization of HKc. The RA treatment (1 nM) of normal HKc, during or immediately following transfection with HPV16 DNA, inhibited immortalization by about 95%. Overall, these results provide a direct biochemical basis for a role of dietary retinoids in the chemoprevention of HPV-induced cancers.

Acknowledgments

This work was supported by Grants 88A05 and 91B38 from the American Institute for Cancer Research. Kim E. Creek is the recipient of a Junior Faculty Research Award from the American Cancer Society.

References

1. R. Doll, R. Peto, The causes of cancer: Quantitative estimates of avoidable risks of cancer in the United States today, *J Natl Cancer Inst* 66:1292 (1981).
2. J.S. Bertram, L.N. Kolonel, F.L. Meyskens, Rationale and strategies for chemoprevention of cancers in humans, *Cancer Res* 47:3012 (1987).
3. R. Peto, R. Doll, J.D. Buckley, M.B. Sporn, Can dietary beta-carotene materially reduce human cancer rates? *Nature* 290:201 (1981).
4. R.C. Moon, L.M. Itri, Retinoids and cancer, *in*: *The Retinoids*, M.B. Sporn, A.B. Roberts, D.S. Goodman, eds., Vol. 2, p. 327, Academic Press, New York (1984).
5. G.L. Peck, Chemoprevention of cancer with retinoids, *Gynecol Oncol* 12:s331 (1981).
6. D.L. Hill, C.L. Grubbs, Retinoids as chemopreventive and anticancer agents in intact animals (review), *Anticancer Res* 2:111 (1982).
7. S.M. Lippman, J.F. Kessler, F.L. Meyskens, Retinoids as preventive and therapeutic anticancer agents (part I), *Cancer Treat Rep* 71:391 (1987).
8. M.B. Sporn, N.M. Dunlop, D.L. Newton, Prevention of chemical carcinogenesis by vitamin A and its synthetic analogs (retinoids), *Fed Proc* 35:1332 (1976).
9. Y. Ito, Effect of an aromatic retinoic acid analog (RO 10-9359) on growth of virus-induced papilloma (Shope) and related neoplasia of rabbits, *Eur J Cancer* 17:35 (1981).
10. H. McMichael, Inhibition of growth of Shope rabbit papilloma by hypervitaminosis A, *Cancer Res* 25:947 (1965).
11. A.K. Verma, B.G. Shapas, H.M. Rice, R.K. Boutwell, Correlation of the inhibition by retinoids of tumor promoter-induced mouse epidermal ornithine decarboxylase activity and of skin tumor promotion, *Cancer Res* 39:419 (1979).
12. E. Bjelke, Dietary vitamin A and human lung cancer, *Int J Cancer* 15:561 (1975).
13. S. Graham, Results of case-control studies of diet and cancer in Buffalo, New York, *Cancer Res* 43:2409s (1983).
14. J.D. Kark, A.H. Smith, B.R. Switzer, C.G. Hames, Serum vitamin A (retinol) and cancer incidence in Evans County, Georgia, *J Natl Cancer Inst* 66:7 (1981).
15. C. Mettlin, S. Graham, M. Swanson, Vitamin A and lung cancer, *J Natl Cancer Inst* 62:1435 (1979).
16. R.B. Shekelle, M. Lepper, S. Liu, C. Maliza, W.J. Raynor, A.H. Rossof, O. Paul, A.M. Shyrock, J. Stamler, Dietary vitamin A and risk of cancer in the Western Electric study, *Lancet* ii:1185 (1981).
17. N. Wald, M. Idle, J. Boreham, A. Bailey, Low serum-vitamin-A and subsequent risk of cancer, *Lancet* ii:813 (1980).
18. A. Bernstein, B. Harris, The relationship of dietary and serum vitamin A to the occurrence of cervical intraepithelial neoplasia in sexually active women, *Am J Obstet Gynecol* 148:309 (1984).
19. T. Hirayama, Diet and cancer, *Nutr Cancer* 1:67 (1979).
20. J.R. Marshall, S. Graham, T. Byers, M. Swanson, J. Brasure, Diet and smoking in the epidemiology of cancer of the cervix, *J Natl Cancer Inst* 70:847 (1983).
21. S.L. Romney, P.R. Palan, C. Duttagupta, S. Wassertheil-Smoller, J. Wylie, G. Miller, N.S. Slagle, D. Lucido, Retinoids and the prevention of cervical dysplasias, *Am J Obstet Gynecol* 141:890 (1981).
22. V. Graham, E.A. Surwit, S. Weiner, F.L. Meyskens, Phase II trial of β-all-*trans*-retinoic acid for cervical intraepithelial neoplasia delivered via a collagen sponge and cervical cap, *West J Med* 145:192 (1986).
23. F.L. Meyskens, V. Graham, M. Chvapil, R.T. Dorr, D.S. Alberts, E.A. Surwit, A phase I trial of β-all-*trans*-retinoic acid delivered via a collagen sponge and a cervical cap for mild or moderate intraepithelial cervical neoplasia, *J Natl Cancer Inst* 71:921 (1983).
24. F.L. Meyskens, E.S. Surwit, Clinical experience with topical tretinoin in the treatment of cervical dysplasia, *J Am Acad Dermatol* 15:826 (1986).
25. Y.M. Peng, D.S. Alberts, V. Graham, E.A. Surwit, S. Weiner, F.L. Meyskens, Cervical tissue uptake of all-*trans*-retinoic acid delivered via a collagen sponge-cervical cap delivery device in patients with cervical dysplasia, *Invest New Drugs* 4:245 (1986).

26. E.A. Surwit, V. Graham, W. Droegemueller, D. Alberts, M. Chvapil, R.T. Dorr, J.R. Davis, F.L. Meyskens, Evaluation of topically applied *trans*-retinoic acid in the treatment of cervical intraepithelial lesions, *Am J Obstet Gynecol* 143:821 (1982).
27. S.A. Weiner, E.A. Surwit, V.E. Graham, F.L. Meyskens, A phase I trial of topically applied *trans*-retinoic acid in cervical dysplasia-clinical efficacy, *Invest New Drugs* 4:241 (1986).
28. H. zur Hausen, Papillomaviruses in human cancers, *Mol Carcinogenesis* 1:147 (1988).
29. H. zur Hausen, Papillomaviruses in anogenital cancer as a model to understand the role of viruses in human cancers, *Cancer Res* 49:4677 (1989).
30. L. Pirisi, S. Yasumoto, M. Feller, J. Doniger, J.A. DiPaolo, Transformation of human fibroblasts and keratinocytes with human papillomavirus type 16 DNA, *J Virol* 61:1061 (1987).
31. L. Pirisi, K.E. Creek, J. Doniger, J.A. DiPaolo, Continuous cell lines with altered growth and differentiation properties originate after transfection of human keratinocytes with human papillomavirus type 16 DNA, *Carcinogenesis* 9:1573 (1988).
32. M. Durst, R.T. Dzarlieva-Petrusevska, P. Boukamp, N.E. Fusenig, L. Gissmann, Molecular and cytogenetic analysis of immortalized human primary keratinocytes obtained after transfection with human papillomavirus type 16 DNA, *Oncogene* 1:251 (1987).
33. C.D. Woodworth, J. Doniger, J.A. DiPaolo, Immortalization of human foreskin keratinocytes by various human papillomavirus DNAs corresponds to their association with cervical carcinoma, *J Virol* 63:159 (1989).
34. P. Kaur, J.K. McDougall, Characterization of primary human keratinocytes transformed by human papillomavirus type 18, *J Virol* 62:1917 (1988).
35. C.D. Woodworth, P.E. Bowden, J. Doniger, L. Pirisi, W. Barnes, W.D. Lancaster, J.A. DiPaolo, Characterization of normal human exocervical epithelial cells immortalized in vitro by papillomavirus types 16 and 18 DNA, *Cancer Res* 48:4620 (1988).
36. R. Schlegel, W.C. Phelps, W.L. Zhang, M. Barbosa, Quantitative keratinocyte assay detects two biological activities of human papillomavirus DNA and identifies viral types associated with cervical carcinoma, *EMBO J* 5:3181 (1988).
37. C.C. Baker, W.C. Phelps, V. Lindgren, M.J. Braun, M.A. Gonda, P.M. Howley, Structural and transcriptional analysis of human papillomavirus type 16 sequences in cervical carcinoma cell lines, *J Virol* 61:962 (1987).
38. E. Schwarz, U.K. Freese, L. Gissmann, W. Mayer, B. Roggenbuck, A. Stremlau, H. zur Hausen, Structure and transcription of human papillomavirus sequences in cervical carcinoma cells, *Nature* 314:111 (1985).
39. C. Yee, I. Krishnan-Hewlett, C.C. Baker, R. Schlegel, P.M. Howley, Presence and expression of human papillomavirus sequences in human cervical carcinoma cell lines, *Am J Pathol* 119:361 (1985).
40. T. Crook, J.P. Morgenstern, L. Crawford, L. Banks, Continued expression of HPV-16 E7 protein is required for maintenance of the transformed phenotype of cells co-transformed by HPV-16 plus EJ-ras, *EMBO J* 8:513 (1989).
41. P. Hawley-Nelson, K.H. Vousden, N.L. Hubbert, D.R. Lowy, J.T. Schiller, HPV16 E6 and E7 proteins cooperate to immortalize human foreskin keratinocytes, *EMBO J* 8:3905 (1989).
42. J.B. Hudson, M.A. Bedell, D.J. McCance, L.A. Laimins, Immortalization and altered differentiation of human keratinocytes in vitro by the E6 and E7 open reading frames of human papillomavirus type 18, *J Virol* 64:519 (1990).
43. M.S. Barbosa, R. Schlegel, The E6 and E7 genes of HPV-18 are sufficient for inducing two-stage in vitro transformation of human keratinocytes, *Oncogene* 4:1529 (1989).
44. C.L. Halbert, G.W. Demers, D.A. Galloway, The E7 gene of human papillomavirus type 16 is sufficient for immortalization of human epithelial cells, *J Virol* 65:473 (1991).
45. W.C. Phelps, C.L. Yee, K. Munger, P.M. Howley, The human papillomavirus type 16 E7 gene encodes transactivation and transformation functions similar to those of adenovirus E1A, *Cell* 53:539 (1988).
46. N. Dyson, P.M. Howley, K. Munger, E. Harlow, The human papilloma virus-16 E7 oncoprotein is able to bind to the retinoblastoma gene product, *Science* 243:934 (1989).

47. K. Munger, B.A. Werness, N. Dyson, W.C. Phelps, E. Harlow, P.M. Howley, Complex formation of human papillomavirus E7 proteins with the retinoblastoma tumor suppressor gene product, *EMBO J* 8:4099 (1989).
48. J.R. Gage, C. Meyers, F.O. Wettstein, The E7 proteins of the nononcogenic human papillomavirus type 6b (HPV-6b) and of the oncogenic HPV-16 differ in retinoblastoma protein binding and other properties, *J Virol* 64:723 (1990).
49. B.A. Werness, A.J. Levine, P.M. Howley, Association of human papillomavirus types 16 and 18 E6 proteins with p53, *Science* 248:76 (1990).
50. K.E. Creek, C.S. Silverman-Jones, L.M. De Luca, Comparison of the uptake and metabolism of retinol delivered to primary mouse keratinocytes either free or bound to rat serum retinol-binding protein, *J Invest Dermatol* 92:283 (1989).
51. S. Yasumoto, J. Doniger, J.A. DiPaolo, Differential early viral gene expression in two stages of human papillomavirus type 16 DNA-induced malignant transformation, *Mol Cell Biol* 7:2165 (1987).
52. U.K. Laemmli, Cleavage of structural proteins during the assembly of the head of bacteriophage T4, Nature 227:680 (1970).
53. S.D. Dunn, Effects of the modification of transfer buffer composition and renaturation of proteins in gels on the recognition of proteins on western blots by monoclonal antibodies, *Anal Biochem* 157:144 (1986).
54. L. Pirisi, A. Batova, G.R. Jenkins, J.R. Hodam, K.E. Creek, Increased sensitivity of human keratinocytes immortalized by human papillomavirus type 16 DNA to growth control by retinoids, *Cancer Res* 52:187 (1992).
55. M. Petkovich, N.J. Brand, A. Krust, P. Chambon, A human retinoic acid receptor which belongs to the family of nuclear receptors, *Nature* 330:444 (1987).
56. V. Giguere, E.S. Ong, P. Segui, R.M. Evans, Identification of a receptor for the morphogen retinoic acid, *Nature* 330:624 (1987).
57. R.M. Evans, The steroid and thyroid hormone receptor superfamily, *Science* 240:889 (1988).
58. L.J. Gudas, Retinoids, retinoid-responsive genes, cell differentiation, and cancer, *Cell Growth Differentiation* 3:655 (1992).
59. L.M. De Luca, Retinoids and their receptors in differentiation, embryogenesis, and neoplasia, *FASEB J* 5:2924 (1991).
60. N. Brand, M. Petkovich, A. Krust, P. Chambon, H. de The, A. Marchio, P. Tiollais, A. Dejean, Identification of a second human retinoic acid receptor, *Nature* 332:850 (1988).
61. D. Benbrook, E. Lernhardt, M. Pfahl, A new retinoic acid receptor identified from a hepatocellular carcinoma, *Nature* 333:669 (1988).
62. A. Krust, P. Kastner, M. Petkovich, A. Zelent, P. Chambon, A third human retinoic acid receptor, hRAR-gamma, *Proc Natl Acad Sci USA* 86:5310 (1989).
63. A. Zelent, A. Krust, M. Petkovich, P. Kastner, P. Chambon, Cloning of murine α and β retinoic acid receptors and a novel receptor gamma predominantly expressed in skin, *Nature* 339:714 (1989).
64. K. Umesono, V. Giguere, C.K. Glass, M.G. Rosenfeld, R.M. Evans, Retinoic acid and thyroid hormone induce gene expression through a common responsive element, *Nature* 336:262 (1988).
65. B. Gloss, H.U. Bernhard, K. Seedorf, G.A. Klock, A DNA segment of the upstream regulatory region of the human papillomavirus-16 exhibits E2 protein-independent enhancer activity in a glucocorticoid hormone-regulated and cell type-specific fashion, *EMBO J* 6:3735 (1987).
66. T. Chong, W.K. Chan, H.U. Bernard, Transcriptional activation of human papillomavirus 16 by nuclear factor I, AP1, steroid receptors and a possibly novel transcription factor, PVF: A model for the composition of genital papillomavirus enhancers, *Nucleic Acids Res* 18:465 (1990).
67. R.C. Nicholson, S. Mader, S. Nagpal, M. Leid, C. Rochette-Egly, P. Chambon, Negative regulation of the rat stromelysin gene promoter by retinoic acid is mediated by an AP1 binding site, *EMBO J* 9:4443 (1990).

68. D. Bartsch, B. Boye, C. Baust, H. zur Hausen, E. Schwarz, Retinoic acid-mediated repression of human papillomavirus 18 transcription and different ligand regulation of the retinoic acid receptor β gene in non-tumorigenic and tumorigenic HeLa hybrids, *EMBO J* 11:2283 (1992).
69. A. Batova, D. Danielpour, L. Pirisi, K.E. Creek, Retinoic acid induces secretion of latent transforming growth factor β1 and β2 in normal and human papillomavirus type 16-immortalized human keratinocytes, *Cell Growth Differentiation* 3:763 (1992).
70. A.B. Glick, K.C. Flanders, D. Danielpour, S.H. Yuspa, M.B. Sporn, Retinoic acid induces transforming growth factor-β2 in cultured keratinocytes and mouse epidermis, *Cell Regulation* 1:87 (1989).
71. G.D. Shiply, M.R. Pittelkow, J.J. Wille, R.E. Scott, H.L. Moses, Reversible inhibition of normal human prokeratinocyte proliferation by type β transforming growth factor-growth inhibitor in serum-free medium, *Cancer Res* 46:2068 (1986).
72. R.J. Coffey, N.J. Sipes, C.C. Bascom, R. Graves-Deal, C.Y. Pennington, B.E. Weissman, H.L. Moses, Growth modulation of mouse keratinocytes by transforming growth factors, *Cancer Res* 48:1596 (1988).
73. M. Reiss, A.C. Sartorelli, Regulation of growth and differentiation of human keratinocytes by type β transforming growth factor and epidermal growth factor, *Cancer Res* 47:6705 (1987).
74. P.A. Dennis, D.B. Rifkin, Cellular activation of latent transforming growth factor β requires binding to the cation-independent mannose 6-phosphate/insulin-like growth factor type II receptor, *Proc Natl Acad Sci USA* 88:580 (1991).
75. C.D. Woodworth, V. Notario, J.A. DiPaolo, Transforming growth factors beta 1 and 2 transcriptionally regulate human papillomavirus (HPV) type 16 early gene expression in HPV-immortalized human genital epithelial cells, *J Virol* 64:4767 (1990).
76. J.A. Pietenpol, R.W. Stein, E. Moran, P. Yaciuk, R. Schlegel, R.M. Lyons, M.R. Pittelkow, K. Munger, P.M. Howley, H.L. Moses, TGF-β1 inhibition of c-*myc* transcription and growth in keratinocytes is abrogated by viral transforming proteins with pRB binding domains, *Cell* 61:777 (1990).
77. C. Agarwal, E.A. Rorke, J.C. Irwin, R.L. Eckert, Immortalization by human papillomavirus type 16 alters retinoid regulation of human ectocervical epithelial cell differentiation, *Cancer Res* 51:3982 (1991).

Chapter 3

Interactions Between Retinoic Acid and Protein Kinase C in Induction of Melanoma Differentiation

RICHARD M. NILES

I. Introduction

Vitamin A is a molecule that plays an indispensable role in many biological processes. Wolbach and Howe[1] were the first to describe the effect of vitamin A deficiency on the proliferation and differentiation of epithelial cells. In spite of the crucial role of this vitamin, we still do not know the exact mechanism by which it performs its biological function.

Retinoids (particularly retinoic acid) are required for normal differentiation of epithelia *in vivo*.[2-5] In addition, retinoids induce the terminal differentiation of certain malignant cells in culture such as F9 teratocarcinoma,[6] HL-60 promyelocytic leukemia,[7-8] erythroleukemia,[9] and neuroblastoma.[10] Retinoids can also act directly on nonneoplastic cells to suppress the process of malignant transformation *in vitro* and carcinogenesis *in vivo*.[11,12] It is thought that the primary action of retinoids is in inhibiting the promotion step of carcinogenesis, since retinoids can counteract the effects of tumor promoters in a variety of *in vivo* and *in vitro* systems.[13,14]

Protein kinase C (PKC), a phospholipid and Ca^{2+}-dependent kinase that phosphorylates substrate proteins on serine and threonine residues, is found in almost all tissues, with the highest activity being observed in brain.[15] PKC is activated by several growth factors, hormones, and neurotransmitters. It is the major—if not the only—receptor for phorbol ester tumor promoters.[16-18] PKC has been shown to phosphorylate the epidermal growth factor receptor,[19] the insulin receptor,[20] and the protein product of the *ras* oncogene (p21).[21] These data suggest that PKC plays an important role in regulating cell proliferation.

Although once considered a single entity, recent cloning and protein separation techniques have revealed the existence of multiple subspecies of PKC. There are three members (PKC-α, -β, and -γ) that meet the strict definition of a Ca^{2+}, phospholipid-activated protein

Richard M. Niles • Department of Biochemistry and Molecular Biology, Marshall University School of Medicine, Huntington, WV 25755 kinase, and others (-δ, -ε, -ζ, and -η) that have similar, but not identical, properties.[22-27] All the PKC subspecies have a single polypeptide chain with conserved and variable regions. The subspecies exhibit tissue-specific expression, with α the most universally expressed, while γ is only found in the brain and spinal cord.

Since PKC is activated by tumor-promoting phorbol esters, it has been postulated that the constant activation of PKC plays a role in tumor progression. Indeed two separate studies in which NIH-3T3 cells[28] and rat-1 cells[29] were transfected with PKC demonstrated that overexpression of this enzyme led to acquisition of a partially transformed phenotype. However, overexpression of PKC-β in HT29 colon cancer cells induced growth inhibition and tumor suppression.[30] Also, introduction of PKC-β into murine erythroleukemia cells accelerated differentiation induced by hexamethylene bisacetamide.[31] Since chronic administration of phorbol esters can lead to downregulation of PKC, investigators have recently hypothesized that depletion of PKC may be the key event in skin carcinogenesis. Indeed, convincing data for this concept was obtained by Hansen *et al.*[32] who found that downregulation of epidermal PKC was associated with and permissive for epidermal hyperplasia and tumor promotion. In addition, Brooks *et al.*[33] found that PKC downregulation, and not transient activation, correlated with melanocyte growth.

Retinoic acid has been shown to inhibit both anchorage-dependent and -independent growth in human and murine melanoma cells.[34-36] It has also been shown to increase tyrosinase and melanin synthesis in both human and murine cell lines.[37] *In vivo* studies demonstrated that retinoids inhibited the development and growth of transplantable murine melanomas[38] and pretreatment of B16 cells with retinoic acid in culture inhibited their subsequent metastatic capability.[39] Because retinoic acid counteracts tumor promotion, and the phorbol tumor promoters are potent activators of PKC, we decided to examine the status of PKC during retinoic acid-induced differentiation of B16 melanoma cells.

II. Material and Methods

A. Cell Culture

B16-F1 cells (obtained from Dr. Fidler, M.D. Anderson Hospital and Tumor Clinic, Houston, TX), were maintained as stock in Dulbecco's Minimal Essential Medium supplemented with 2 mM L-glutamine, 2 mM sodium pyruvate, 50 U/ml penicillin G, 50 μg/ml streptomycin sulfate, and 10% newborn supplemented calf serum (Sterile Systems, Logan, UT). Every 2 months new cultures were initiated from frozen stock in order to minimize the changes in cell phenotype that can occur with prolonged *in vitro* culture.

B. Transfections

B16 cells were transfected with 19 μg PKC-α expression plasmid DNA and also with 1 μg of a plasmid (pSV-neo) expressing the gene for neomycin resistance. DNA was introduced into the cells using the $CaPO_4$ precipitation of DNA onto the cells for 4 hr, followed by a 2 min glycerol shock. Cells were then re-fed with normal growth medium and incubated for 2-3 days, after which they were split into media containing neomycin (1 mg/ml). Resistant clones were isolated via a cloning cylinder subcultured and expanded for subsequent testing of overexpression of PKC.

C. Western Blot Analysis of PKC

Total cell protein from wild type B16 cells and stable transfectants was separated on 10% SDS-PAGE gels. Gels were transferred to nitrocellulose sheets in transfer buffer containing 20 mM Tris, 150 mM glycine (pH 8.3), 10% methanol, and 10 μg/ml SDS, at 250 mA overnight at 4°C. The blots were incubated for 4 hr in blocking solution (50 mM Tris (pH 7.4), 150 mM NaCl, 5% nonfat dry milk) to saturate the nonspecific sites, and then incubated with the appropriate titer of the anti-PKC monoclonal antibodies overnight at 4°C. The blots were washed four times with blocking buffer containing 0.1% Triton X-100, and then incubated with blocking buffer plus 1:5000 titer of rabbit anti-mouse IgG conjugated with peroxidase for 2 hr at room temperature. The blots were washed once with blocking buffer and then incubated with solutions #1 and #2 of the Amersham Chemiluminesence kit. This step was carried out in the dark room and the blot exposed to X-ray film for 30 sec to 30 min depending upon the level of antigen present. After developing, the relative amount of PKC-α, -β, and -γ in the different cells was assessed by densitometry, using the immunoreactivity of β-tubulin as an internal standard (the amount of tubulin does not change with retinoic acid treatment of in the PKC overexpressing clones).

D. RNA Extraction

Since polyA$^+$ RNA was required for all experiments in order to clearly visualize the PKC mRNAs, we developed a new method of RNA extraction in which the isolation of total RNA is bypassed and polyA$^+$ RNA is directly isolated. The advantage of this procedure is that it takes about 4 hr to obtain polyA$^+$ RNA vs. 4 days using older methodology.

Plates of cells were washed three times with cold PBS and then scraped into lysis buffer containing freshly added proteinase K [0.2 M NaCl, 0.2 M Tris-HCl (pH 7.5), 1.5 mM $MgCl_2$, 2% SDS, and 200 µg/ml proteinase K]. The DNA in the cell lysate was sheared by passage through a 22-gauge needle about ten times, and then the lysate was incubated for 1.5 hr at 45°C with intermittent agitation. While the lysate was incubating, the oligo-dT resin (5 mg/10^7 cells) was hydrated in 3 ml of elution buffer [0.01 M Tris-HCl (pH 7.5), 1.0 mM EDTA, and 0.2% SDS]. The resin was then equilibrated by washing twice with 5 volumes of binding buffer [0.5 M NaCl, 0.01 M Tris-HCl (pH 7.5), 1.0 mM EDTA, and 0.5% SDS]. The lysate was adjusted to 0.2 M NaCl, added to the oligo-dT resin and this mixture was incubated for 1 hr at room temperature with intermittent agitation. The resin was then washed with 5 volumes of binding buffer and loaded into a 1 ml syringe. The polyA$^+$ RNA was eluted with 2 × 0.4 ml aliquots of elution buffer. The polyA$^+$ RNA was precipitated with 2 volumes of 100% ethanol and 0.08 volumes of 5 M NaCl at −20°C overnight, and the pellet was resuspended in sterile water. The amount of polyA$^+$ RNA was estimated by hybridization to 3H-polyU, where a standard curve of known amount of polyA$^+$ RNA (1-100 ng) hybridizing to the 3H-polyU was established.

E. DNA Probes and ^{32}P-Labeling

Probes containing DNA sequences corresponding to PKC-α, -β, and -γ were obtained from Dr. A. Ullrich (Genentech, Palo Alto, CA) and Dr. Ohno (Yokohama University, Yokohama, Japan). Mouse β-tubulin cDNA was obtained from Dr. Farmer (Boston University School of Medicine, Boston, MA). The inserts were cut out using the appropriate restriction enzyme, and isolated by electrophoresis in low-melt agarose. The isolated inserts were labeled with dCTP^{32} using the random primer method.

F. Northern Blots

Equal amounts of polyA$^+$ RNA from each sample (usually 5 µg) were separated by electrophoresis in a 1% agarose, 2.2 M formaldehyde gel containing ethidium bromide. Following electrophoresis, the gels were photographed under UV light, soaked in sterile water for 2-4 hr and the RNA transferred by capillary blotting onto Hybond-N (nylon,

Amersham) membranes. The membranes were exposed to UV light for 5 min and then baked for 3 hr in a vacuum oven at 60°C to fix the RNA onto the blot.

Blots were prehybridized (5X SSC, 50% formamide, 5× Denhardt's solution, 10% dextran sulfate and 2.0 µg of Herring sperm DNA) for 24 hr at 42°C. They were then hybridized in the same solution containing 1.0×10^6 DPM per lane of the appropriate denatured probe. After 48 hr at 42°C, the blots were washed twice at room temperature with 2X SSC, once at 60°C with 0.6X SSC + 0.1% SDS for 30 min, and once at room temperature with 0.1X SSC for 30 min. The blots were partially dried, wrapped in Saran wrap, and exposed to Kodak XAR-5 film at −80°C for 24-72 hr. RNA from bovine brain served as a positive control for PKC, since all isotypes are expressed in brain tissue. It should be noted that between species such as mouse, bovine, and human there is greater than 90% homology within isotypes for PKC.

G. Monolayer Growth Rates, Soft Agar Colony Formation, and Melanin Production

Wild-type B16 cells treated ± 10 µM retinoic acid and various stable transfectants were seeded in triplicate 60-mm dishes at 5×10^4/dish. Cells counts were performed every day for 5 days, with medium changes every 48 hr, and growth curves established. Colony formation in soft agarose was determined by pouring a 1% agarose base in 60-mm dishes and overlaying with a 0.3% agarose mixture containing 1000 wild type cells or various clones of stable transfectants. One set of wild type dishes also contained 10 µM retinoic acid. This top layer of agarose (made in DME + 10% serum) was overlaid with a small amount of additional DME + 10% serum. After 10 days (the liquid phase was replenished every 48 hr) of incubation, the number of colonies containing greater than 25 cells was counted. Each group had 3 dishes. For analysis of melanin production wildtype cells ± 10 µM retinoic acid and various stable transfectants were seeded at 2×10^5 in 100-mm dishes. After 4 days in culture, the cells were re-fed with medium that did not contain phenol red. An aliquot of this medium (1.5 ml) was centrifuged (2000 × g, 15 min) exactly 24 hr later and melanin content measured by the optical density (OD) at 492 nm. The cell layers were collected by scraping into 1 ml of 1N KOH; an aliquot was removed for DNA quantitation (diphenylamine) and the remainder was hydrolyzed at 80°C for 1 hr. After clarification by centrifugation (12,000 × g, 30 min), the OD 492 nm of the supernatant was determined.

H. Transcription Run-on Analysis

In vivo transcription was performed by the method of Greenberg and Ziff.[40] Briefly, 2×10^7 nuclei were isolated from both untreated cells and cells treated for 24 hr with retinoic acid. The nuclei were immediately placed at −70°C in glycerol storage buffer until they were needed. DNA probes (15 µg) were slot-blotted onto nitrocellulose using a Schleicher and Schuell slot-blot apparatus. The frozen nuclei were thawed and immediately added to an equal volume of reaction buffer containing 1 mM each of ATP, GTP, and CTP, and 130 mM cold UTP, in addition to 150 mCi of (α-^{32}P) UTP (3000 Ci/mmol), (Dupont-NEN, Boston, MA). Endogenous transcription was allowed to proceed for 25 min at 30°C and the newly synthesized transcripts were isolated by trichloroacetic acid (TCA) precipitation and phenol chloroform extraction. The radiolabeled transcripts were hybridized with the slot-blotted probes at 65°C for 72 hr, and subsequently washed for 2 hr at 65°C in 2X SSC, and 30 min at 37°C in 2X SSC containing 20 mg/ml of RNAase A. The nitrocellulose was then exposed on film for 10-14 days, depending on the experiment.

III. Results

A. Stimulation of PKC Activity by Retinoic Acid

We have previously determined that melanocyte stimulating hormone and retinoic acid arrest growth of B16 melanoma cells in G1.[41] Since PKC has been implicated in regulating the growth of various cell types,[28,29] we measured its activity in cells treated with the above agents and also in cells growth-arrested by serum deprivation. Table 1 shows that all of the agents inhibited growth relative to the control cells, however, only retinoic acid stimulated PKC activity. The degree of stimulation varied between experiments, but was generally in the range of four- to eight-fold. The enzyme activity from both control and treated cultures was proportional to protein concentration, and was predominantly (>95%) in the 100,000 × g soluble fraction (data not shown).

Table 1. PKC Activity in B16 Mouse Melanoma Cells Treated with Various Growth Inhibitors[a]

Treatment	Specific activity		Cell number
	per cell	per mg protein	
Control	9.5 ± 1.2	109 ± 14	3.1×10^6
10 µM retinoic acid	35.1 ± 3.5	589 ± 59	1.7×10^6
2 µg/ml MSH	7.2 ± 0.2	95 ± 3	1.6×10^6
0.2% serum	11.8 ± 0.6	115 ± 6	1.2×10^6

[a] B16 cells were seeded at 2×10^5/100-mm dish in Dulbecco's medium plus 10% supplemented calf serum (Sterile Systems, Logan, UT). One day after seeding, the cultures were re-fed with the appropriate additives. Following an additional 48 hr incubation, cells were harvested, sonicated, and (following chromatography on DEAE-cellulose) assayed for PKC activity as described in "II. Material and Methods." Specific activity of PKC is presented as pmol ^{32}P transferred to histone/min per 10^6 cells or per mg cell protein. The data is presented as the mean ±SEM, where n = 3.

The relationship between the time of retinoic acid treatment and stimulation of PKC activity was examined. Figure 1 shows that by 24 hr of treatment with 10 µM retinoic acid, B16 melanoma cells have a four-fold higher specific activity of PKC compared to untreated cells. The absolute specific activity of PKC from retinoic acid-treated cells increased three-fold during the second 24 hr of treatment, however, since the untreated cells also had higher PKC activity at this timepoint, the relative difference was still about four-fold. It was technically difficult to measure time points beyond 48 hr of treatment because the large amount of the melanin produced by these cells became cytotoxic.

B. Western Blot Analysis of PKC Levels

The increase in the amount of PKC was verified by a second approach in which we measured the steady-state levels of PKC by the Western blot technique. B16 cells were incubated with and without 10 µM retinoic acid for 48 hr. Equal protein aliquots from the crude 100,000 × g supernatant fractions and a 150 mM NaCl eluate of a DEAE-cellulose column were separated by SDS-PAGE. After blotting the proteins onto nitrocellulose, PKC

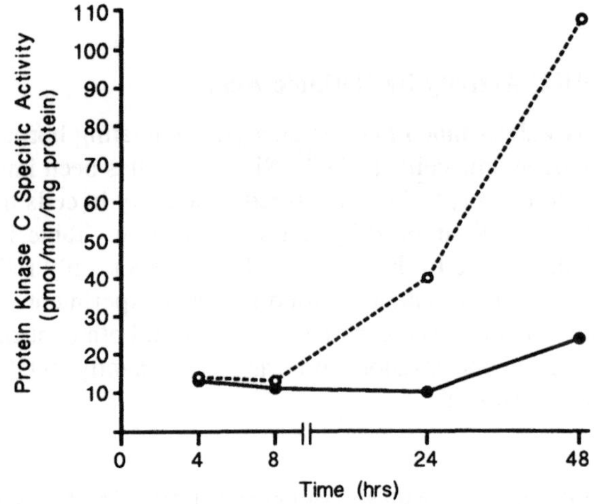

Figure 1. Time-dependent increase in PKC activity in retinoic acid-treated B16 mouse melanoma cells. B16 cells were treated with or without 10 µM retinoic acid. At the indicated timepoints, cells were harvested and assayed for PKC activity as described in the text. Closed circles: control; open circles: +10 µM retinoic acid. The bars above and below the data points represent the SEM. ($n = 3$)

was detected by incubation with either a mouse monoclonal PKC antibody (Amersham) or chicken polyclonal antibodies prepared against purified rat brain PKC. Detection of antibody binding was accomplished by incubation with ^{125}I protein A.

The Western blot in Figure 2 illustrates two points. First, the antibody binds to several different concentrations of purified rat brain PKC (kindly provided by Dr. Curtis Ashendel, Purdue University, West Lafayette, IN) (lanes 1 and 2: 80,000 MW protein) and, second, that absorption and elution of the crude 100,000 × g supernatant fractions from control and 48 hr retinoic acid-treated cells on DEAE-cellulose columns results in enrichment of immunoreactive PKC (compare lanes 4 and 6). When lanes 5 and 6 were quantitated by densitometry scanning, there was a 20- to 25-fold greater amount of immunoreactive PKC in samples from retinoic acid-treated cells. Using the more potent chicken anti-PKC antisera, we could detect an increase in immunoreactive PKC as early as 16 hr of retinoic acid treatment (data not shown).

C. PKC Isotype Expression in B16 Cells

Northern analysis demonstrated that B16 mouse melanoma cells express only PKC-α RNA (Figure 3A). When the cells were treated with 10 µM retinoic acid, an induction of PKC-α mRNA was observed between 4 and 8 hr after the retinoic acid was added. PKC-α mRNA expression continued to rise until 48 hr. The ethidium bromide-stained gel was photographed to show that equal amounts of RNA were loaded in each lane, except for the last lane, which contained bovine brain RNA (Figure 3B). It should be noted that the sizes of the PKC-α mRNA are somewhat higher than those reported by Coussens *et al.*[23] This may be due to different methods of estimating RNA sizes. We used molecular size markers obtained from Bethesda Research Laboratories to obtain our kilobase pair (Kbp) estimates for all the PKC mRNA species. This photograph was scanned with a computerized densitometer and the induction of PKC-α was normalized to this scan and graphed (Figure 3C). By 48 hr the induction was 10- to 15-fold over control level (Figure 3C).

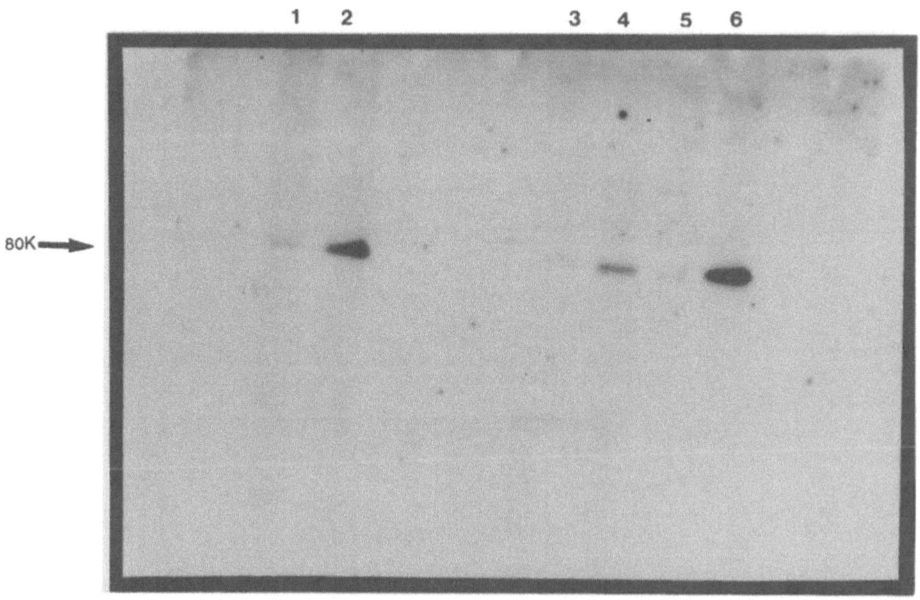

Figure 2. Western blot analysis of PKC levels in control and 48 hr retinoic acid-treated B16 mouse melanoma cells. B16 cells were treated with or without 10 μM retinoic acid for 48 hr. Cells were harvested and 100 μg of protein from each treatment group directly loaded onto SDS-gels. Alternatively, protein extracts were applied to DEAE-cellulose columns and 100 μl of the dialyzed 150 mM NaCl eluate loaded onto the SDS-gel. Following fractionation by SDS-PAGE, proteins were electrophorectically transferred to nitrocellulose and incubated with a monoclonal anti-PKC antibody as described in the text. Lanes 1 and 2: 20 μl and 100 μl respectively of purified rat brain PKC. Lanes 3 and 5: crude extract and DEAE-cellulose purified fraction respectively from control cells. Lanes 4 and 6: crude extract and DEAE-cellulose purified fraction respectively from retinoic acid-treated cells. This autoradiogram represents a 72 hr exposure at –70°C of the blot in a cassette with double reflector screens.

D. Mode of Regulation of PKC-α Expression by Retinoic Acid

In order to determine whether the effect of retinoic acid on PKC-α mRNA expression was direct or indirect, cycloheximide was used to inhibit new protein synthesis, and Northern analysis was performed to determine whether or not the induction of PKC-α mRNA was abolished. B16 cells were treated with or without retinoic acid and/or 5 mg/ml of cycloheximide for 8 hr. An 8 hr timepoint was used because induction of PKC-α mRNA was not apparent before this time, and cycloheximide proved to be toxic after this point. A 1 hr pulse with [^{35}S]-methionine showed that protein synthesis was inhibited by greater than 90% (data not shown). Figure 4 shows that the retinoic acid induction of PKC-α mRNA is indirect, since cycloheximide inhibited the effect of retinoic acid. Cycloheximide in the absence of retinoic acid had no effect on PKC-α mRNA levels.

To determine if retinoic acid increased PKC-α transcription, we performed a nuclear run-on analysis. The cells were treated with or without retinoic acid for 24 hr and the nuclei were isolated and stored at –70°C. Nuclei were thawed and transcription was allowed to continue. Figure 5 shows that the PKC-α message is induced only two- to three-fold (as determined by laser densitometry), which is not enough to account for the 10- to 12-fold increase in steady-state mRNA levels. The tubulin message is essentially the same in both

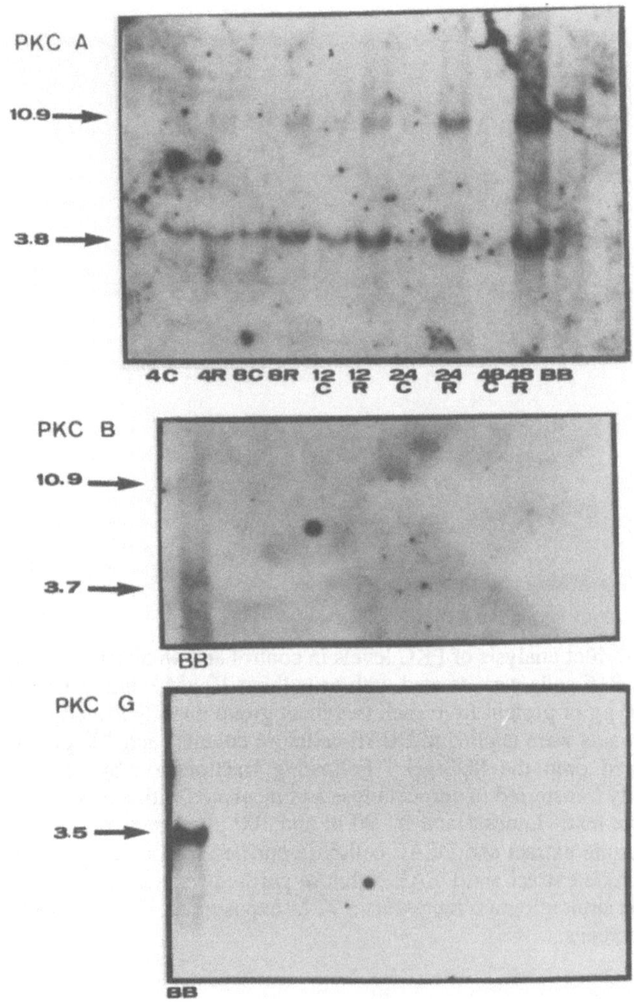

Figure 3. Northern analysis of PKC RNA levels in control and retinoic acid-treated cells. B16 mouse melanoma cells were treated with 10 μM retinoic acid for 4, 8, 12, 24, and 48 hr. Total RNA was extracted from control and treated cells at each time point and 40 μg was loaded in each lane. Bovine brain (BB) RNA was used as a positive control. (A) The top panel was probed with ^{32}PdCTP-labeled PKC-α (PKC A) insert; the middle panel with PKC-β (PKC B) insert, and the bottom panel with PKC-γ (PKC G) insert. The blots were washed under high stringency conditions, 1 hr with 2X SSC at 42°C and 1 hr with 0.6X SSC, 0.1% SDS at 65°C. The autoradiograms were exposed for 7-10 days. (B) The ethidium bromide stained gel from the top panel of Figure 3A was photographed to show the relative amounts of RNA loaded in each lane. (C) The two PKC-α messages in the autoradiogram in the top panel of Figure 3A were scanned at each timepoint using a Molecular Dynamics computing densitometer; the relative ODs were normalized to the scan of the ethidium-stained gel, and then plotted as a function of time with and without retinoic acid treatment. The 10.9 Kb message was not detected by the densitometer unless retinoic acid was present and therefore this line was omitted from the graph.

the control and retinoic acid-treated nuclei (less than 2% change). This is consistent with the northern analysis, which showed that the tubulin message was not affected by retinoic acid treatment. Conversely, there is a small decrease in the level of c-*myc* transcript. This

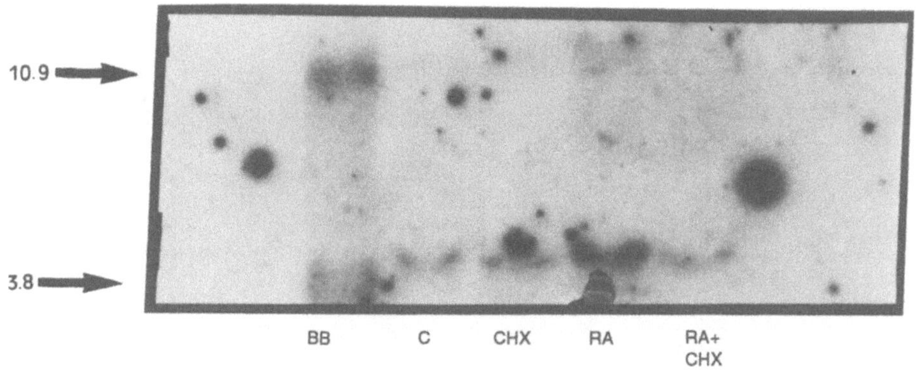

Figure 4. The effect of cycloheximide (CHX) on retinoic acid induction of PKC-α mRNA. B16 cells were treated as follows: Lane 1: untreated; Lane 2: 5 μg/ml of CHX for 8 hr; Lane 3: 10 μM retinoic acid for 8 hr, and Lane 4: both CHX and retinoic acid for 8 hr. RNA was then extracted from each group and 40 μg was loaded into each lane. The level of PKC-α RNA was determined by Northern analysis. The autoradiogram was exposed for 3 weeks.

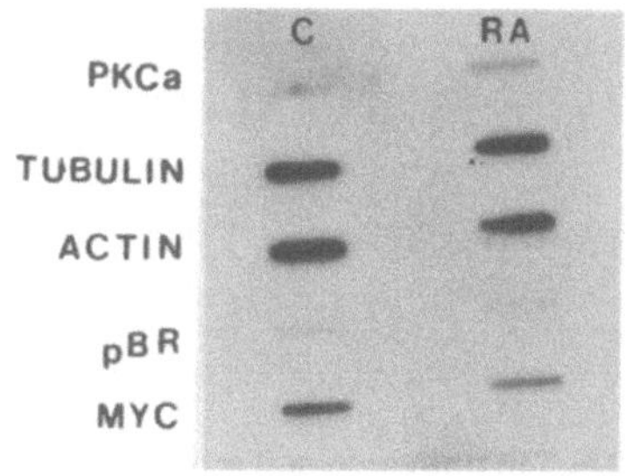

Figure 5. Transcription run-on analysis of PKC-α in retinoic acid-treated and control nuclei. Nuclear run-ons were performed with nuclei isolated from control and 24 hr retinoic acid-treated cells. 10 μg of each plasmid was slot-blotted onto nitrocellulose and subsequently hybridized with the ^{32}P-RNA transcripts for 3 days. The autoradiogram was exposed for 2 weeks.

decrease in transcription rate could partially explain our previous finding that the steady-state level of this message is decreased in retinoic acid-treated B16 cells (data not shown). Nonspecific hybridization (pBR322) was negligible.

Figure 6 shows the results of a message stability study performed on cells that had been treated for 24 hr with or without retinoic acid. When actinomycin D was added for 0, 2, and 8 hr (4 μg/ml), no change in message stability was found. The PKC-α message is apparently very stable and the addition of retinoic acid has no effect. The half-life for the 3.8 Kb PKC-α message is obviously longer than 8 hr, but treatment with actinomycin D for longer periods of time was not possible due to the toxic effects of the drug. The 10.9 Kb PKC-α message was less stable than the smaller message; this finding was consistent in both control and treated cells. The blot was reprobed with c-*myc*, which has a very short half-life (less than 30 min)[42] to show that the actinomycin D treatment was effective. The level of c-*myc* RNA was high at 0 hr and was undetectable by 2 hr. The c-*myc* RNA level at 0 hr was less in the retinoic acid-treated cells, which confirms previous data that c-*myc* RNA levels decrease in retinoic acid-treated B16 cells (unpublished data). Therefore, the major control of PKC-α mRNA level by retinoic acid is probably posttranscriptional, either due to accelerated RNA processing or increased transport out of the nucleus.

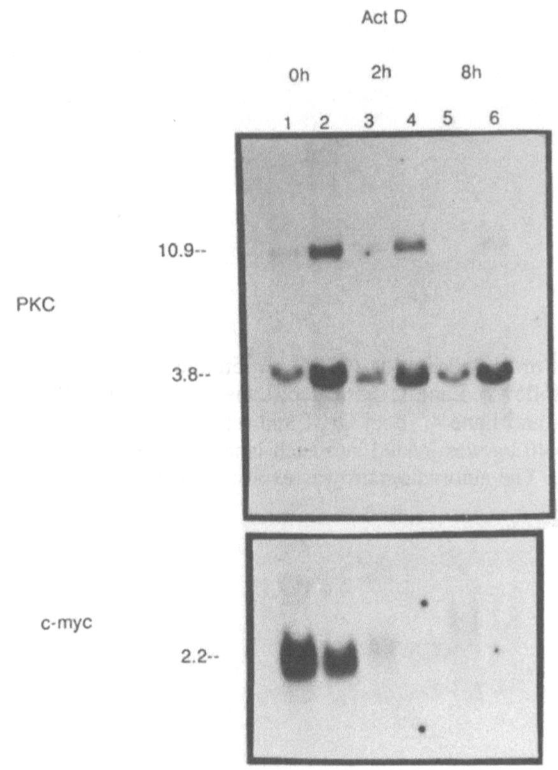

Figure 6. Stability of PKC-α mRNA in untreated and retinoic acid-treated B16 cells. B16 cells were treated with or without 10 μM retinoic acid for 24 hr. Cells were then incubated with actinomycin D (4 μg/ml) for 2 or 8 hr. At the end of each timepoint RNA was extracted and polyA$^+$ RNA was isolated by oligo-dT chromatography. The amount of PKC-α mRNA was then estimated by Northern analysis using 10 μg of polyA$^+$ RNA from each sample. The autoradiogram was exposed for 1 week. The blot was reprobed with c-*myc*, a message that has a very short half-life (less than 30 min). This autoradiogram was exposed for 2 days.

E. Role of Increased PKC-α in Retinoic Acid-Induced B16 Melanoma Differentiation

Since this increase in PKC-α occurs relatively early during the differentiation program, the question arises as to whether this enzyme mediates some of the phenotypic changes induced by retinoic acid (i.e., decreased monolayer growth rate, elimination of anchorage-independent growth, and increased melanin production). This question was addressed by establishing and characterizing stable transfectants of B16-F1 cells that overexpress PKC-α in the absence of exogenous retinoic acid.

We transfected early passage B16 cells with a PKC-α cDNA expression vector as described in "II. Material and Methods." Thirty-two G418-resistant clones were screened, yielding two consistently positive clones, designated as H and K, respectively. These clones were determined to overexpress PKC-α based on both Western blot analysis (Figure 7) and enzyme activity (Figure 8). Densitometric analysis showed that clones H and K contained approximately 3.3- and 2.3-times more PKC-α than wild type (WT) cells or cells transfected with only the gene for neomycin resistance (Neo). This agreed well with the increased enzyme activity found in clones H and K.

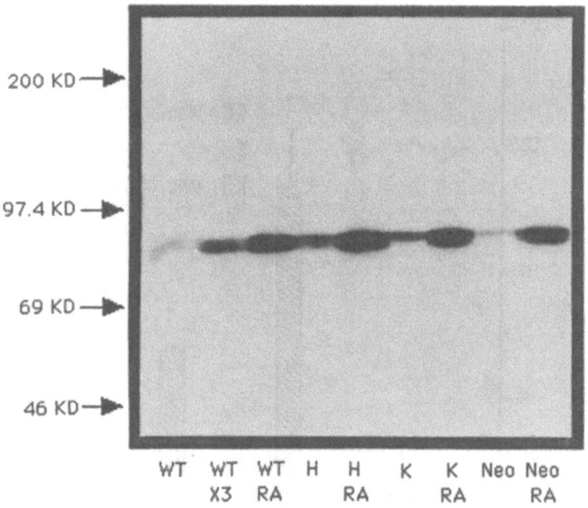

Figure 7. Western blot analysis of PKC-α overexpressing clones. Cells were harvested by trypsinization and lysed as described in materials and methods. Crude protein (50 µg) was loaded in all lanes except for lane 2 which contained 150 µg protein for calibration purposes. PKC-α overexpressing clones (H and K), WT cells, and cells transfected with pSV-Neo only (Neo).

In monolayer growth rate experiments, we found that cells from clones H and K had slower growth rates than untreated WT or Neo cells (Figure 9). As expected, retinoic acid treatment depressed the monolayer growth of WT and Neo cells. In comparing the growth of the two overexpressing clones, we found that clone H had fewer cells per dish than clone K ($p < 0.05$) by 48 hr. This finding correlated with the PKC levels and activities in these clones. When clones H and K were treated with retinoic acid, their respective growth rates did not significantly decrease (data not shown). At the light microscopy level, we were unable to detect obvious morphology differences between PKC-α overexpressing clones and control cells.

The ability of PKC-α overexpressing clones to form viable colonies in soft agarose was significantly lower than that of untreated WT and Neo cells (Figure 10). Cells from clone H formed significantly fewer colonies than those from clone K ($p < 0.001$). As with the monolayer growth rate data, the difference in soft agarose growth between clones H and K correlated with their respective PKC levels. As expected, retinoic acid treatment of WT cells decreased their ability to form viable colonies in a concentration-dependent fashion.

As an estimate of the degree of differentiation, we analyzed the cells for melanin production (Figure 11). Cells from clones H and K contained and secreted more melanin than untreated control cells. Retinoic acid treatment of WT cells resulted in melanin levels comparable to those from clones H and K.

In light of the large reduction in soft agarose colony formation by the PKC-α overexpressing clones, we decided to study their tumorigenicity. When syngeneic C57B1/6 mice were injected subcutaneously with the different cell types, we observed statistically significant differences in tumor latencies ($p < 0.01$) and tumor weights ($p < 0.01$) between the two PKC-α overexpressing clones and their WT and Neo counterparts (Table 2). Tumors originating from control cells weighed more, had shorter latencies, and commonly traversed the peritoneal cavity. Peritoneal involvement was much less common in tumors arising from clones H and K.

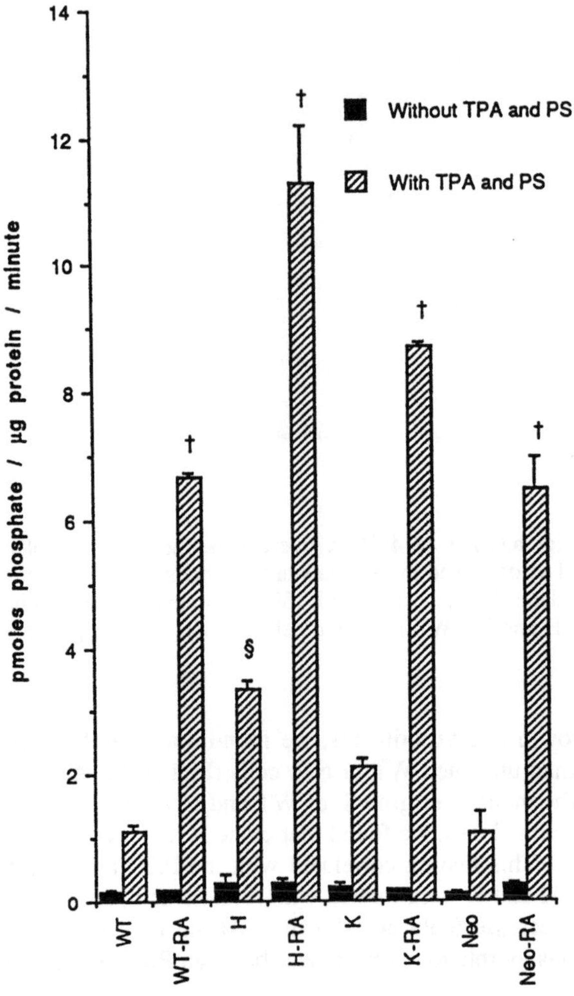

Figure 8. *In vitro* enzyme activities of partially purified PKC from PKC-α overexpressing clones. Cells were harvested by trypsinization and lysed. PKC enriched fractions were obtained by ion-exchange chromatography. These fractions were further concentrated by microfiltration. Equal amounts of protein were assayed using a commercially available PKC assay system (Amersham) in the presence of co-factors phosphatidyl serine and phorbol ester (hatched bars) and in the absence of these co-factors (black bars). Wild type (WT). PKC-α overexpressing clones (H and K), and cells transfected with pSV-Neo only (Neo). The data are presented as the mean ± SD where n = 2. Means were determined to be significantly different from WT by ANOVA followed by Newman-Keuls multiple comparisons ($p^* < 0.05$).

F. Involvement of Retinoic Acid Receptor-β in the Induction of PKC-α

B16 melanoma cells constitutively express mRNA for retinoic acid receptors (RAR) -α and -γ, but very little RAR-β. Addition of retinoic acid results in a rapid increase in RAR-β mRNA (Figure 12). In contrast to the retinoic acid induction of PKC-α, this increase in RAR-β mRNA is not blocked by cycloheximide (data not shown). Since the actions of

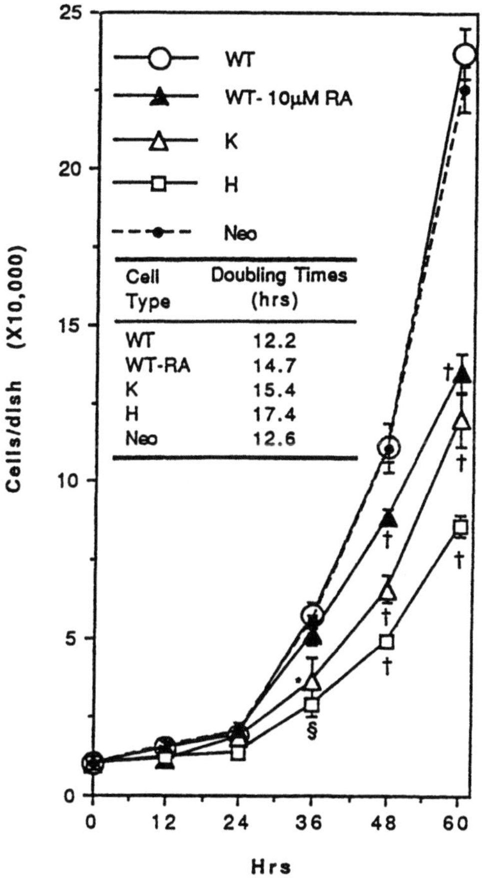

Figure 9. Monolayer growth curves and doubling times of PKC-α overexpressing clones. Cells were seeded at equal densities and counted at 6-hr intervals. PKC-α overexpressing clones (H and K), WT cells, and cells transfected with pSV-Neo only (Neo). Each point is presented as the mean of triplicate plates ±SD.

Table 2. Tumorigenicity of WT and Transfected Clones of B16 Cells[a]

Cell type	Latency (days) median/range	Tumor weight (gs) median/range	Tumor incidence
WT	5/3–9	2.369/1.379–5330	8/8
H	9/7–13[b]	1.057/0.208–1.470[b]	8/8
K	9/7–13[b]	0.915/0.081–1.299[b]	8/8
Neo	5/3–7	2.807/1.687–4.890	8/8

[a] Each mouse received 6×10^5 viable cells suspended in 0.1 ml of phosphate-buffered saline (PBS) injected subcutaneously into the right flank. Mice were sacrificed 14 days following injection. WT, PKC-α overexpressing clones (H and K), and cells transfected with pSV-Neo only (Neo).

[b] Significantly different from WT by Dunnetts nonparametric multiple comparisons ($p < 0.01$).

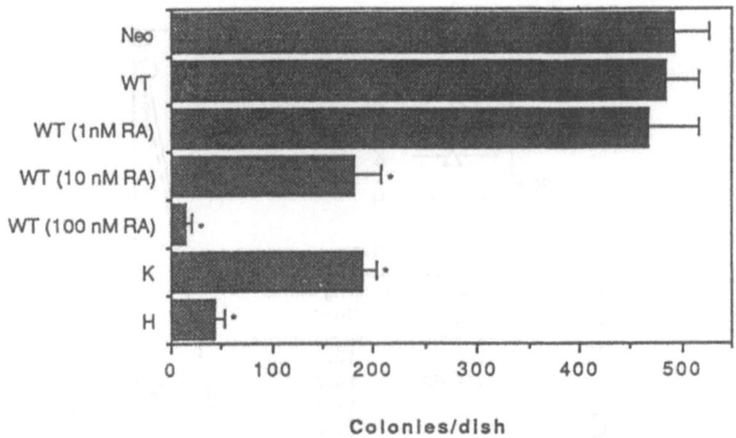

Figure 10. Colony formation in soft agarose of PKC-α overexpressing clones. Five thousand viable cells in 0.3% agarose were seeded into each dish. Dishes were re-fed with their respective media every 48 hr. Ten days after seeding, colonies equal to or greater than 300 μm in diameter were counted. WT and transfected (H, K, and Neo) B16 cells were used in these experiments. The data are presented as the mean ±SD of triplicate plates. (*Mean is significantly different from WT, $p < 0.001$; Newman-Keuls).

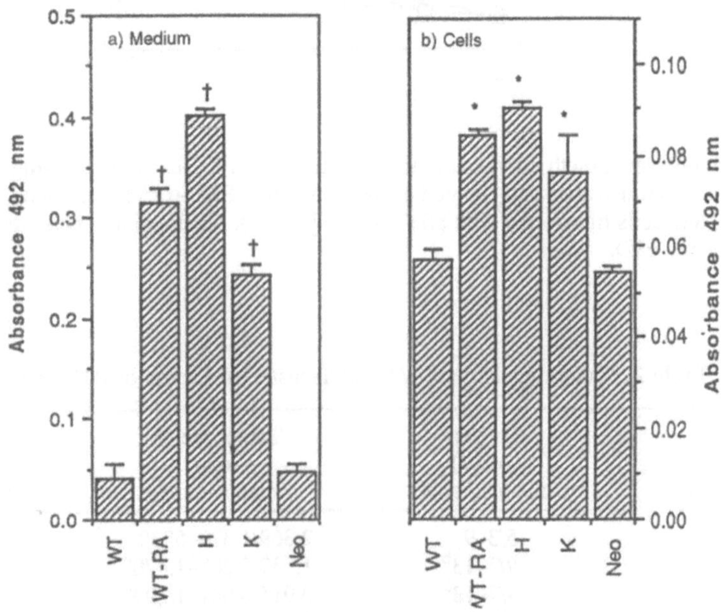

Figure 11. Relative melanin content of medium (a) and cells (b). (a) Equal amounts of medium were collected from each plate and centrifuged at 12,000 × g. The absorbance at 462 nm was determined for each supernatant. (b) An equal number of cells was harvested and lysed in 1 N KOH at 80°C for 1 hr. The hydrolysate was centrifuged at 12,000 × g, and the adsorbance at 462 nm was determined for WT, PKC-α overexpressing clones (H and K), and cells transfected with pSV-Neo only (Neo). The data are presented as the means ±SD from duplicate plates.

retinoic acid are mediated through the RARs, we investigated the role of RAR-β in mediating the induction of PKC-α, by transfecting B16 cells with full-length RAR-β cDNA under the control of an SV40 promoter. Sixty-five neomycin-resistant clones were screened for RAR-β expression by Northern blots (Figure 13). Clones 7 and 19 were found to consistently overexpress RAR-β. These clones were then compared to both WT and Neo clones in terms of their expression of PKC-α. Western blot analysis using an isotype-specific PKC-α monoclonal antibody revealed that both clones had significantly more PKC-α than any of the control cells (Figure 14). In other experiments not shown here, we found that both RAR-β overexpressing clones responded faster to retinoic acid induction of PKC-α mRNA than WT or Neo clones and that the induction was not blocked by cycloheximide.

IV. Discussion

In this report, we have shown that retinoic acid can induce PKC mRNA and protein, that the isotype induced is PKC-α, and that overexpression of PKC-α can mimic the action of retinoic acid, and we provided preliminary evidence that RAR-β may mediate the induction of PKC-α by retinoic acid. This effect of retinoic acid is specific and not simply due to the growth state of the cells, since other agents that inhibited proliferation did not increase the activity of PKC.

It was previously shown that retinoic acid induced differentiation of the human promyelocytic cell line HL-60 was accompanied by increased PKC activity.[43] Other agents such as DMSO and vitamin D_3, which induced granulocytic differentiation of HL-60, also increased the PKC activity.[44] In the case of vitamin D_3-induced differentiation of HL-60, there was also an increase in phorbol ester receptors.[45] Makowske *et al.*,[46] using antipeptide antibodies that specifically recognized the α, β and γ isozymes of PKC, showed that all three isozymes increased about three-fold in abundance after a 96 hr of treatment of HL-60 cells with 1 μM retinoic acid. DMSO, which also induced differentiation, also increased the amount of all three isozymes, although there was a relatively greater increase in the α and β forms of PKC. In addition to HL-60, the human monoblastoid cell line U937 also exhibited increased PKC activity when induced to differentiate by treatment with vitamin D_3.[47] It is not clear from these studies whether the increase in PKC is a result of differentiation or the inhibition of cell proliferation that accompanies this process.

The results presented here demonstrate that B16 melanoma cells only express PKC-α mRNA, and that this was the major RNA species induced by retinoic acid. The effect of retinoic acid on PKC-α mRNA expression was indirect, since inhibition of protein synthesis by cycloheximide abolished the induction of PKC-α mRNA. Although retinoic acid did not change the stability of either of the PKC-α mRNAs, it is interesting to note that the 10.9 Kb PKC-α message is more unstable than the 3.8 Kb mRNA. The difference between the two messages resides in the 3′ untranslated region, where the larger message has an alternate polyadenylation signal further downstream.[48] The 10.9 Kb message has been shown to contain four AUUUA sequences, whereas the 3.8 Kb message contains only one such sequence.[48] This sequence has been proposed by Shaw and Kamen[49] to be associated with message instability, and thus may explain the decreased stability of the 10.9 Kb PKC-α message.

The induction of PKC-α mRNA appears to be primarily posttranscriptional, since retinoic acid treatment did not alter the stability of the PKC-α mRNAs, and increased the rate of transcription by only two- to three-fold, which is not enough to account for the 10- to 12-fold increase in steady-state mRNA levels. The small increase in transcription is specific, however, since the tubulin transcription remains the same after retinoic acid treatment (Figure 5). However, the primary method of PKC-α regulation must be posttranscriptional, either enhanced RNA processing or increased transport out of the nucleus.

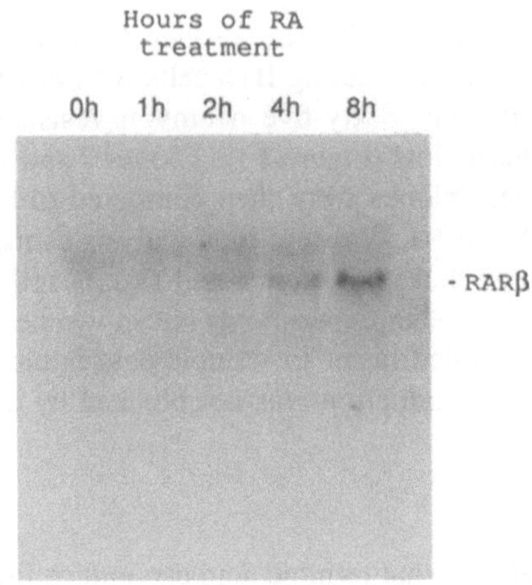

Figure 12. Induction of RAR-β mRNA by retinoic acid. B16 cells were treated ± 10 μM all-*trans* retinoic acid for the indicated times. Cells were then harvested, RNA extracted, and 20 μg of total RNA analyzed by Northern blots using the full length RAR-β cDNA as a probe. The autoradiogram represents an 18 hr exposure of the blot.

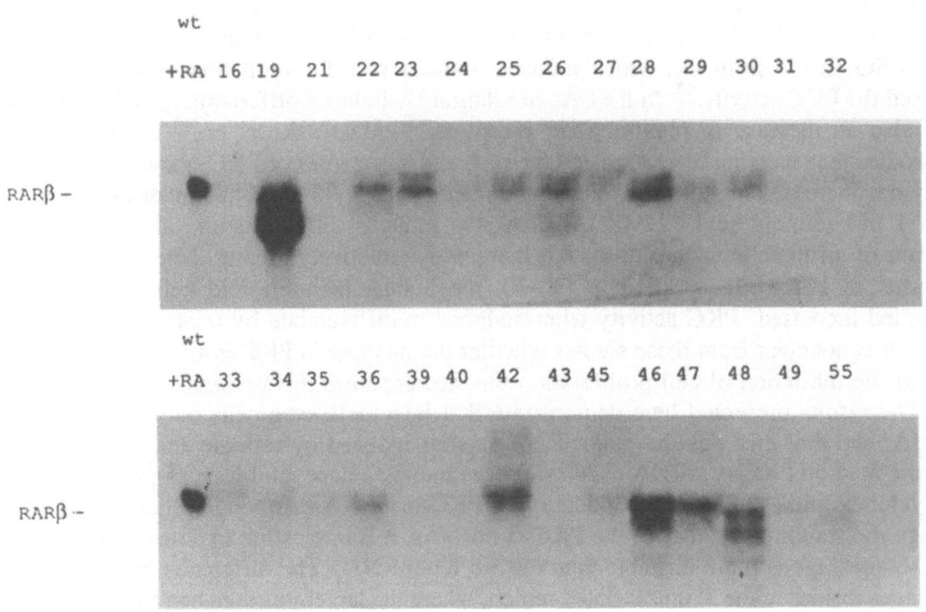

Figure 13. Screening of B16 melanoma-RAR-β transfectants. B16 cells were transfected with full length RAR-β cDNA driven by an SV-40 promoter as described in "II. Material and Methods." Sixty-five neomycin-resistant clones were isolated and expanded. RNA was extracted and the amount of RAR-β mRNA was determined by Northern blotting. The autoradiogram represents a 36 hr exposure of the blot.

This work examined the expression of the conventional PKCs. In light of the recent finding that the novel subtype, PKC-η[50] is expressed predominantly in the skin, it would be interesting to look at the expression of this subtype in the B16 melanoma cells, and whether retinoic acid might effect the level of its expression.

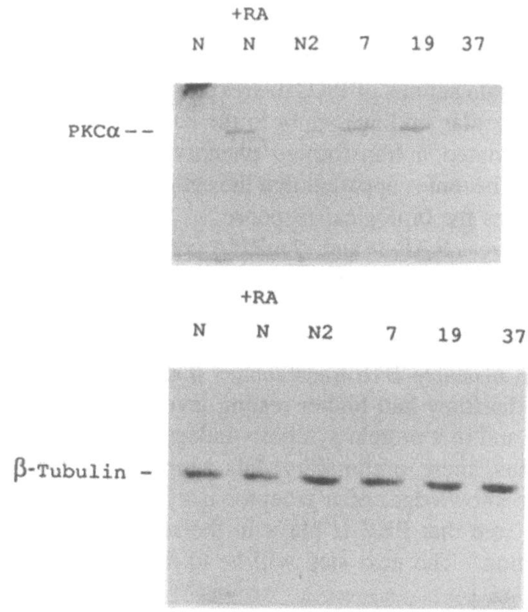

Figure 14. PKC-α expression in stable transfectants that express RAR-β. Stable transfectant clones that express the Neo gene only (N,N2) or both the Neo gene and the RAR-β gene (clones 7, 19, 37), were examined for the level of PKC-α protein by Western blotting. One of the clones that only expresses the Neo gene was treated with 10 μM retinoic acid for 24 hr to induce PKC-α (positive control). Immunoreactivity was detected using the chemiluminesence kit from Amersham. To ensure that there was equal protein loading and transfer for each sample, the blots were reprobed with anti-β tubulin, which we have previously shown does not change in amount during retinoic acid-induced melanoma differentiation.

When stable transfectants overexpressing PKC-α were examined, it was found that they exhibited phenotypes very similar to those of WT cells treated with retinoic acid. It is important to note that the level of PKC-α overexpression in both clones was lower than that induced by retinoic acid treatment of WT or Neo cells. The reason for this could be explained in that PKC-α is apparently associated with diminished growth. Therefore, we might have been selecting cells that overexpress an apparently negative growth regulator. If a cell should overexpress very large amounts of PKC-α, it might grow so slowly that no colony would be formed. Therefore, we believe that inducing higher levels of PKC-α in this system would require transfection with PKC-α under the control of an inducible promoter.

Other laboratories have transfected various cells with other PKC isozymes.[28-30] In two of these reports, the investigators transfected nonmalignant fibroblasts and obtained clones that overexpressed PKC-γ and PKC-β1, respectively, at very high levels. In both cases, the PKC overexpressors assumed some characteristics of transformed cells.[28,29] When malignant HT29 cells (human colon carcinoma) cells were transfected with PKC-β1, the overexpressing cells (11- to 15-fold) showed evidence of diminished malignancy.[30] Also, when PKC-β protein was introduced into erythroleukemia cells, they differentiated at a faster rate.[31] An important difference between the HT29 study[30] and that reported here is that we did not have to treat our overexpressing clones with phorbol esters in order to induce a phenotypic change. This might be explained by the presence of factors in the cell culture medium (i.e., calf serum or autocrine factors) that could stimulate a constant diacylglycerol production, thus activating the enzyme in the absence of additional co-factors such as phorbol esters. This would suggest that in B16 melanoma cells the limiting factor is the amount of PKC-α enzyme and not the activators of the enzyme.

Although the disparate actions of PKC overexpression on the phenotypes of the host cell still seem an enigma, similar findings apply to the *ras* oncogene. Some cells, when transfected with v-*ras*, adopted a transformed phenotype,[51-53] while others are induced to differentiate.[54] Thus, it becomes apparent that the cellular milieu in which certain regulatory proteins function dictates the biological response.

B16 melanoma express RAR-α and -γ mRNA constitutively. RAR-β mRNA levels are dramatically induced by retinoic acid in a protein synthesis-independent manner. Since retinoic acid induction of PKC-α mRNA does require new protein synthesis, we hypothesized that RAR-β was the mediator of the induction of PKC by retinoic acid. This was tested by the isolation of stable B16 transfectants that overexpressed RAR-β. Analysis of these clones showed that they had higher resting levels of PKC-α than WT cells and that they responded faster and in a protein synthesis-independent manner to induction of PKC-α by retinoic acid. Thus, these preliminary data support our hypothesis and are the first demonstration, to our knowledge, of a receptor isotype-specific function. These data all point to the essential role that PKC-α plays in the retinoic acid-induced pathway of B16 melanoma differentiation. The next step will be to determine the downstream events that are regulated by PKC-α.

V. Summary

Retinoic acid treatment of B16 mouse melanoma cells induces a differentiated phenotype. This is accompanied by a decrease in monolayer growth rate, loss of the ability to form colonies in soft agarose, increased production of melanin and other melanocyte-specific markers. In addition, retinoic acid treatment of these cells decreases their tumorigenicity when injected subcutaneously into mice. Our laboratory has found that an early biochemical change after the addition of retinoic acid is a large increase in PKC. PKC is an enzyme whose activity is activated by diacylglycerol and calcium and has been shown to be an important mediator of substances that stimulate growth or differentiation. Since PKC is a multi-gene family, it was important for us to determine which isotype(s) was expressed in B16 cells and which type was induced by retinoic acid. We found that only PKC-α is expressed in these cells, and this is the form that is induced by retinoic acid. The retinoic acid-induced increased in PKC-α is found at both the RNA and protein level. The mechanism of induction is not yet clear since there is only a small increase in the transcription rate and no change in the stability of the mRNA for PKC-α in treated cells. In addition, the induction of PKC by retinoic acid can be blocked by inhibitors of protein synthesis, suggesting that the induction requires the synthesis of new protein(s). In order to determine the role of increased PKC-α in the retinoic acid-induced differentiation, we transfected full-length PKC-α cDNA in mammalian expression vectors into B16 cells. Two clones that stably overexpressed PKC-α to different levels were isolated. The phenotype of these clones resembled WT cells treated with retinoic acid, i.e. they had longer doubling times, decreased ability to form colonies in soft agar, increased melanin production, and decreased tumorigenicity in mice. Recent data suggest a role for the RAR-β in mediating the effect of retinoic acid on PKC induction. B16 cells express a very low amount of RAR-β mRNA. The level is increased drastically by retinoic acid treatment without any requirement for protein synthesis. When B16 cells were transfected with and overexpressed RAR-β, they also expressed more PKC-α mRNA and protein, and the induction of PKC by retinoic acid was not blocked by protein synthesis inhibitors. In summary, these finding suggest a key role for PKC-α in the pathway by which retinoic acid induces B16 mouse melanoma differentiation.

Acknowledgments

This work was supported in part by grant 92A58 from the American Institute for Cancer Research and grant RO1 CA59530 from the National Cancer Institute. The author would like to thank Barbara Loewy, Susan Rosenbaum, Joachim Gruber and Yong-Hong Xiao, who were involved in various aspects of these studies, Dr. Shigeo Ohno (Yokohama University School of Medicine, Yokohama, Japan) for providing the PKC expression vectors and Dr. P. Chambon (CNRS, Strasbourg, France) for providing the RAR plasmids.

References

1. S.B. Wolbach, P.R. Howe, Tissue changes following deprivation of fat soluble vitamin A, *J Exp Med* 62:753 (1925).
2. M.B. Sporn, G.H. Clamon, N.M. Dunlop, D.L. Newton, J.M. Smith, U. Saffiotti, Activity of vitamin A analogues in cell cultures of mouse epidermis and organ cultures of hamster trachea, *Nature* 253:47 (1975).
3. D.L. Newton, W.R. Henderson, M.B. Sporn, Structure-activity relationships of retinoids in hamster tracheal organ culture, *Cancer Res* 40:3413 (1980).
4. Y. Wong, R. Buck, An electron microscopic study of metaplasia of the rat tracheal epithelium in vitamin A-deficiency, *Lab Invest* 24:55 (1971).
5. R. Wu, M.M.J. Wu, Effect of retinoids on human bronchial epithelial cells: Differential regulation of hyaluronate synthesis and keratin protein synthesis, *J Cell Physiol* 127:73 (1986).
6. S. Strickland, K.K. Smith, K.R. Marotti, Hormonal induction of differentiation in teratocarcinoma stem cells: Generation of parietal endoderm by retinoic acid and dibutyryl cAMP, *Cell* 21:347 (1980).
7. T.R. Breitman, S.E. Selonick, S.J. Collins, Induction of differentiation of the human promyelocytic leukemia cell line (HL-60) by retinoic acid, *Proc Natl Acad Sci USA* 77:2936 (1980).
8. T.R. Breitman, S.J. Collins, B.R. Keene, Terminal differentiation of human promyelocytic leukemic cells in primary culture in response to retinoic acid, *Blood* 57:1000 (1981).
9. L.C. Garg, J.C. Brown, Friend erythroleukemia cell differentiation: Induction by retinoids, *Differentiation* 25:79 (1983).
10. N.J. Sidell, Retinoic acid-induced growth inhibition and morphologic differentiation of human neuroblastoma cells in vitro, *J Natl Cancer Inst* 68:589 (1981).
11. M.B. Sporn, A.B. Roberts, Role of retinoids in differentiation and carcinogenesis, *Cancer Res* 43:3034 (1983).
12. M.B. Sporn, D.L. Newton, Prevention of chemical carcinogenesis by vitamin A and its synthetic analogs (retinoids), *in*: *Inhibition of Tumor Induction and Development*, M.S. Zedeck, M. Lipkin, eds., Plenum Press, NY, p. 71 (1981).
13. F.L. Meyskens, Studies of retinoids in the prevention and treatment of cancer, *J Am Acad Dermatol* 6:824 (1982).
14. J. Abb, F. Deinhardt, Retinoids inhibit the mitogenic activity of tumor-promoting phorbol esters on human lymphocytes, *Int J Cancer* 25:267 (1980).
15. Y. Takai, A. Kishimoto, Y. Iwasa, Y. Kawahara, T. Mori, Y. Nishizuka, Calcium-dependent activation of a multifunctional protein kinase by membrane phospholipids, *J Biol Chem* 254:3692 (1979).
16. Y. Nishizuka, The role of protein kinase C in cell surface transduction and tumor promotion, *Nature* 308:693 (1984).
17. Y. Nishizuka, Studies and perspectives of protein kinase C, *Science* 233:305 (1986).

18. C. Ashendel, The phorbol ester receptor: A phospholipid regulated protein kinase, *Biochem Biophys Acta* 822:219 (1984).
19. T. Hunter, N. Ling, J.A. Cooper, Protein kinase C phosphorylation of EGF receptor at a threonine residue close to the cytoplasmic face of the plasma membrane, *Nature* 311:480 (1984).
20. G.E. Bollag, R.A. Roth, J. Beaudoin, D. Mochly-Rosen, D.E. Koshland, Jr., Protein kinase C directly phosphorylates the insulin receptor in vitro and reduces its protein-tyrosine kinase activity, *Proc Natl Acad Sci USA* 83:5822 (1986).
21. A.Y. Jeng, S.K. Srivastava, J.C. Lacal, P.M. Blumberg, Phosphorylation of ras oncogene product by protein kinase C, *Biochem Biophys Res Comm* 145:782 (1987).
22. Y. Nishizuka, Studies and perspectives of protein kinase C, *Science* 233:305 (1986).
23. L. Coussens, P.J. Parker, L. Rhee, T.L. Yang-Feng, E. Chen, M.D. Waterfield, U. Francke, A. Ullrich, Multiple distinct forms of bovine and human protein kinase C suggest diversity in cellular signaling pathways, *Science* 233:859 (1986).
24. G.M. Housey, C.A. O'Brien, M.D. Johnson, P. Kirschmeier, I.B. Weinstein, Isolation of cDNA clones encoding protein kinase C: Evidence for a protein kinase C-related gene family, *Proc Natl Acad Sci USA* 84:1065 (1987).
25. Y. Ono, U. Kikkawa, K. Ogita, T. Fujii, T. Kurokawa, Y. Asaoka, K. Sekiguchi, K. Ase, K. Igarashiu, Y. Nishizuka, Expression and properties of two types of protein kinase C: Alternative splicing from a single gene, *Science* 236:1116 (1987).
26. Y. Ono, T. Fujii, K. Ogita, U. Kikkawa, K. Igarashi, Y. Nishizuka, Identification of three additional members of rat protein kinase C family, *FEBS Lett* 226:125 (1987).
27. Y. Ono, T. Fujii, K. Ogita, U. Kikkawa, K. Igarashi, Y. Nishizuka, The structure, expression, and properties of additional members of the protein kinase C family, *J Biol Chem* 263:6927 (1988).
28. D.A. Persons, W.O. Wilkison, R.M. Bell, O.J. Finn, Altered growth regulation and enhanced tumorigenicity of NIH 3T3 fibroblasts transfected with protein kinase C-1 cDNA, *Cell* 52:447 (1988).
29. G.M. Housey, M.D. Johnson, W.L.W. Hsiao, C.A. O'Brien, J.P. Murphy, P. Kirschmeier, I.B. Weinstein, Overproduction of protein kinase C causes disordered growth control in rat fibroblasts, *Cell* 52:343 (1988).
30. P.M. Choi, K.-M. Tchou-Wong, I.B. Weinstein, Overexpression of protein kinase C in HT29 colon cancer cells causes growth inhibition and tumor suppression, *Mol Cell Biol* 10:4650 (1990).
31. E. Melloni, S. Pontremoli, B. Sparatore, M. Patrone, F. Grossi, P.A. Marks, R.A. Rifkind, Introduction of the β isozyme of protein kinase C accelerates induced differentiation of murine erythroleukemia cells, *Proc Natl Acad Sci USA* 87:4417 (1990).
32. L.A. Hansen, N.A. Monteiro-Riviere, R.C. Smart, Differential down-regulation of epidermal protein kinase C by 12-O-tetradecanoylphorbol-13-acetate and diacylglycerol: Association with epidermal hyperplasia and tumor promotion, *Cancer Res* 50:5740 (1990).
33. G. Brooks, R.E. Wilson, T.P. Dooley, M.W. Goss, I.R. Hart, Protein kinase C down-regulation, and not transient activation, correlates with melanocyte growth, *Cancer Res* 51:3281 (1991).
34. F.L. Meyskens, Jr., S.E. Salmon, Inhibition of human melanoma colony formation by retinoids, *Cancer Res* 39:4055 (1979).
35. R. Lotan, D. Lotan, A. Kadouri, Comparison of retinoic acid effects on anchorage-dependent growth, anchorage-independent growth and fibrinolytic activity of neoplastic cells, *Exp Cell Res* 141:79 (1982).
36. R. Lotan, T. Stolarsky, D. Lotan, A. Ben-Zeev, Retinoic acid restores shape-dependent growth control in neoplastic cells cultured on poly(2-hydroxyethyl methacrylate)-coated substrate, *Int J Cancer* 33:115 (1984).
37. F.L. Meyskens, Jr., B.B. Fuller, Characterization of the effects of different retinoids on the growth and differentiation of human melanoma cell lines and selected subclones, *Cancer Res* 40:2194 (1980).

38. E.L. Felix, B. Loyd, M.H. Cohen, Inhibition of the growth and development of a transplantable murine melanoma by vitamin A, *Science* 189:886 (1975).
39. D.M. Morre, T.M. Kloppel, A.L. Rosenthal, P.C. Fink, Chemoprevention of tumor development and metastasis of transplantable hepatocellular carcinomas in rats by vitamin A, *J Nutr* 110:1629 (1980).
40. M.E. Greenberg, E.B. Ziff, Stimulation of 3T3 cells induces transcription of the c-*fos* proto-oncogene, *Nature* 311:433 (1984).
41. R. Niles, Retinoic acid-induced arrest of mouse melanoma cells in G1 without inhibition of protein synthesis, *In Vitro* 23:803 (1987).
42. C. Dani, J.M. Blanchard, M. Piechaczyk, S. El Sabouty, L. Marty, P. Jeanteur, Extreme instability of myc mRNA in normal and transformed human cells, *Proc Natl Acad Sci USA* 81:7046 (1984).
43. J.P. Durham, D.A. Emler, F.R. Butcher, J.A. Fontana, Calcium activated, phospholipid-dependent protein kinase activity and protein phosphorylation in HL-60 cells induced to differentiate by retinoic acid, *FEBS Lett* 185:157 (1985).
44. E. Zylber-Katz, R.I. Glazer, Phospholipid and Ca++-dependent protein kinase activity and protein phosphorylation patterns in the differentiation of human promyelocytic leukemia cells HL-60, *Cancer Res* 45:5159 (1985).
45. R.E. Martell, R.U. Simpson, J.M. Taylor, 1,25-Dihydroxyvitamin D_3 regulation of phorbol ester receptors in HL-60 leukemia cells, *J Biol Chem* 262:5570 (1987).
46. M. Makowske, R. Ballester, Y. Cayre, O.M. Rosen, Immunochemical evidence that three protein kinase C isozymes increase in abundance during HL-60 differentiation induced by dimethyl sulfoxide and retinoic acid, *J Biol Chem* 263:3402 (1988).
47. D.K. Ways, R.C. Dodd, T.E. Bennett, T.K. Gray, H.S. Earp, 1,25 Di-hydroxyvitamin D_3 enhances phorbol ester-stimulated differentiation and protein kinase C-dependent substrate phosphorylation activity in the U937 human monoblastoid cell, *Endocrinology* 121:1654 (1987).
48. S. Rose-John, A. Dietrich, F. Marks, Molecular cloning of mouse protein kinase C (PKC) cDNA from Swiss 3T3 fibroblasts, *Gene* 74:465 (1988).
49. G. Shaw, R. Kamen, A conserved AU sequence from the 3′ untranslated region of GM-CSF mediates selective mRNA degradation, *Cell* 46:659 (1986).
50. N. Bacher, Y. Zisman, E. Berent, E. Livneh, Isolation and characterization of PKC-L, a new member of the protein kinase C-related gene family specifically expressed in lung, skin and heart, *Mol Cell Biol* 11:126 (1991).
51. G.H. Yaokum, J.F. Lechner, E.W. Gabrielson, B.E. Korba, L. Malan-Shibley, J.C. Willey, M.G. Valerio, A.M. Shamduddin, B.F. Trump, C.C. Harris, The v-ras gene can transform primary cultures of human tracheal epithelial cells, *Science* 227:1174 (1985).
52. J.S. Rhim, G. Jay, P. Arnstein, F.M. Price, K.K. Sanform, S.A. Aaronson, Neoplastic transformation of human epidermal keratinocytes by AD12-SV-40 Kirsten sarcoma viruses, *Science* 227:1250 (1985).
53. C.D. Scher, E.M. Scolnick, R. Siegler, Introduction of erythroid leukaemia by Harvey and Kirsten sarcoma viruses, *Nature* 256:225 (1975).
54. M. Noda, M. Kuo, A. Ogura, D. Liu, T. Amaon, T. Takano, Y. Ikawa, Sarcoma viruses carrying ras oncogenes induce differentiation-associated properties in a neuronal cell line, *Nature* 318:73 (1985).

Chapter 4

Racial and Ethnic Issues in Diet and Cancer Epidemiology

SHIRIKI KUMANYIKA

I. Introduction

This article describes differences in patterns of diet-related cancers among racial and ethnic groups. The objective is to demonstrate the potential insights about the possible role of diet in cancer etiology that might be gained from appropriately designed racial and ethnic comparisons. The examples used relate to racial and ethnic groups within the United States, although the issues are also relevant internationally. A comprehensive consideration of racial/ethnic differences in cancer occurrence from a global perspective has been published elsewhere.[1]

The four major racial/ethnic categories used by the U.S. Census bureau are African Americans, Hispanic Americans, Asian and Pacific Islander Americans, and American Indians and Alaskan Natives.[2,3] Together these groups comprise more than 20% of the U.S. population (Table 1), but it is inappropriate to consider minorities as a single group. Minority populations differ, within and across subgroups, with respect to ethnic and cultural origins, lifestyles, socioeconomic factors, region of residence, general health profiles,[2-7] and, as described here, there are marked differences in cancer risk among minority populations and in comparison to whites.[2,8-11]

Hispanic Americans include several Spanish-speaking or Spanish-origin subgroups: Mexican Americans, Puerto Ricans, Cuban Americans, Central Americans, South Americans, and others.[6,7] Asian and Pacific Islanders are even more diverse. This category aggregates people form diverse countries in East Asia (China, Japan, and Korea), Southeast Asia (the Philippines, Vietnam, Cambodia, Laos, Thailand, Malaysia, Singapore, and Indonesia), the Indian Subcontinent (India, Pakistan, and Bangladesh, Sri Lanka, and Burma), and the Pacific Islands (Hawaii, Guam, Samoa, Tonga, Fiji, and other Micronesian Islands).[3] American Indians and Alaska Natives include more than 500 federally recognized tribes and bands who live throughout the United States.[12]

Shiriki Kumanyika • Center for Biostatistics and Epidemiology, College of Medicine, Pennsylvania State University, Hershey, PA 17033

Diet and Cancer: Markers, Prevention, and Treatment
Edited by M.M Jacobs, Plenum Press, New York, 1994

Table 1. U.S. Racial and Ethnic Composition, 1988[a]

Racial category	Estimated number (in millions)	Estimated increase from 1980 to 1988	Percent of population in 1988
Black American	30.2	+13%	12
Hispanic American	19.8	+34%	8[b]
Asian and Pacific-Islander American	6.5	+70%	3
American Indian and Alaskan Natives	1.7	+19%	1
White American	207.4	+6%	84

[a] Data are from Reference 2.
[b] Hispanics may be of any race and are included, for example, in the estimates for blacks and whites. Therefore, this 8% overlaps with the estimates for the other categories, which total to 100%.

Since these racial and ethnic groups have different cancer incidence rates,[2,8-11] and since some aspects of the diets of these groups also differ,[13-18] the extent to which dietary differences contribute to the differences in cancer risks becomes an obvious issue for consideration by epidemiologists studying patterns of cancer occurrence. There are two possible ways in which a contribution of diet to racial and ethnic differences in cancer risk can occur: (1) racial and ethnic groups can differ in the prevalence of dietary factors that cause or protect against cancer; and (2) racial and ethnic groups can differ in the susceptibility to develop cancer when exposed to a given dietary risk factor or protective factor. This latter possibility is of direct relevance to the discussion of biological markers, since racial/ethnic groups do differ in at least their superficial biological characteristics as well as in some documented health-related biological characteristics.[19]

It is tempting to postulate that biological differences are the explanation for some of the more striking differences in cancer risk that are observed. It may be difficult to imagine that environmental risk factors could be so differentially distributed among groups sharing essentially the same environment as to produce the observed variations in patterns of cancer occurrence. However, while race/ethnicity as defined in the United States can in no sense be considered biological variables,[20-22] the clustering of risk or protection in racial and ethnic groups can be used as one rubric for focusing studies that probe for biological differences in cancer susceptibility. Any factors found to be more common in certain racial/ethnic groups than in others may provide important leads to etiology. Having said this, however, it must also be acknowledged that any explanations for racial/ethnic differences based on ideas about genetically determined biological differences will be extremely difficult to confirm in the face of a host of other plausible explanations based either on methodological factors or on racial/ethnic differences in environmentally determined cancer risks.[23]

II. Cancer Incidence in Racial and Ethnic Minority Populations in the United States

Data from cancer registries in the Surveillance, Epidemiology, and End Results (SEER) Program have been used to develop race- or ethnicity-specific estimates in which rates for minority populations can be compared with rates in reference data from white populations.[2,8,9] Figures 1 through 6 show racial/ethnic comparisons of cancer incidence rates for several types of cancer that have been associated with dietary causes. Rates are averaged over 3- to 5-year periods to smooth out possible artifacts associated with the calculation of rates based on small numbers. Geographic-specific comparisons were used, when available

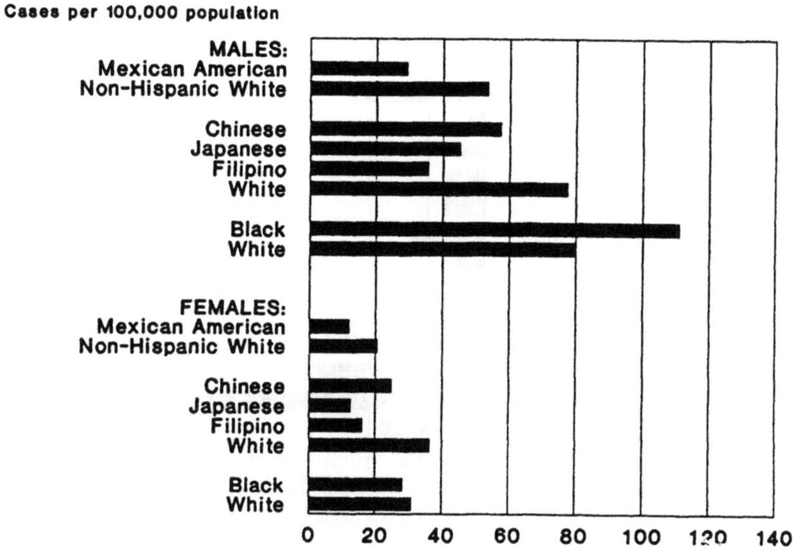

Figure 1. Average annual age-adjusted incidence of lung and bronchus cancer, United States, 1977-83, by sex and race/ethnicity, from selected SEER registries. Data for white and minority populations are geographic specific. Source: Ref. 2.

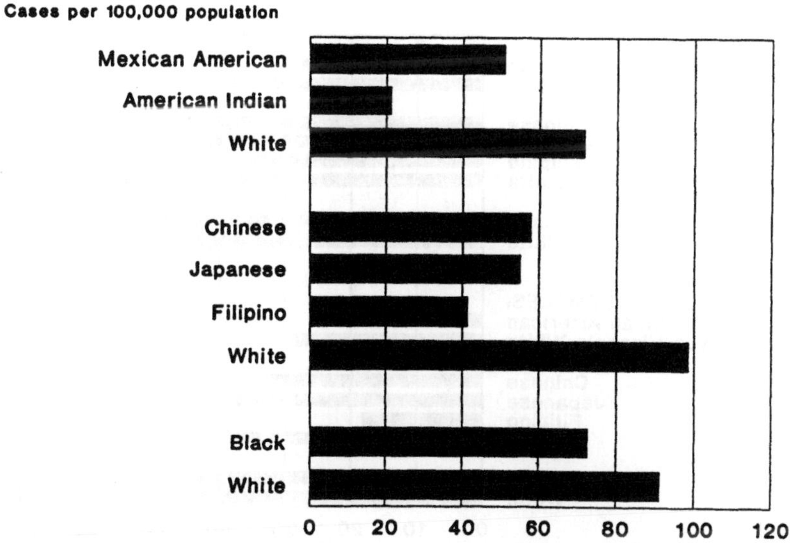

Figure 2. Average annual age-adjusted incidence of breast cancer among females, United States, 1977-83, by race/ethnicity, from selected SEER registries. Data for white and minority populations are geographic specific. Source: Ref. 2.

(Figures 1-4). The greater validity of such within-region comparisons is evident in the amount of variation in the incidence of cancer of the same site among white populations in different regions in Figures 1, 2 and 4. Comparisons of minority populations in one region with whites in the United States as a whole may be less precise. All comparisons in Figures 1 through 6 are adjusted for age differences across populations. However, none has been

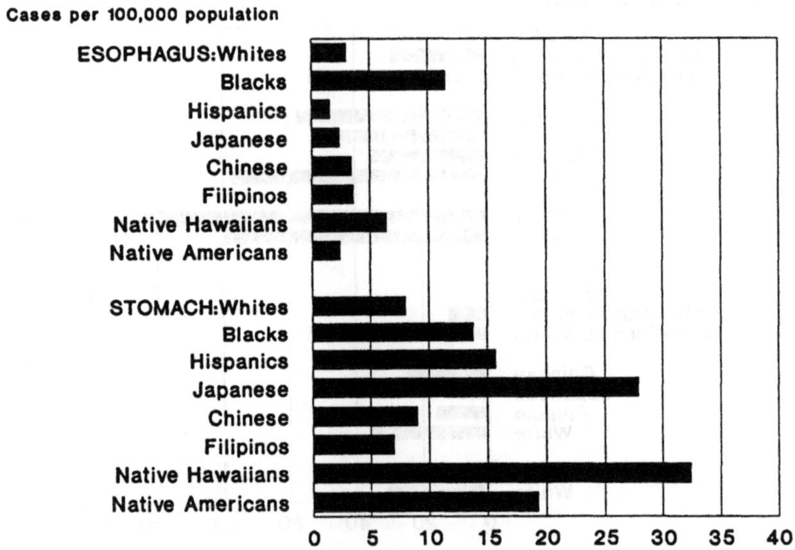

Figure 3. Average annual age-adjusted incidence of prostate cancer among males, United States, 1977-83, by race/ethnicity, from selected SEER registries. Data for white and minority populations are geographic specific. Source: Ref. 2.

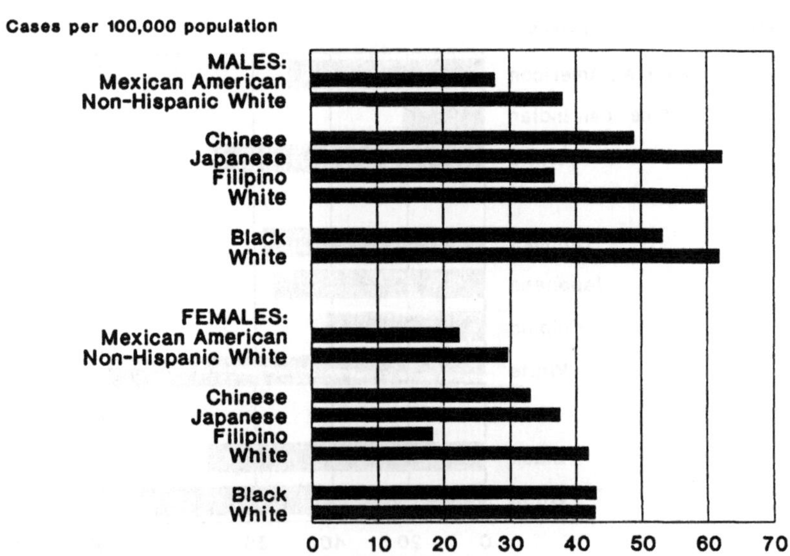

Figure 4. Average annual age-adjusted incidence of colon and rectum cancer, United States, 1977-83, by sex and race/ethnicity, from selected SEER registries. Data for white and minority populations are geographic specific. Source: Ref. 2.

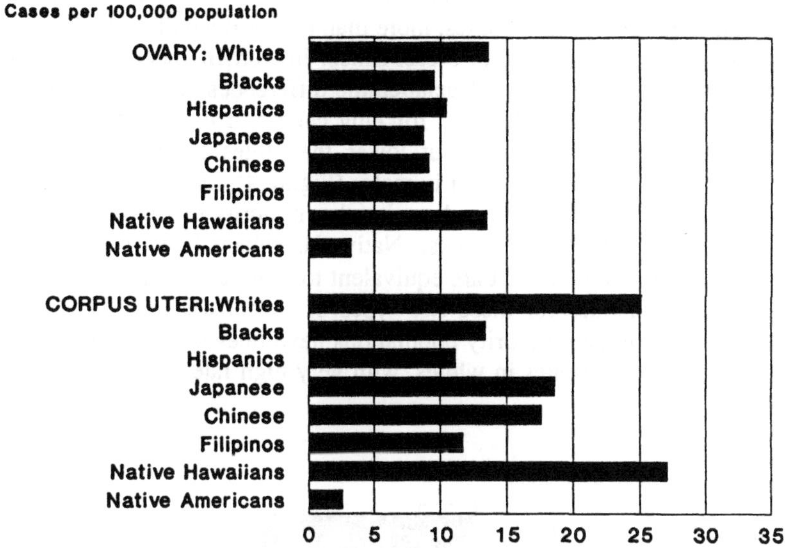

Figure 5. Average annual age-adjusted incidence of cancer of the ovary and corpus uteri, United States, 1978-81, by race/ethnicity, from selected SEER registries. Source: Ref. 9.

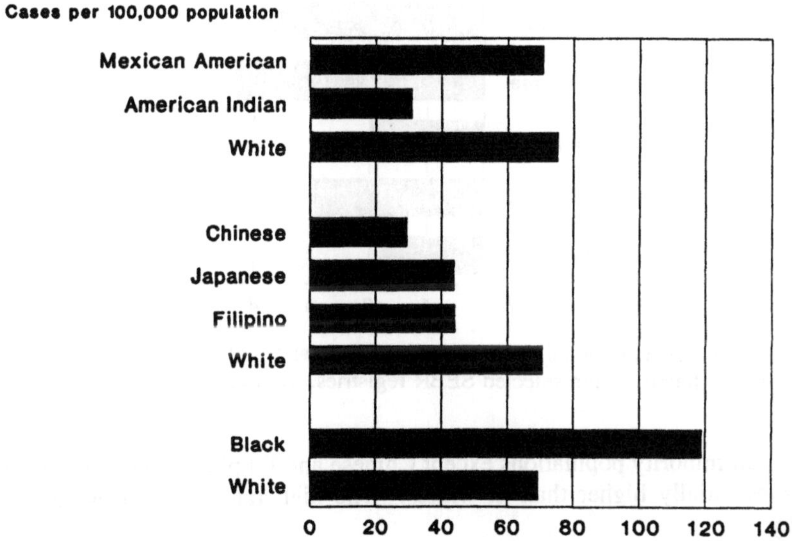

Figure 6. Average annual age-adjusted incidence of cancer of the esophagus and stomach, United States, 1978-81, by race/ethnicity, from selected SEER registries. Source: Ref. 9.

adjusted for socioeconomic status differences among the groups being compared. Effects of socioeconomic factors on comparisons of this type are discussed in Section IV of this chapter.

Several striking racial/ethnic differences can be noted in these data. Among males the incidence of lung cancer is notably higher in blacks than whites, whereas for the other minority populations the rates in the white comparison population are higher by as much as two-fold in both sexes (Figure 1). Breast cancer incidence is higher in white females than in females in minority groups in all cases, more than two or three times higher in some cases (Figure 2). However, the incidence of breast cancer in women under 40 years is higher for blacks than for whites[9] (not shown). Racial/ethnic differences in prostate cancer (Figure 3) follow a similar pattern to that observed for lung cancer in that rates for black males are notably higher than for white males but lower than rates for males in other minority populations. For colon and rectal cancer in males (Figure 4), rates in Japanese males exceed those in white males but are lower than those in other minority group males and for minority group females in all groups except blacks. Native Hawaiians have rates of ovarian cancer and cancer of the uterine corpus that are equivalent to or greater than rates in white females (Figure 5). Otherwise, rates for these two types of cancer are lower for minority group females. In Figure 6, several minority populations have rates of esophageal cancer that are equivalent to or higher than rates in whites, with very high rates in blacks. As also shown

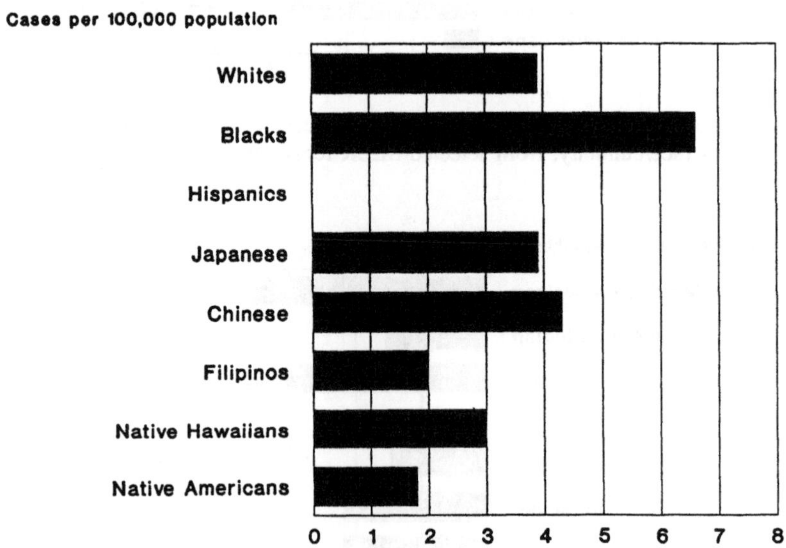

Figure 7. Average annual age-adjusted mortality from cancer of the corpus uteri, United States, 1978-81, by race ethnicity, from selected SEER registries. Source: Ref. 9.

in Figure 6, all minority populations except Chinese and Filipinos have stomach cancer rates that are substantially higher than in whites. The high rates in Japanese are particularly noteworthy.

In addition to these differences in cancer incidence, there are also racial/ethnic differences in tumor characteristics such as histological type, site of occurrence, and survival rates. For example, blacks appear to have more aggressive types of tumors for some cancer sites compared to whites.[9] The proportion of adenocarcinoma of the stomach that occurs in the gastric cardia is much lower for black males than for white males.[24,25] The 5-year relative survival rates for corpus uteri cancer were 88% for whites and 57% for blacks for 1973-1981.[9]

The uterine corpus cancer mortality rates in Figure 7, which show much higher mortality in blacks than whites, are a clear example of the importance of using cancer incidence rather than cancer mortality data when attempting to identify the role of potential etiological factors such as diet, since the data in Figure 5 indicate that black females are actually less likely to develop uterine corpus cancer than are white females. Survival factors that may be unrelated to etiology intervene to reverse the relative risks in the mortality data.

III. Racial or Ethnic Comparisons of Diet and Cancer

Using recognized diet and cancer associations[26] as a guideline, these comparative cancer incidence data can be used to develop hypotheses about dietary factors that might be more or less prevalent in certain racial/ethnic groups in association, respectively, with higher or lower than average rates of cancer. For example, do the high rates of stomach cancer in several minority populations, and especially in Japanese Americans (Figure 6), occur in conjunction with a high prevalence of associated dietary risk factors such as intakes of smoked, salted, and pickled foods and cured meat products containing nitrates? For this association the answer appears to be "yes,"[13-15] and further parallel studies within these populations might be expected to indicate that there is no difference in stomach cancer risk in Japanese compared to whites after accounting for these dietary factors, adjusting for possible confounding variables. Another plausible hypothesis would be that the racial/ethnic differences in ovarian cancer rates are related to the lower exposure to dietary galactose among minority populations due to a combination of higher rates of lactose maldigestion and lower dairy product consumption among minorities.[27] Provocative evidence that dairy products, especially those such as yogurt, in which free galactose levels are higher, increase the risk for ovarian cancer has been reported by Cramer and co-workers.[28,29] The high incidence of prostate cancer in black men (Figure 3) could be related to the high intakes of retinol and β-carotene-containing fruits and vegetables in the black population.[13,30,31] This follows the line first suggested by Graham *et al.*[32], i.e., that high vitamin A intake may increase the risk of prostate cancer even though it appears to decrease risks for cancers at other sites.

In contrast, the evidence indicating higher rates of lung or esophageal cancer in blacks than whites (Figure 1 and Figure 6) in the context of higher than average intakes of protective foods such as cruciferous and β-carotene-containing vegetables,[13,31] might lead to a hypothesis that these foods are less protective for blacks than would appear to be the case based on studies in whites. Similarly, the observation that black women have rates of uterine corpus cancer, for which obesity is an important risk factor,[26] that are 50% of rates in white women (Figure 5) while the prevalence of obesity in black women is twice that in white women[2] suggests that the sensitivity to obesity as a risk factor for this type of cancer differs in black and white women.

Overall, numerous discrepancies between the incidence of cancer and the population prevalence of associated risk factors can be identified to suggest that differences in the prevalence of dietary risk factors do not explain, or do not fully explain, the differences in cancer risks. This is also true for certain other diet-related chronic diseases.[33,34] However, population-level or "ecological" comparisons, in which the population is the unit of analysis and in which the risk factor data are collected separately from the data on cancer incidence, are useful primarily for generating hypotheses, e.g., in identifying possible patterns to be studied in more detail.[35] Data in which risk factor occurrence is associated with disease at the individual level provide a more appropriate basis for actually determining whether a given dietary factor has the same effect in one racial/ethnic group as in another. For example, a study by Correa *et al.*[36] indicates some food variables that appear to have differing associations with gastric cancer risk in blacks and whites. However, there are relatively few studies of diet and cancer across racial/ethnic groups that will support valid comparisons of this type. The remainder of this chapter argues that more such studies should be done, but done with attention to several potential interpretational problems.

IV. Methodological Issues

Racial/ethnic comparisons of diet and cancer associations involve heterogeneity that can be both beneficial and problematic. The benefits include the broader range of dietary exposures in diverse cultural groups that may include both lower or higher intakes of certain foods or nutrients compared to other populations, as well as a broader range of intakes within populations, e.g., if the population includes migrants or persons who retain different degrees of adherence to traditional diets. Different cultural groups also have different sources of the same nutrients.[36,37] In the data in Table 2, for example, beef and sausage provide a higher proportion of total fat intake in black than in white Americans; liver contributes a third of total vitamin A in Mexican Americans and one-fifth of total vitamin A in black Americans but makes no substantial contribution to vitamin A in white Americans. Greens (particularly collard and mustard greens) contribute substantially to vitamin A intakes in black Americans but not in other groups. Both black Americans and Mexican Americans consume more of the total vitamin C from fruit-flavored beverages than do white Americans. Similar examples could be given for racial/ethnic differences in sources of dietary fiber or other nutrients.[13,38] For example, whole-grain products contribute less and greens and legumes more to fiber intakes in black than in white Americans.[38] These differences in nutrient sources can be useful for differentiating risks associated with foods vs. risks associated with nutrients contained in those foods.

Table 2. Differences in the Percent Contribution of Selected Foods to Intakes of Total Fat and Vitamins A and C[a]

	Black Americans	White Americans	Mexican Americans
Total fat intake			
Beef cuts	13.3	9.6	12.3
Sausage	5.1	1.7	2.5
Fried chicken	4.2	3.2	1.8
Vitamin A intake			
Liver	19.8	[b]	33.7
Carrots	7.0	16.5	2.2
Greens	11.1	1.2	
Sweet potatoes, pumpkin, (including pies)	8.7		3.5
Vitamin C intake			
Oranges, orange juice	30.6	33.9	14.9
Fruit-flavored beverages	14.7	6.6	25.8

[a] Data are from Reference 30.

[b] Blanks indicate that the food contributed less than 0.5% to intake of this nutrient in this racial or ethnic group.

Racial/ethnic groups in the United States differ on aspects of cancer epidemiology such as the slope and direction of trends in cancer incidence, the prevalence of nondietary risk factors such as smoking; alcohol use, sedentary behavior, and occupational and regional exposure to carcinogens; and the prevalence of other chronic conditions such as obesity and diabetes mellitus[1-4,6,8,39] that may affect cancer.[26] Also, as noted previously, there are racial and ethnic differences in tumor histological characteristics and anatomical sites[9] that may increase the ability to differentiate effects of various dietary factors. Racial/ethnic differences in survival may provide a context for studying effects of diet on tumor recurrence.

As indicated in Table 3, however, many of these same aspects that may render racial/ethnic comparisons of diet and disease beneficial in providing for greater contrast or for studying associations across a broader range or more varied combination of exposures are problematic when attempting to determine whether apparent differences in susceptibility to diet-related cancers are artifactual, are explained by the sum total of risk factor differences between populations, or are in fact evidence of underlying biological differences.

Table 3. Problems in Estimating Racial or Ethnic Differences in Susceptibility to Dietary Factors

Dietary assessment	Socioeconomic factors
Food vs. food constituents	Occupational factors
Other lifestyle differences	Regional factors
Definitions of race	Secular trends
Tumor characteristics	Survival differences

As reviewed elsewhere,[34] the complexities of nutritional epidemiology involving minority populations are added to the already numerous methodological problems in this field.[40,41] Differences in nutrient sources place an extra burden on dietary assessment in a cross-cultural context in that intake of a nutrient may be underascertained in populations for whom important food sources are omitted or underspecified in food-frequency questionnaires.[30] Analyses based on food groups will be difficult to interpret cross culturally because, as noted earlier, the same food will contribute different proportions of the intakes of different dietary constituents in different populations.

The potential confounding of cross-cultural diet and cancer studies by socioeconomic status differences between groups is significant, since socioeconomic factors are strongly associated with cancer incidence, survival, and risk factors and with race/ethnicity,[42] to such an extent that some racial differences in cancer risk are reversed when socioeconomic factors are taken into account.[43] Racial/ethnic differences in exposure to geographic cancer risk factors can be potentially controlled by making comparisons within regions (as in Figures 1-4). However, subarea factors such as urban vs. rural residence will still be differential between minorities and nonminorities. Further, occupational exposures, both type of occupations and differences in exposure based on job assignment, vary with race/ethnicity and socioeconomic status.[8]

The potential confounding of diet and cancer comparisons across race/ethnicity by secular trends in both dietary patterns and cancer is also substantial because of the long latency periods from exposure to disease emergence. Current eating patterns in a population may be less representative of cancer risks than eating patterns 10 or 20 years previously, as considered by Hargreaves *et al.* in their consideration of diet and cancer in black Americans.[14] Where cancer incidence is changing over time, current (i.e., cross-sectional) differences in cancer risk across populations may give a misleading picture. The effects of diet may be reflected in *future* risks as indicated by an upward secular trend.[39]

Finally, as noted in Section I, the definition of race or ethnicity can be problematic. Our sociopolitical designations are useful for identifying populations with certain common characteristics, but should not be considered genetic classifications. A population such as black Americans may be equally or more homogenous behaviorally than biologically in relation to certain risks, allowing the behavioral factor to masquerade as a genetic variable. For example, black-white differences in the presentation of tobacco-related cancers could be due to underlying biological differences between blacks and whites in the susceptibility of tissues to tobacco smoke or could be found to relate to marked differences in smoking patterns (the number of cigarettes smoked, tar and nicotine levels, menthol content of brands chosen),[44] if these factors could be tied to the specifics of cancer causation. Chakraborty and co-workers[22] have proposed the use of multiple unique alleles as a more scientific basis for estimating racial admixture from a parental population than the demographic characterizations commonly used. Without this type of approach, the utility of racial/ethnic comparisons in elucidating diet and cancer associations will of necessity be limited to inferences based on environmental factors because of the methodological infeasibility of ever ruling out confounding by unmeasured or inadequately measured environmental differences.

V. Conclusion

Many racial or ethnic differences in diet-related cancers can probably be explained on the basis of differences in environmental risk factors, although this has been insufficiently demonstrated. More relevant to the issue of biological markers is that there may be racial/ethnic differences in susceptibility to diet-related cancers that may lead to insights into mechanisms of dietary effects and/or to identification of biological markers for cancer risk that are associated with genetic differences. However, conclusions about genetic differences can only be made if biological markers for racial admixture are used to define race/ethnicity variables. The measurement of environmental exposures will never be sufficiently precise to support the attribution of racial/ethnic differences to genetic factors whenever differences cannot be explained environmentally. Even apparent differences in susceptibility may be artifacts introduced by the noncomparable measurements obtained in heterogeneous cultural and biological contexts. However, appropriately designed racial/ethnic comparisons offer considerable potential for generating insights into the role of diet in cancer prevention and treatment that would not arise from studies of homogeneous groups.

References

1. A.P. Polednak, *Racial and Ethnic Differences in Disease*, New York, Oxford, p. 168 (1989).
2. National Center for Health Statistics, *Health, United States, 1990*, U.S. Department of Health and Human Services, Public Health Service, Centers for Disease Control, DHHS pub. no. (PHS) 91-1232, Washington, D.C., U.S. GPO (1991).
3. U.S. Department of Health and Human Services, *Report of the Secretary's Task Force on Black and Minority Health, Volume II, Crosscutting Issues in Minority Health*, Washington, D.C., U.S. GPO (1985).
4. U.S. Department of Health and Human Services, *Report of the Secretary's Task Force on Black and Minority Health, Volume I, Executive Summary*, Washington, D.C., U.S. GPO (1985).
5. J.S. Lin-Fu, Population characteristics and health care needs of Asian Pacific Americans, *Publ Health Rep* 103:18 (1988).
6. Hispanic Americans, An emerging group, Part 1, Magnitude, ethnicity, and geographic location, *Statist Bull* 69(4):2 (1988).

7. Hispanic Americans, An emerging group, Part 2, Profile of diverse characteristics, *Statist Bull* 69(4):8 (1988).
8. U.S. Department of Health and Human Services, *Report of the Secretary's Task Force on Black and Minority Health, Volume III, Cancer*, Washington, D.C., U.S. GPO (1986).
9. C.R. Baquet, K. Ringen, *Cancer among Blacks and Other Minorities, Statistical Profiles*, National Cancer Institute, NIH pub. no. 86-2785, Washington, D.C., U.S. GPO (1986).
10. P.E. Wolfgang, P.A. Semeiks, W.S. Burnett, Cancer incidence in New York City Hispanics, *Ethnic Dis* 1:263 (1991).
11. A.P. Lanier, L.R. Bulkow, B. Ireland, Cancer in Alaskan Indians, Eskimos, and Aleuts, 1969-83, Implications for etiology and control, *Pub Health Rep* 104:658 (1989).
12. U.S. Department of Commerce, Bureau of the Census, *We, the First Americans*, Superintendent of Documents (SuDoc No C3.2:AM3.9), Washington, D.C., U.S. GPO (1986).
13. S. Kumanyika, D.L. Helitzer, Nutritional status and dietary patterns of racial minorities in the United States, in: U.S. Department of Health and Human Services, *Report of the Secretary's Task Force on Black and Minority Health, Volume II, Crosscutting Issues in Minority Health*, Washington, D.C., U.S. GPO, pp. 118-190 (1985).
14. M.K. Hargreaves, C. Baquet, A. Gamshadzahi, Diet, nutritional status, and cancer risk in American blacks, *Nutr Canc* 12:1 (1989).
15. P.G. Kittler, K. Sucher, *Food and Culture in America*, New York, Van Nostrand Reinhold (1989).
16. D. Sanjur, Ethnicity and food habits, in: *Social and Cultural Perspectives in Nutrition*, Englewood Cliffs, NJ, Prentice-Hall, p. 233 (1982).
17. B.F. Harland, S.A. Smith, R. Ellis, R. O'Brien, E.R. Morris, Comparison of the nutrient intakes of blacks, Siouan Indians, and whites in Columbus County, North Carolina, *J Am Diet Assoc* 92:348 (1992).
18. E.D. Nobmann, T. Byers, A.P. Lanier, J.H. Hankin, M.Y. Jackson, The diet of Alaska Native adults, 1987-1988, *Am J Clin Nutr* 55:1024 (1992).
19. A.P. Polednak, Physical anthropology of racial/ethnic groups, in: *Racial and Ethnic Differences in Disease*, New York, Oxford, p. 32 (1989).
20. R.S. Cooper, A note on the biologic concept of race and its application in epidemiologic research, *Am Heart J* 108:715 (1987).
21. A. Weissman, "Race-ethnicity," A dubious scientific concept, *Publ Health Rep* 105:102 (1990).
22. R. Chakraborty, M. Ilyas Kamboh, and R.E. Ferrell, "Unique" alleles in admixed populations, A strategy for determining "hereditary" population differences of disease frequencies, *Ethnic Dis* 1:245 (1991).
23. A.P. Polednak, Interpretation of racial/ethnic differences in disease, in: *Racial and Ethnic Differences in Disease*, New York, Oxford, pp. 17-31 (1989).
24. W.J. Blot, S.S. Devesa, R.W. Kneller, J.F. Fraumeni, Rising incidence of adenocarcinoma of the esophagus and gastric cardia, *JAMA* 265:1287 (1991).
25. P.C. Yang, S. Davis, Epidemiologic characteristics of adenocarcinoma of the gastric cardia and distal stomach in the United States 1973-1982, *Int J Epidemiol* 12:293 (1988).
26. National Research Council, *Diet and Health, Implications for Reducing Chronic Disease Risk, Report of the Committee on Diet and Health*, Food and Nutrition Board, Commission on Life Sciences, Washington, D.C., National Academy Press (1989).
27. N.S. Scrimshaw, E.B. Murray, The acceptability of milk and milk products in populations with a high prevalence of lactose intolerance, *Am J Clin Nutr* 48:supplement to October (1988).
28. D.W. Cramer, Lactase persistence and milk consumption and determinants of ovarian cancer, *Am J Epidemiol* 130:904 (1989).
29. D.W. Cramer, B.L. Harlow, W.C. Willett, W.R. Welch, et al., Galactose consumption and metabolism in relation to the risk of ovarian cancer, *Lancet* 2:66 (1989).
30. L.G. Borrud, S. McPherson, M.Z. Nichaman, P.C. Pillow, G.R. Newell, Development of a food frequency instrument: Ethnic differences in food sources, *Nutr Canc* 12:201 (1989).

31. B.H. Patterson, G. Block, Food choices and the cancer guidelines, *Am J Publ Health* 78:282 (1988).
32. S.B. Graham, B. Haughey, J. Marshall, R. Priore, T. Byers, T. Rzepka, C. Mettlin, J.E. Pontes, Diet in the epidemiology of carcinoma of the prostate gland, *J Natl Canc Inst* 70:687 (1983).
33. S. Kumanyika, Diet and chronic disease issues for minority populations, *J Nutr Educ* 22:89 (1990).
34. S.K. Kumanyika, Diet and nutrition as influences on the morbidity/mortality gap, *Ann Epidemiol* 3:154 (1993).
35. J.L. Kelsey, W.D. Thompson, A.S. Evans, Methods in Observational Epidemiology, New York, Oxford (1986).
36. P. Correa, E. Fontham, L.W. Pickle, V. Chen, Y.P. Lin, W. Haenszel, Dietary determinants of gastric cancer in South Louisiana inhabitants, *J Natl Canc Inst* 75:645 (1985).
37. L.G. Borrud, P.C. Pillow, P.K. Allen, R.S. McPhearson, M.Z. Nichman, G.R. Newell, Food group contributions to nutrient intake in whites, blacks, and Mexican Americans in Texas, *J Am Diet Assoc* 8:1061 (1989).
38. G. Block, E. Lanza, Dietary fiber sources in the United States by demographic group, *J Natl Canc Inst* 79:83 (1987).
39. S.K. Kumanyika, P.M. Golden, Cross-sectional differences in health status in U.S. racial/ethnic minority groups, Potential influence of temporal changes, disease, and life-style transitions, *Ethnic Dis* 1:50 (1991).
40. W. Willett, Nutritional epidemiology, Issues and challenges, *Int J Epidemiol* 16(suppl):312 (1987).
41. G.G. Rhoads, N. Kurinij, Epidemiological studies in nutrition, utility and limitations, *J Nutr* 118:240 (1988).
42. H.P. Freeman, Cancer in the socioeconomically disadvantaged, *Ca-A Cancer Journal for Clinicians* 39:266 (1989).
43. C.R. Baquet, J.W. Horm, T. Gibbs, P. Greenwald, Socioeconomic factors and cancer incidence among blacks and whites, *J Natl Canc Inst* 83:551 (1991).
44. G.C. Kabat, A. Morabia, E.L. Wynder, Comparison of smoking habits of blacks and whites in a case-control study, *Am J Publ Health* 81:1483 (1991).

Chapter 5

Dietary Modulation of Oxidative DNA Damage

ZORA DJURIC, SILVANA MARTINO,
LANCE K. HEILBRUN and RONALD W. HART

I. Introduction

Breast cancer is a major cause of death and disability among women in the United States. One in nine women will develop breast cancer at some point in their lives. Over the years, treatment modalities have improved, but mortality from breast cancer has changed little since 1930.[1] This observation dictates a need for greater efforts toward disease prevention. One important new prevention strategy to reduce the incidence of breast cancer is dietary intervention.

Epidemiological and case-control studies have suggested a relationship between dietary fat and breast cancer risk,[2-4] although not all studies have indicated such a relationship.[5-7] In order to examine the relationship between breast cancer and fat intake, a wide range of fat intake among subjects is desirable. One difficulty in determining the influence of fat intake on breast cancer risk in epidemiological studies is that within any given cultural group, the range of fat intake may be too narrow to observe an effect.[8] This may be one reason why studies comparing breast cancer incidence in the United States and Japan have shown more consistent associations of fat intake and breast cancer risk.[2,9] Similarly, in an Italian study where the diet varied considerably due to regional differences in culinary practices, increased fat intake was associated with increased breast cancer risk.[10]

Caloric content of the diet also may be a key factor, and increased body weight, which may reflect caloric intake, has been associated with increased cancer risk.[11,12] The independent effects of fat and caloric intake may, however, be difficult to separate in human epidemiological studies. The controversy between the relative importance of fat and calories remains unsettled. In the absence of controlled clinical trials, it is doubtful that more epidemiological studies will yield a resolution of the issues.[4]

Zora Djuric, Silvana Martino, and Lance K. Heilbrun • Department of Internal Medicine, Wayne State University, Detroit, MI 48201; Ronald W. Hart • National Center for Toxicological Research, Jefferson, AR 72079

Diet and Cancer: Markers, Prevention, and Treatment
Edited by M.M Jacobs, Plenum Press, New York, 1994

Controlled clinical trials to examine the effects of a caloric restriction on breast cancer incidence have not been reported. Low-fat interventions have been instituted, but low-fat diets often result in decreased caloric intake, making it difficult to separate the effects of fat reduction and reduced caloric intake from each other. There was a national low-fat intervention trial initiated in 1987, the Women's Health Trial, which randomized women into either non intervention or low-fat diets. One year after randomization, the percent of calories from dietary fat decreased from 39% to 22% in the intervention group and from 39% to 37% in the control group. There were concomitant decreases in caloric intake of 25% and 7% in the intervention and control groups, respectively.[13] The trial was terminated in 1988, but breast cancer incidence results from Seattle participants in the Women's Health Trial are very encouraging and indicate a 40% reduction in cancer incidence 1 to 5 years after randomization.[14]

The importance of both caloric intake and dietary fat on mammary tumorigenesis in laboratory rodents was first demonstrated by Tannenbaum in 1945.[15] Subsequently, numerous animal studies have shown that both reduced dietary fat and decreased caloric intake are protective against the development of mammary gland tumors in rodents.[2,16,17] A review of 100 animal studies has indicated that both higher caloric intake and higher fat intake independently increase mammary tumor incidence in rats and mice.[16]

One mechanistic explanation for the effects of dietary fat and calories on mammary gland tumorigenesis may include the dietary modulation of oxidative DNA damage levels. Caloric restriction has been shown to induce enzymes that detoxify oxygen free radicals as well as DNA repair enzymes.[18-22] Fats can be oxidized, leading to production of reactive oxygen species. These reactive oxygen species, and perhaps lipid radicals, can oxidize DNA bases.[23] There is strong evidence that oxidative DNA damage plays a role in first-stage promotion, initiation, and progression of tumors.[24-27] One therefore might expect that the levels of oxidized DNA bases can be utilized as an index of the carcinogenic effect of fat in the diet.

Numerous reports have examined the role of oxidative DNA damage in carcinogenesis. Hydroxylated DNA bases have been shown to be mutagenic when incorporated into the genome.[28-31] Oxidants and 5-hydroxymethyluracil may have a role in gene expression.[32,33] Oxidative DNA damage also can induce aberrant oncogene expression and the malignant transformation of cells.[25-27] Later in the carcinogenic process, oxidant-induced DNA damage may be part of the inherent genetic changes associated with tumor cell growth. Endogenous production of hydrogen peroxide by tumor cells has been associated with tumor cell proliferation and may confer a growth advantage to tumor cell populations via mutation that favors heterogeneity, invasion, and metastasis.[34-36] High levels of oxidative DNA damage have been found in surgically obtained breast cancer tissues.[37]

Of the various roles oxidative DNA damage can have in carcinogenesis, promotion is perhaps the most closely tied to dietary fat and calories. Promotion is thought to be reversible and may be influenced by intervention strategies such as diet.[27,38] In addition, the effects of a promoter may be evident in many tissues other than the target organ, including peripheral blood.

Reports in the literature indicate that endogenous DNA damage and oxidative stress are increased in leukocytes of breast cancer patients and high-risk individuals. DNA repair in response to hydrogen peroxide was shown to be compromised in leukocytes of patients with cancer of various sites,[39] as was DNA repair in leukocytes of individuals with a family history of cancer.[40] DNA repair ability in lymphocytes from breast cancer patients and from high-risk women also was found to be decreased relative to control women.[41,42] Lipid peroxidation was increased in women with mammographic dysplasia relative to control women, as determined by urinary malonaldehyde levels.[43] Thus, individuals at increased cancer risk may have increased levels of oxidative stress and/or compromised DNA repair. Both of these factors should be reflected in the levels of oxidative DNA damage in peripheral leukocytes.

We have examined 5-hydroxymethyluracil levels in DNA as a marker of oxidative stress and oxidative DNA damage to assess the possible beneficial effects of low-fat and calorie-restricted diets over a short period of time. This type of DNA damage was examined in peripheral nucleated blood cells of humans and mammary gland and liver of rats after dietary change.

II. Methods

A. Clinical Intervention

1. Patients. Women, aged 18-65, were eligible for the dietary intervention program if they satisfied one or more of the following criteria: (1) at least one first-degree relative had been diagnosed with breast cancer; (2) a P2 or DY mammogram; or (3) atypia or florid papillomatosis detected by breast examination. In addition, dietary fat intake upon entry to the program had to be at least 25% of the total caloric intake. All subjects gave their informed consent for participation in the study. Of the women in the program, a consecutive series of 21 were selected for analysis of oxidative DNA damage.

2. Dietary Intervention. The women were randomized onto either nonintervention or low-fat diets after stratification into three age groups. The nonintervention diet women followed their own usual diet. The women in the low-fat diet group were instructed by a dietitian on how to follow the Low Fat Eating Plan (target of 15% of calories from fat) developed by the American Health Foundation and the University of Minnesota Nutrition Coordinating Center. Three-day food records and 24-hr recall data for both groups were collected every 3 months by a registered dietitian and analyzed for nutritional content by the University of Minnesota Nutrition Coordinating Center.[44,45]

3. Blood Samples. The blood samples were drawn into heparinized tubes after a 14-hour overnight fast, and the women were between days 17 and 24 of the menstrual cycle. The blood was refrigerated immediately after drawing and subsequently stored at −70° until analysis. Nuclei were prepared from 5–10 ml of heparinized blood by the method of Ciulla *et al.*,[46] and the DNA was isolated using organic extractions as previously described using phenol, chloroform, and isoamyl alcohol.[47]

4. Statistical Analyses. Square root transformation of the DNA damage levels (in 5-hydroxymethyluracil/10^4 thymines) was necessary to better normalize the data. Differences between the two diet groups in mean (square root of) DNA damage levels and means of other variables were evaluated by two-sample *t*-tests. The relationship of total fat intake (g/day) and (square root of) DNA damage level was assessed via simple linear regression.

B. Animal Diets

1. Animals. Female, Fischer 344 rats were weaned at 21 days of age and maintained on NIH-31 diets. At 38 days of age, the rats were placed on the control diet (see "Diets," below). Food intake was monitored such that the amount of food to be fed to the calorically restricted rats could be determined. At an age of 45 days, the animals were weighed and divided arbitrarily into the four diet groups with 12 animals per group. The animals were maintained on the diets for 2 weeks with daily food changes.

The animals were housed singly in hanging, mesh-bottom cages fitted with trays to facilitate the collection of spilled food for the determination of total intake. Food intake, body weight, and the general health of the animals were monitored on a daily basis. The diets were presented to the animals in jars, and food intake was calculated from the amount of food left in the jar the next day plus the amount of food spilled.

The lights in the animal room were kept on a schedule of 12 hr on and 12 hr off. The food was presented to the animals immediately before the lights were turned off. All animals thus consumed their diets during the same time period.

After 2 weeks, the animals were sacrificed by carbon dioxide inhalation and decapitation. The mammary gland tissue and livers were removed. Whole livers were frozen. The mammary gland epithelium was isolated from fresh mammary gland tissue by the method of Moon *et al.*[48] Briefly, the tissue was minced and incubated with collagenase, and the fat was separated from the epithelial cells by centrifugation. The epithelial cells were frozen until DNA extractions could be performed.

2. Diets. The diets consisted of ingredients (casein, dextrose, cellulose, AIN76 vitamin and mineral mixes, D,L-methionine, and choline bitatrate) that were obtained from Teklad Premier Laboratory Diets (a division of Harlan Sprague Dawley Inc., Madison, WI). The diets prepared were: (1) 3% corn oil fed *ab libitum* (low fat), (2) 5% corn oil fed *ab libitum* (control), (3) 20% corn oil fed *ab libitum* (high fat), and (4) 5% corn oil fed at 40% caloric restriction relative to the *ab libitum* control diet (CR). The dextrose content of the diets was modified to accommodate the desired percentage of corn oil but the other ingredients were kept constant.[49] The percentages of all ingredients, except dextrose, were increased in the formulation of the CR diet such that when fed at 40% restriction, the calorically restricted rats consumed the same amount of nutrients as control. The diets were stored in a refrigerator.

The total energy values of the diets were calculated using the standard Atwater value estimates for physiological fuel values.[50] Carbohydrates, protein, and cellulose were calculated at 4 kcal/g and fat at 9 kcal/g. Cellulose is often considered to be an inert filler; however, it can be fermented in the hindgut to a limited extent, ranging from 12% to 20%.[51] In our calculation of caloric intake, 12% digestibility of cellulose was used.

3. DNA Extraction. The mammary gland epithelium and liver tissue were homogenized in 1% sodium dodecyl sulfate and 1 mM EDTA, and the DNA was extracted as previously described.[47] The procedure involved incubation of the homogenates with protease followed by successive organic extractions with phenol, chloroform/isoamyl alcohol/phenol (24:1:25) and chloroform/isoamyl alcohol (24:1), treatment of the semipurified nucleic acid with RNases A and T1 and another series of the same organic extractions. The DNA was recovered by precipitation with ethanol and quantified spectrophotometrically.

4. Statistical Methods. Dependent variables were analyzed to determine the effect of the four experimental diets on mean DNA damage level, nutrient intake, and animal weight. Nutrient intakes were calculated based on a simple mean food intake over the 14 days of study. DNA damage level was assessed at day 14 only. Due to the heavily nonnormal distribution of DNA damage level and nutrient intake, nonparametric methods of statistical analysis were used. For each dependent variable, separate analyses were performed using the *k*-sample Kruskal-Wallis rank sum test to compare mean levels by diet group.[52]

To perform the multiple comparisons of interest (all pairwise comparisons of diet groups), a nonparametric rank sum method was used, based on a large sample approximation for balanced designs. This allowed us to maintain the experimentwise Type I error rate at 0.05 while performing all pairwise comparisons of diet groups.[53] There were 12 rats for each diet group, yielding a completely balanced design, and there were no missing data. Rats were consecutively and arbitrarily assigned to diet groups. The data were viewed as resulting from separate (1 × 4) complete one-way layout experimental designs.

C. DNA Damage Analysis

The gas chromatography-mass spectrometry (GC-MS) methods of Dizdaroglu *et al.*[54] were modified in our laboratory to allow quantitation of 5-hydroxymethyl-2′-deoxyuridine and thymidine in enzymatically hydrolyzed DNA.[55] The enzymatic hydrolysis avoided problems with analyte instability that can be encountered with acid hydrolysis of the DNA. The purified DNA samples were hydrolyzed to nucleosides enzymatically with DNase I, snake venom phosphodiesterases I and II, and alkaline phosphatase. The protein and buffer salts in the hydrolysates were removed with use of Sep-paks, and trimethylsilyl derivatives of the nucleosides were prepared for analysis by GC-MS as previously described.[55]

The amount of thymine in each sample was determined using thymine-d_4 as an internal standard. This controlled for the amount of DNA injected onto the column and is important since spectrophotometrically quantitation of DNA is hampered by difficulties in pipetting viscous DNA solutions. In addition, quantitation of thymine serves as a monitor of DNA hydrolysis efficiency. The amount of 5-hydroxymethyluracil in each sample was determined using 5-hydroxymethyluracil-2-^{13}C,5-d_2,6-d as an internal standard. The isotopically labeled internal standards controlled for possible variations in the derivatization efficiency of each sample.

The GC-MS analyses were carried out with a Hewlett-Packard 5971 mass spectrometer operated in the EI single-ion mode, and positive ions were monitored. The Hewlett-Packard 5890 GC was operated in the splitless injection mode and injections were made with a Hewlett-Packard 7673 autosampler.

III. Results

A. Clinical Study

Oxidative DNA damage was analyzed in the nucleated blood cells of 21 women who had been in the dietary intervention program between 3 and 24 months. The level of fat intake was calculated from dietary records obtained at the same time blood was drawn for analysis of DNA damage. The mean total fat intake in the intervention group was decreased by almost 50% relative to the nonintervention diet group.[56] Polyunsaturated, monounsaturated, and saturated fats all were decreased by the same amount. The mean level of 5-hydroxymethyluracil in the nuclear DNA of peripheral nucleated blood cells from women on the low-fat diet was two-thirds lower relative to the nonintervention diet group (Table 1).

Our preliminary data indicate that changes in other nutrients besides fat occurred in women who consumed a low-fat diet, but these changes were not significantly different between the diet groups. It is interesting to note that mean vitamin C and β-carotene levels were increased 84% and 60%, respectively, in the women on the low-fat diet relative to control women, whereas mean vitamin E levels were decreased 44%. Mean % body fat differed significantly ($p = 0.013$) by diet arm: 26.8% vs. 32.4% for women on the low-fat and nonintervention diet arms, respectively. Fiber intake increased significantly in the low-fat group as well.

Among these 21 women, there was a significant linear regression relationship of daily total fat intake with the square root of DNA damage level, irrespective of diet arm (Figure 1). The fitted regression model (square root DNA damage level = 0.93 + 0.03 × total fat intake) had a significant slope ($p = 0.0003$) and a significant intercept ($p = 0.0196$). Further adjustment for % body fat and fiber intake, in a trivariate regression model, did not reduce the strength of the association between total fat intake and DNA damage level.

Table 1. Selected Parameters in 21 Women on Low-Fat and Standard Diets

Parameter	Diet group[a]		
	Low-fat diet	Nonintervention diet	p Value[b]
DNA damage[c]	3.0 ± 0.6	9.3 ± 1.9	0.004
Fat intake (g/day)	33 ± 7	57 ± 28	0.013
% calories from fat	19 ± 5	36 ± 10	0.0001
Caloric intake (kcal/day)	1615 ± 236	1390 ± 489	0.218
Age (years)	40 ± 12	46 ± 7	0.21
Body weight (pounds)	144 ± 44	147 ± 20	0.87
% body fat	27 ± 4	32 ± 5	0.013
Vitamin C intake (mg/day)	567 ± 770	307 ± 321	0.363
Vitamin E intake (IU/day)	58 ± 107	131 ± 267	0.405
β-Carotene (IU/day)	4017 ± 4146	2506 ± 3337	0.366
Fiber (g/day)	18 ± 7	12 ± 6	0.0336
Saturated fat (g/day)	12 ± 1	20 ± 3	0.018
Monounsaturated fat (g/day)	11 ± 1	20 ± 3	0.005
Polyunsaturated fat (g/day)	7 ± 1	13 ± 2	0.029

[a] Daily dietary intakes were calculated from 3-day food records collected at the time the blood was drawn (mean ± standard deviation). There were 9 women in the low-fat group and 12 women in the nonintervention group.

[b] The p values are for the difference between the two diet groups by the two sample t-test.

[c] DNA damage is expressed as 5-hydroxymethyluracil/10^4 thymine residues.

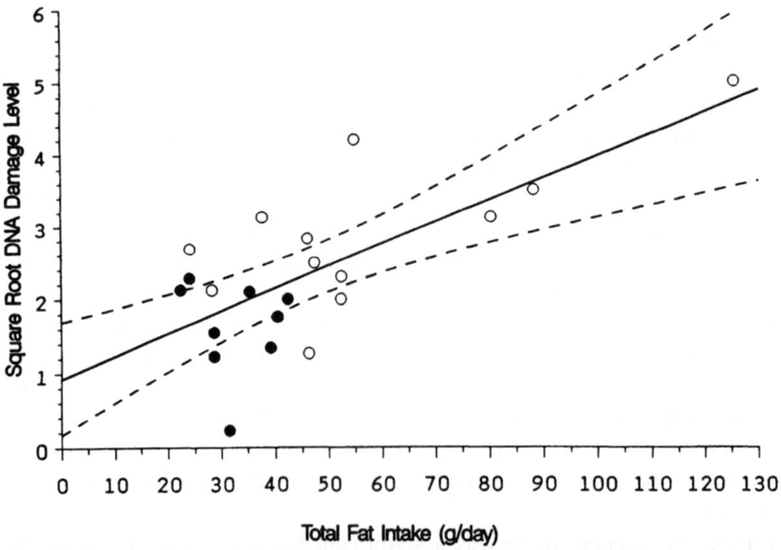

Figure 1. Relationship of (square root of) oxidative DNA damage level in peripheral nucleated blood cells to total dietary fat intake. Fat content of the diet was analyzed from dietary records completed at the same time the blood sample was drawn for analysis of DNA damage. Closed circles indicate values for women on the low-fat diet and open circles indicate values for women on the nonintervention diet. The dashed lines represent the 95% confidence limits about the predicted mean DNA damage level.

B. Oxidative DNA Damage in Rats Fed Low-Fat or Calorie-Restricted Diets

In order to develop an animal model and examine whether diet-induced changes in oxidative DNA damage levels occur in mammary gland, we utilized rats fed varying diets.[49] Female, Fischer 344 rats were either given diets with varying levels of fat or a diet restricted to 60% of *ab libitum* for 2 weeks. The low-fat, control, and high-fat diets contained 3, 5, and 20% corn oil, respectively. The calorie-restricted diet provided 40% fewer calories than the control diet, but the fat intake of the rats was the same.

Mean food intake was significantly different by diet group (p<0.0001) because the mean for the CR rats was significantly different from that of all the other diets (Table 2). With regard to mean daily caloric intake, the significant difference between diet groups again was due to the decreased caloric intake of the CR group. There were no significant differences in food or caloric intake between the control, high-fat, and low-fat groups.

Table 2. Dietary Intakes and Animal Weights

	Diet group[a]				*p* Value[b]
Variable	Control	High-fat	Low-fat	Calorie restricted	
Food intake (g/day)	12.10 ± 1.07 c	10.67 ± 0.54 d	11.84 ± 0.54 c	7.14 ± 0.07 c c d	<0.0001
Calories (kcal/day)	46.81 ± 4.13 c	49.27 ± 3.64 c	44.62 ± 2.04 d	226.77 ± 0.27 c c d	<0.0001
Fat intake (g/day)	0.61 ± 0.05 d e	2.13 ± 0.16 c e d	0.36 ± 0.02 c d e	0.60 ± 0.01 d e	<0.0001
Day 1 body weight (g)	105.3 ± 7.7	111.3 ± 5.9	106.4 ± 6.1	109.5 ± 10.5	0.245
Day 14 body weight (g)	137.7 ± 9.7 e	146.9 ± 9.2 c	136.6 ± 9.5	113.7 ± 8.1 c e e	<0.0001

[a] Table entries are the mean ± the standard deviation. There were 12 rats in each diet group. For the dietary variables, the mean over the 14 days of study was first calculated for each rat; the mean of those 12 animal-specific means was then calculated for each diet group shown in this table. Data taken from ref. 49 with permission.

[b] For simultaneous comparison of the four diet group means, by the Kruskal-Wallis rank sum text.

[c-e] Pairs of means significantly different by Miller's multiple comparisons procedure with: ^{c}p <0.001; $^{d}p < 0.05$; $^{e}p < 0.01$.

The impact of reducing caloric intake was evident in the body weight differentials among groups. Initially the mean animal weight did not differ significantly by diet group. After 2 weeks, mean body weight did not differ significantly as a function of dietary fat, but it was significantly lower in the CR group as compared with the other three groups (Table 2).

Daily fat intake differed significantly among the diet groups in the expected fashion: the means for the high-fat and low-fat rats each differed significantly from that of the other three diet groups (Table 2). Mean fat intake of the high-fat group was three- to six-fold higher than that in any of the other diet groups. Conversely, the daily fat intake of the low-fat group was lower than that in any of the other groups. The daily fat intake of the CR and control groups was virtually the same.

As shown in Table 3, the mean liver DNA damage level differed significantly across the four diet groups ($p = 0.002$). As revealed by the rank sum multiple comparisons analysis, this difference was entirely due to only two pairs of diet groups. Mean DNA damage levels in the liver of CR rats differed significantly from that of control rats ($p<0.01$) and rats in the high-fat group ($p<0.05$). Mean 5-hydroxymethyluracil levels in the liver DNA of rats fed high-fat and low-fat diets were decreased relative to control, but this was not significant.

Table 3. DNA Damage Levels in Rats after 2 Weeks of Dietary Modification

	Diet group[a]				p Value[b]
Organ	Control	High-fat	Low-fat	Calorie restricted	
Liver	3.28 ± 1.32	2.59 ± 0.66	2.30 ± 0.68	1.86 ± 0.36	0.002
	c			c	
		d		d	
Mammary gland	1.76 ± 0.49	1.12 ± 0.27	1.19 ± 0.40	1.10 ± 0.26	0.001
	c			c	
	d	d			
	d		d		

[a] Table entries are the mean (5-hydroxymethyluracil/10^4 thymine residues) ± the standard deviation. There were 12 rats in each diet group. Data taken from ref. 49 with permission.
[b] For simultaneous comparison of the four diet group means, by the Kruskal-Wallis rank sum text.
[c,d] Pairs of means significantly different by Miller's multiple comparisons procedure with: $^c p <0.01$; $^d p < 0.05$.

In the mammary gland, control rats likewise had the highest mean level of DNA damage. Mean DNA damage level in mammary gland differed significantly across the four diet groups ($p = 0.001$). Multiple comparisons analysis confirmed that this was due to significantly different mean DNA damage levels for each of the three diet groups when compared to the mean level in control rats ($p<0.05$ in each case).

IV. Discussion

Both dietary fat and reduced calorie intake have been suggested to exert their tumor-promoting effects via modulation of oxygen free radical levels.[57,58] In this work, we have examined the effects of such dietary modification on the levels of oxidative DNA damage.

We quantified 5-hydroxymethyluracil levels in the DNA of nucleated peripheral blood cells of women on standard and low-fat diets. In addition, we quantified 5-hydroxymethyluracil levels in the DNA of rat mammary gland and liver after changes in either dietary fat or calories.

In the clinical dietary intervention program, the significantly lower fat intake in the low-fat diet group relative to the nonintervention group was not accompanied by significant differences in caloric intake. The significant decrease in 5-hydroxymethyluracil levels in DNA of nucleated blood cells therefore is closely linked to the fat content of the diet (Table 1). Irrespective of the diet arm, there was also a significant linear relationship between fat intake and DNA damage level (Figure 1). This relationship was not weakened by considering percent body fat, fiber, or vitamin intakes as co-variates. Since DNA damage was measured in the blood, these results indicate that decreased fat intake decreases systemic oxidative stress. This may be one mechanism by which low-fat diets can reduce breast cancer risk and incidence.

Since animal models are useful for more detailed mechanistic studies, we examined whether or not dietary change also could influence oxidative DNA damage in rats. In the animal studies, the high-fat diet containing 20% corn oil was chosen based on the published tumor-promoting effects of such diets during mammary gland carcinogenesis (as reviewed[16]). The rats on this diet did not consume a statistically higher amount of calories than those on the control diet, but the fat intake was significantly higher (Table 2). Similarly, the caloric intake of animals on the low-fat diet was not significantly different from control, but fat intake was significantly decreased. The low-fat diet used is similar to that which can be clinically achieved in humans, i.e. about 40% reduction in daily fat intake.[56] The comparison of the control, high-fat, and low-fat diets therefore should not be obscured by differences in caloric intake.

Both the high-fat and low-fat groups exhibited decreased 5-hydroxymethyluracil levels in the liver and mammary gland DNA. Both organs were examined in order to determine whether or not fat intake would specifically affect DNA damage levels in mammary gland. The decreases in DNA damage detected in liver were not significant, but in mammary gland the decreases were statistically significant (Table 3). This finding is of interest in that mammary gland tumorigenesis in particular is known to be modulated by dietary fat.[16]

The decreased 5-hydroxymethyluracil levels in the high-fat diet group could be due to compensatory mechanisms that can be induced by oxidative stress. In male, Sprague-Dawley rats, a 20% corn oil diet has been shown to increase glutathione levels in liver and kidney relative to a 5% corn oil diet.[59] Diets very high in fat thus may induce detoxification systems (such as glutathione) in response to oxidative stress. Due to the possible induction of such antioxidant defenses, the effects of oxidative stress on the levels of oxidative DNA damage may be difficult to predict and may depend on the severity or duration of the stress. Using ethane exhalation as an index of oxidative stress, rats on both low-fat (2% sunflower seed oil) and high-fat (25% sunflower seed oil) exhibited decreased ethane exhalation relative to control (12.5% sunflower seed oil) after 10 weeks of feeding. Increased ethane exhalation in the high-fat group was obtained only after 20-27 weeks of dietary change.[60] This indicates that low-fat diets may lead to decreased oxidative stress in rats, whereas high-fat diets do not appear to increase oxidative damage unless feeding is continued for longer periods of time.

The effect of reduced caloric intake had a stronger influence on 5-hydroxymethyluracil levels than altered fat intake (Table 3), and 5-hydroxymethyluracil levels were significantly decreased in both liver and mammary gland DNA of calorically restricted rats. In another study, 8-hydroxy-2′-deoxyguanosine levels were decreased by caloric restriction in liver DNA.[61] Caloric restriction in one human subject also has been shown to decrease urinary oxidized DNA bases.[62] These decreases in oxidative DNA damage during caloric restriction could be due to decreased production of oxygen free radicals as a result of increased metabolic efficiency, free radical detoxification, and/or DNA repair.[18-22] Although the relative contributions of decreased oxygen free radical production and DNA repair are unknown, decreased caloric intake significantly decreased 5-hydroxymethyluracil levels in the DNA of both liver and mammary gland.

In summary, our results indicate that both fat and caloric intake can modulate the levels of oxidative DNA damage in rats after only 2 weeks of dietary modulation. This change in oxidative DNA damage levels may be due to an early change in cellular redox balance that may contribute to the known effects of such diets on tumor development. Since low-fat diets decrease oxidative DNA damage in both humans and rats, it may be possible to suggest clinical intervention strategies using the rat model. Such issues as type of fat and degree of fat and/or calorie reduction that are optimal for breast cancer risk reduction remain to be resolved.

Acknowledgments

This work was supported by the American Institute for Cancer Research Grant 91B53, Public Health Service grant CA-22453 and the Wayne State University Ben Kasle Trust Fund for Cancer Research. The authors acknowledge D.A. Luongo, M.H. Lu, S.M. Lewis, X.W. Chen, P.H. Duffy, B.A. Reading, F.A. Valeriote, A. Boomer, and J. Depper for their contributions to the studies described.

References

1. U.S. Department of Health and Human Services, National Cancer Institute, Cancer rates and risks, NIH Publication No. 85-691 (1985).
2. A. Schatzkin, P. Greenwald, D.P. Byar, C.K. Clifford, The dietary fat-breast cancer hypothesis is alive, *JAMA* 261:3284-3287 (1989).
3. G.R. Howe, T. Hirohata, T.G. Hislop, J.M. Iscovich, J.-M. Yuan, K. Katsouyanni, F. Lubin, E. Marubi, B. Modan, T. Rohan, P. Toniolo, Y. Schunzhang, Dietary factors and risk of breast cancer: Combined analysis of 12 case-control studies, *J Natl Cancer Inst* 82:561-569 (1990).
4. P. Greenwald, Issues raised by the Women's Health Trial, *J Natl Cancer Inst* 80:788-790 (1988).
5. W.C. Willett, D.J. Hunter, M.J. Stampfer, G.A. Colditz, J.-A.E. Manson, D. Spiegelman, B. Rosner, C.H. Hennekens, F.E. Spiezer, Dietary fat and fiber in relation to risk of breast cancer, an eight year follow-up, *JAMA* 268:2037-2081 (1992).
6. G.R. Howe, C.M. Friedenreich, M. Jain, A.B. Miller, A cohort study of fat intake and risk of breast cancer, *J Natl Cancer Inst* 83:336-340 (1991).
7. D.A. Evans, The dietary fat-breast cancer hypothesis: Is it really alive? *JAMA* 263:237-238 (1990).
8. E.L. Wynder, S.D. Stellman, The "over-exposed" control group, *Am J Epid* 135:459-461 (1992).
9. R.T. Chlebowski, D.W. Nixon, G.L. Blackburn, P. Jochinsen, E.F. Scanlon, W. Insull, I.M. Buzzard, R. Elashoff, R. Butrum, E.L. Wynder, A breast cancer adjuvant study (NAS): Protocol design and initial patient adherence, *Breast Cancer Res Treat* 10:21-29 (1987).
10. P. Toniolo, E. Riboli, F. Protta, M. Charrel, A.P.M. Cappa, Calorie-providing nutrients and risk of breast cancer, *J Natl Cancer Inst* 81:278-286 (1989).
11. D. Albanes, Caloric intake, body weight and cancer: A review, *Nutr Cancer* 9:199-217 (1987).

12. D. Kritchevsky, Nutrition and breast cancer, *Cancer* 66:1321-1325 (1990).
13. S.L. Gorbach, A. Morill-LaBrode, M.N. Woods, J.T. Dwyer, W.D. Selles, M. Henderson, W. Insull, Jr., S. Goldman, D. Thompson, C. Clifford, L. Sheppard, Changes in food patterns during a low-fat dietary intervention in women, *J Am Diet Assoc* 90:802-809 (1990).
14. M.H. Henderson, E. White, R.S. Thompson, Cancer incidence in Seattle Women's Health Trial participants by group and time since randomization, *J Natl Cancer Inst* 83:1260-1261 (1991).
15. A. Tannenbaum, The dependence of tumor formation on the composition of the calorie-restricted diet as well as on the degree of restriction, *Cancer Res* 5:616-625 (1945).
16. L.S. Freeman, C. Clifford, M. Messina, Analysis of dietary fat, calories and body weight, and the development of mammary tumors in rats and mice: A review, *Cancer Res* 50:5710-5719 (1990).
17. C.W. Welsch, Relationships between dietary fat and experimental mammary tumorigenesis: A review and critique, *Cancer Res* 52:2040s-2048s (1992).
18. J.M. Lipman, A. Turturro, R.W. Hart, The influence of dietary restriction on DNA repair in rodents: A preliminary study, *Mech Ageing Develop* 48:135-143 (1989).
19. R.J. Feuers, P.H. Duffy, J.A. Leakey, A. Turturro, R.A. Mittelstaedt, R.W. Hart, Effect of chronic caloric restriction on hepatic enzymes of intermediary metabolism in the male Fischer 344 rat, *Mech Ageing Development* 48:179-189 (1989).
20. A. Koizumi, R. Weindruch, R.L. Walford, Influence of dietary restriction and age on liver enzyme activities and lipid peroxidation in mice, *J Nutr* 117:361-367 (1987).
21. N. Weraarchakul, R. Strong, W.G. Wood, A. Richardson, The effect of aging and dietary restriction on DNA repair, *Experimental Cell Res* 181:197-204 (1989).
22. B.P. Yu, S. Langaniere, J.-W. Kim, Influence of life-prolonging food restriction on membrane lipoperoxidation and antioxidant status, *Basic Life Sciences* 49:1067-1073 (1988).
23. B.N. Ames, R.L. Saul, Cancer, ageing, and oxidative damage, *in*: *Theories of Carcinogenesis*, O.H. Iversen, ed., Hemisphere, Cambridge, pp. 203-270 (1988).
24. P.A. Cerutti, Prooxidant states and tumor promotion, *Science* 227:375-381 (1987).
25. P.A. Cerutti, B.F. Trump, Inflammation and oxidative stress in carcinogenesis, *Cancer Cells* 3:1-7 (1991).
26. C.G. Cochrane, Cellular injury by oxidants, *Amer J Med* 91 (suppl. 3C):23s-30s (1991).
27. K. Frenkel, Carcinogen-mediated oxidant formation and oxidative DNA damage, *Pharm Ther* 53:127-166 (1992).
28. T.J. McBride, B.D. Preston, L.A. Loeb, Mutagenic spectrum resulting from DNA damage by oxygen radicals, *Biochemistry* 30:207-213 (1991).
29. T.M. Reid, L.A. Loeb, Mutagenic specificity of oxygen radicals produced by human leukemia cells, *Cancer Res* 52:1082-1086 (1992).
30. S. Shibutani, M. Takeshita, A.P. Grollman, Insertion of specific bases during DNA synthesis past the oxidation damaged base 8-oxodG, *Nature* 349:431-434 (1991).
31. L. Shirname-More, T.G. Rossman, W. Troll, G.W. Teebor, K. Frenkel, Genetic effects of 5-hydroxymethyl-2′-deoxyuridine, a product of ionizing radiation, *Mutat Res* 178:177-186 (1987).
32. R.G. Allen, V.S. Venkatraj, Oxidants and antioxidants in development and differentiation, *J Nutr* 122:631-635 (1992).
33. R.J. Boorstein, L.N. Chiu, G.W. Teebor, Phylogenetic evidence of a role for 5-hydroxymethyluracil-DNA glycosylase in the maintenance of 5-methylcytosine in DNA, *Nucleic Acids Res* 17:7653-7661 (1989).
34. R.H. Burdon, V. Gill, C. Rice-Evans, Oxidative stress and tumour cell proliferation, *Free Rad Res Commun* 11:65-76 (1990).
35. T. Galeotti, L. Masotti, S. Borrello, E. Casali, Oxy-radical metabolism and control of tumour growth, *Xenobiotica* 21:1041-1051 (1991).
36. T.P. Szatrowski, C.F. Nathan, Production of large amounts of hydrogen peroxide by human tumor cells, *Cancer Res* 51:794-798 (1991).

37. D.C. Mallins, R. Haimanot, Major alterations in the nucleotide structure of DNA in cancer of the female breast, *Cancer Res* 51:5430-5432 (1991).
38. H.C. Pitot, D. Beer, S. Hendrich, Multistage carcinogenesis: The phenomenon underlying the theories, *in*: *Theories of Carcinogenesis*, O.H. Iverson, Hemisphere, Cambridge, pp. 159-177 (1988).
39. R.W. Pero, D.B. Johnson, D.G. Miller, E. Zang, M. Markowitz, G.A. Doyle, M. Lund-Pero, L. Salford, P. Sordillo, N. Raskin, E.J. Beattie, Adenosine diphosphate ribosyltransferase responses to a standardized dose of hydrogen peroxide in the mononuclear leukocytes of patients with a diagnosis of cancer, *Carcinogenesis* 10:1657-1664 (1989).
40. R.W. Pero, D.B. Johnson, M. Markowitz, G. Doyle, M. Lund-Pero, J. Seidegard, M. Halper, D.G. Miller, DNA repair synthesis in individuals with and without a family history of cancer, *Carcinogenesis* 10:693-697 (1989).
41. E. Kovacs, A. Almendral, Reduced DNA repair synthesis in healthy women having first degree relatives with breast cancer, *Eur J Cancer Clin Oncol* 23:1051-1057 (1987).
42. E. Kovacs, D. Stucki, W. Weber, H.J. Muller, Impaired DNA-repair synthesis in lymphocytes of breast cancer patients, *Eur J Cancer Clin Oncol* 22:863-869 (1986).
43. N.F. Boyd, V. McGuire, Evidence of lipid peroxidation in premenopausal women with mammographic dysplasia, *Cancer Lett* 50:31-37 (1990).
44. A. Boomer, S. Martino, G. Krag, D. Liang, J. Bacus, B. Reading, L. Heilbrun, Compliance to a 15% fat diet in women at risk for breast cancer, *Breast Cancer Res Treat* 14:179 (1989).
45. I.M. Buzzard, E.H. Asp, R.T. Chlebowski, A.P. Boyar, R.W. Jeffery, D.W. Nixon, G.L. Blackburn, P.R. Jochimsen, E.F. Scanlon, W. Insull, R.M. Elashoff, R. Butrum, E.L. Wynder, Diet intervention methods to reduce fat intake: Nutrient and food group composition of self-selected low-fat diets, *J Am Diet Assoc* 90:42-53 (1990).
46. T.A. Ciulla, R.M. Sklar, S.L. Hauser, A simple method for DNA purification from peripheral blood, *Anal Biochem* 174:485-488 (1988).
47. Z. Djuric, E.K. Fifer, Y. Yamazoe, F.A. Beland, DNA binding by 1-nitropyrene and 1,6-dinitropyrene *in vitro* and *in vivo*: Effects of nitroreductase induction, *Carcinogenesis* 9:357-364 (1988).
48. R.C. Moon, D.H. Janss, S. Young, Preparation of fat cell 'free' rat mammary gland, *J Histiochem Cytochem* 17:182-186 (1969).
49. Z. Djuric, M.H. Lu, S. Lewis, D.A. Luongo, X.W. Chen, L.K. Heilbrun, B.A. Reading, P.H. Duffy, R.W. Hart, Oxidative DNA damage in rats fed low-fat, high fat or calorie-restricted diets, *Toxicol Appl Pharmacol* 115:156-160 (1992).
50. H.A. Harper, Calorimetry: Elements in nutrition, *in*: *Review of Physiological Chemistry*, 17th ed., H.A. Harper, V.W. Rodwell, P.A. Mayes, eds., Lange Medical Publications, Los Altos, CA, pp. 596-612 (1979).
51. P.J. VanSoest, *Nutritional Ecology of the Ruminant*, O&B Books, Corvallis, OR (1982). R. Weindruch, R.L. Walford, *The Retardation of Aging and Disease by Dietary Restriction*, Charles C. Thomas, Springfield, IL (1988).
52. M. Hollander, D.A. Wolfe, *Nonparametric Statistical Methods*, Wiley Inc., New York, pp. 114-119 (1973).
53. R.G. Miller, Jr., *Simultaneous Statistical Inference*, 2nd Ed., SpringerVerlag, Inc., New York, p. 166 (1981).
54. M. Dizdaroglu, Application of capillary gas chromatography-mass spectrometry to chemical characterization of radiation-induced base damage of DNA: Implications for assessing DNA repair processes, *Anal Biochem* 144:593-603 (1985).
55. Z. Djuric, D.A. Luongo, D.A. Harper, Quantitation of 5-(hydroxymethyl)uracil in DNA by gas chromatography-mass spectroscopy, *Chem Res Toxicol* 4:687-691 (1991).
56. Z. Djuric, L.K. Heilbrun, B.A. Reading, A. Boomer, F.A. Valeriote, S. Martino, Effects of a low-fat diet on levels of oxidative damage to DNA in human peripheral nucleated blood cells, *J Natl Cancer Inst* 83:766-769 (1991).
57. B.N. Ames, Dietary carcinogens and anticarcinogens. Oxygen radicals and degenerative

disease, *Science* 221:1256-1264 (1983).
58. R. Weindruch, R.L. Walford, *The Retardation of Ageing and Disease by Dietary Restriction*, Charles C. Thomas, Springfield, IL (1988).
59. V.K. Nonavinakere, C.L. Black, S. Curtis, R.A. Potmis, J.L. Early, Effects of dietary fat and selenium (Se) on weight gain; and liver, kidney, colon glutathione (GSH) of 1,2-dimethylhydrazine dihydrochloride (DMH) treated rats, *Toxicologist* 12:409 (1992).
60. E. Hietanen, M. Ahotupa, J.-C. Bereziat, A.-M. C. Bussacchini, H. Bartsch, Elevated lipid peroxidation in rats induced by dietary lipids and *N*-nitrosodimethylamine and its inhibition by indomethacin monitored via ethane exhalation, *Toxicologic Pathol* 15:93-96.
61. M.H. Chang, H. Kasai, S. Nishimura, B.P. Yu, Protection of DNA damage by dietary restriction, *Free Radical Biol Med* 12:523-525 (1992).
62. M.G. Simic, D.S. Bergtold, Dietary modulation of DNA damage in human, *Mutat Res* 250:17-24 (1991).

Chapter 6

Carbohydrate Tumor Marker: Basis for a Simple Test for Colorectal Cancer

ABULKALAM M. SHAMSUDDIN and KOSAKU SAKAMOTO

I. Introduction

Large intestinal cancer (LIC) is one of the commonest cancers in the industrialized world. It is ranked among the major causes of cancer death in the United States and other Western countries.[1] Because of the magnitude of the health problem and associated cost in life and materials, control of this cancer is vital. Prevention is one of the methods of cancer control, and detection of the cancer at the very early stage of the disease is fundamental to prevention. Early detection in its turn is dependent on screening the population (or those at risk) for the disease. Without screening, a 50-year-old person at average risk has approximately a 530 in 10,000 chance of developing invasive colorectal cancer during the rest of her/his life.[2] A host of currently available diagnostic assays have been recommended and are in use for LIC screening.[3] The common ones are the fecal occult blood tests (FOBT), barium enema X-rays, and endoscopic visualization.[3] The cost-effectiveness of these vary tremendously; their use as screening assays is therefore seriously in question.[2-7] On one hand, FOBTs are relatively inexpensive[2,3] (approximately $10) but are not cost-effective due to their inaccuracies. While the cost of barium enemas and colonoscopies can range from $250-800, they are highly accurate, although marred by the discomfort they cause.

The cost-effectiveness of screening assays depends not only on their actual expense, but also on the sensitivity and specificity of the assays. (*Sensitivity* is the proportion of diseased subjects who have a positive test and *specificity* is the proportion of nondiseased subjects who yield a negative test result.[3]) Perhaps because of their relative simplicity and low price, the FOBTs have become part of the screening strategy for colorectal cancer despite their high inaccuracy. The sensitivity and specificity of the FOBTs are so poor that, "Occult blood testing is, at best an imperfect approach to the screening of colorectal cancer" decries Ahlquist.[8] Table 1 shows the difference in net saving in terms of cost per year of life gained from screening with assays that have different sensitivities.

Abulkalam M. Shamsuddin • Department of Pathology, University of Maryland School of Medicine, Baltimore, MD 21201; Kosaku Sakamoto • First Department of Surgery, Gunma University School of Medicine, Maebashi, Gunma 371, Japan

Diet and Cancer: Markers, Prevention, and Treatment
Edited by M.M Jacobs, Plenum Press, New York, 1994

Table 1. Effect of Sensitivity of FOBT on Cost-Effectiveness[a]

	Cost per year of life gained from screening FOBT		
Screening regimen	25% Sensitivity	40% Sensitivity	$ Saving (%)
FOBT only[b]	$43,167	$35,054	$8,113 (19%)
FOBT + Sig[c]	$48,338	$42,509	$5,829 (12%)

[a] Adapted from Congressional Office of Technology Assessment, 1990.
[b] Once a year.
[c] Sigmoidoscopy every 5 years.

The U.S. Congressional Office of Technology Assessment (OTA) estimates that a mere increase in sensitivity of FOBT from 25% to 40% would reduce the cost per year of life gained by nearly 20%. Table 1 also shows that the cost as estimated could always be reduced by enhancing the sensitivity even when using various screening regimens utilizing FOBT with or without flexible sigmoidoscopic examination at different schedules. The fundamental problem with the FOBTs are that they are based on the faulty premise of blood in stool being a marker of the cancer. Thus, assays that enjoy sensitivity rates higher than FOBT are most likely to reduce that cost even more and make screening cost-effective. These assays must be based on the markers that are expressed not only by the cancer but also by the precancerous lesions, so that the disease can be detected at a rather early stage.

Recently, a host of markers have been identified that are claimed to represent cancer and have potential use in early diagnosis and screening. Alterations associated with neoplastic transformation have resulted in the discovery of some of these markers, while the discoveries of others have been rather serendipitous. These include morphological abnormalities (light microscopic and ultrastructural features), allelic deletion of a particular chromosome, altered gene expression and enzyme activities, abnormal accumulation of a particular metabolite, altered cell proliferation, and immunogenicity.[3] Although all of these markers have potential for being utilized in a screening assay for LIC,[9,10] we shall only discuss the use of the less exotic ones, such as simple carbohydrates in rectal mucus and an easy and inexpensive assay for their detection.

Phenotypic alterations such as mucin histochemical changes associated with malignancy or premalignant lesions of the large intestine have been studied both in human tissues as well as in experimental models.[11-14] An approach that compares and correlates the data obtained from *in vitro* and *in vivo* models using both experimental animals and human tissues and cells was utilized.

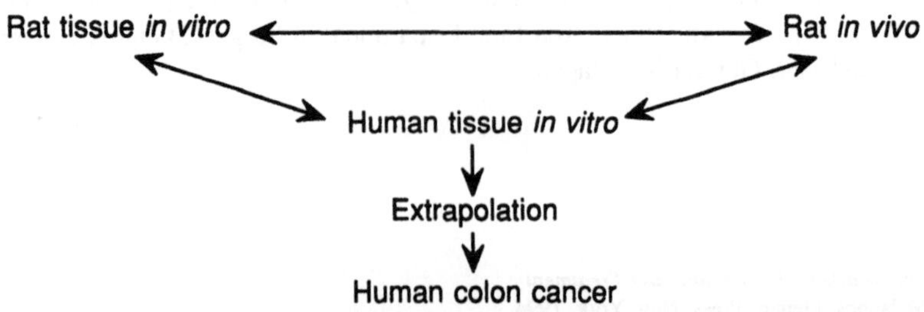

These changes are recognized as an alteration of antigenicity of glycoconjugate molecules on the cell membrane. The phenotypic changes are categorized as (a) the expression of newly synthesized antigens, (b) deletion of antigens that should normally be expressed, and (c) an ectopic expression of antigens that are not normally expressed.[15] The majority of the molecules responsible for these phenotypic changes consist of short carbohydrate side chains (oligosaccharides) distributed on the cell membrane and/or secreted mucous glycoprotein (mucin).

In this chapter, we will review some of the basic knowledge of carbohydrate tumor markers and describe in detail a simple method that detects an alteration in the marker with a very simple assay.

II. Structure, Synthesis and Expression of Glycoconjugates

Most phenotypes of a particular cell, as well as tumor-associated antigens, are determined by a carbohydrate structure that is covalently linked (conjugated) with either lipid (glycolipid) or protein (glycoprotein). Glycolipid molecules are predominantly inserted into the lipid bilayer of a cell membrane, leaving its carbohydrate moiety outside the cell. Glycoproteins contain oligosaccharide side chains conjugated to protein by either *O*-glycosidic linkages for mucin-type glycoproteins or *N*-glycosidic bonds for serum glycoproteins. The mucopolysaccharides in the gastrointestinal (GI) tract mucin are also glycoproteins. The oligosaccharide side chains of these glycoconjugates consist of up to 10 monosaccharide units. They are commonly branched in structure and serve as antigenic determinants when immunized to allogeneic animals. Because of the variety of chain length, branching, saccharide sequence, and anomeric linkage of the individual monosaccharide units, the carbohydrate side chains have enormous potential antigenicity. The majority of the resulting antibodies recognize carbohydrate epitopes that are usually related to the human blood group antigens. The glycoconjugates are synthesized at the endoplasmic reticulum, where the oligosaccharide side chains are conjugated to a polypeptide core chain, and transferred to the Golgi, lysosome, or plasma membrane. They are liberated into the extracellular fluid including the mucin in the GI tract.

A. Mucin Histochemical Markers

The mucin in the GI tract consists of mucopolysaccharides, or glycoproteins. The characteristic feature of mucopolysaccharides is that the terminal moiety of the oligosaccharide side chain consists of *N*-acetylneuraminic acid (NANA, or sialic acid), and is negatively charged. The sialic acid residue is transferred to the terminal galactose or penultimate *N*-acetylgalactosamine (GalNAc) through the action of specific sailosyltransferases.

Mucopolysaccharides are traditionally classified as either neutral or acidic. One can differentially stain these mucopolysaccharides with alcian blue (AB) and periodic acid-Schiff (PAS) sequence; all the acidic mucin is first stained with AB and then the remaining neutral mucin is visualized by PAS. Acidic mucin is differentiated into either sulfomucin or sialomucin, and the latter is further subdivided into either *N*-acetyl or *O*-acetyl groups.

Histochemical examination of a normal human large intestine indicates that the mucus in the crypts of the ascending colon consists of a mixture of acidic and neutral mucin with a predominance of the latter. In contrast, the mucus in the rectal crypts is almost exclusively acidic in nature.[16] These topological differences in the mucin histochemistry are altered with malignant transformation, the phenomenon presumably attributed to the quantitative and/or qualitative composition of terminal sialic acid in the mucus glycoprotein.[17] Reid *et al.*[18] and Culling *et al.*[19] demonstrated the reduction or loss of *O*-acetylated sialic acids in patients

Histochemical Classification of Mucin

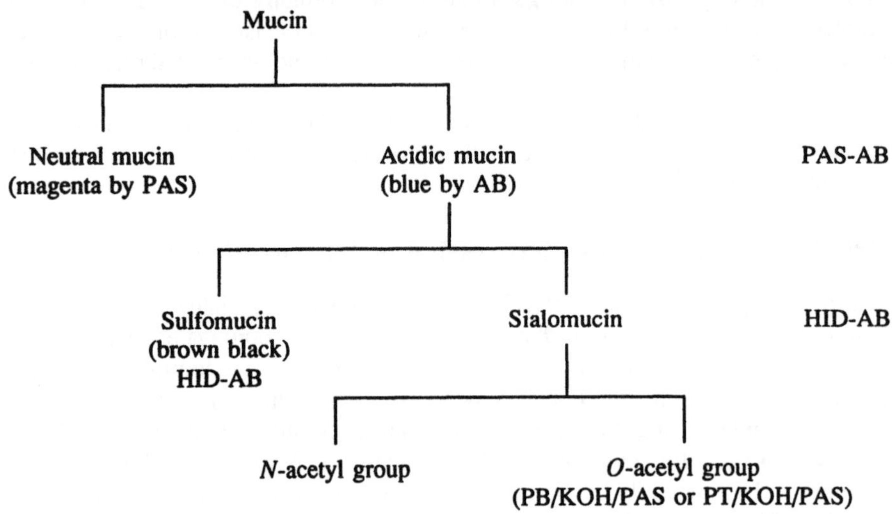

at high risk of colon cancer or in colon tumors themselves. Shamsuddin and Trump demonstrated in rats that, in contrast to the presence of sulfomucin in normal colon mucosa, abnormal sialomucin was detected both *in vivo*[12] and *in vitro*[13] shortly after treatment with carcinogens. A shift from normal sulfomucin to abnormal sialomucin was also demonstrated in extensive comparative studies of human colon.[14] The altered expression of the colonic mucin is observed not only in the carcinomas but also in the crypts of morphologically normal-appearing mucosa adjacent to and distant from the carcinomas both in humans[14] and experimental animals treated with azoxymethane[12] or dimethylhydrazine.[20] These observations were first reported by Filipe and colleagues[11,21] but the reason and the exact significance of the abnormal mucin expression in the normal-appearing mucosa remained unknown until Shamsuddin established the concept of the "field effect" of the carcinogens on the large intestinal epithelium.[3,14] The concept of field effect will be discussed later in the chapter. Suffice it to say here that the normal-appearing mucosa that is remote from the cancer may sporadically harbor a wide variety of progressive changes which include alteration of mucin phenotype (emergence of sialomucin), presence of dilated or distorted crypts, various degrees of atypia/dysplasia, and even foci of carcinoma *in situ*. Figure 1 shows the spectrum of these changes in mucosa away from cancers in the human colon.

Shamsuddin rationalized that (a) the presence of cancer in the large intestine implies previous exposure of the host to carcinogens, (b) most carcinogens act by way of the field effect where the entire target tissue is subjected to the carcinogenic stimuli, (c) carcinogens induce multifocal changes throughout the entire target tissue viz. colorectal mucosa, (d) of the many initiated sites, only some of them may be promoted to a recognizable carcinoma. Thus, the alterations in the normal-appearing, initiated but not promoted mucosa may express some of the markers of cancer and precancer. Since mucin is secreted by the colorectal mucosa and can easily be sampled from the rectum, Shamsuddin thus embarked on exploiting this fact, in conjunction with the altered mucus of cancer in developing screening assays. But before we describe the alterations in the mucin, let's review their common structures and characteristics.

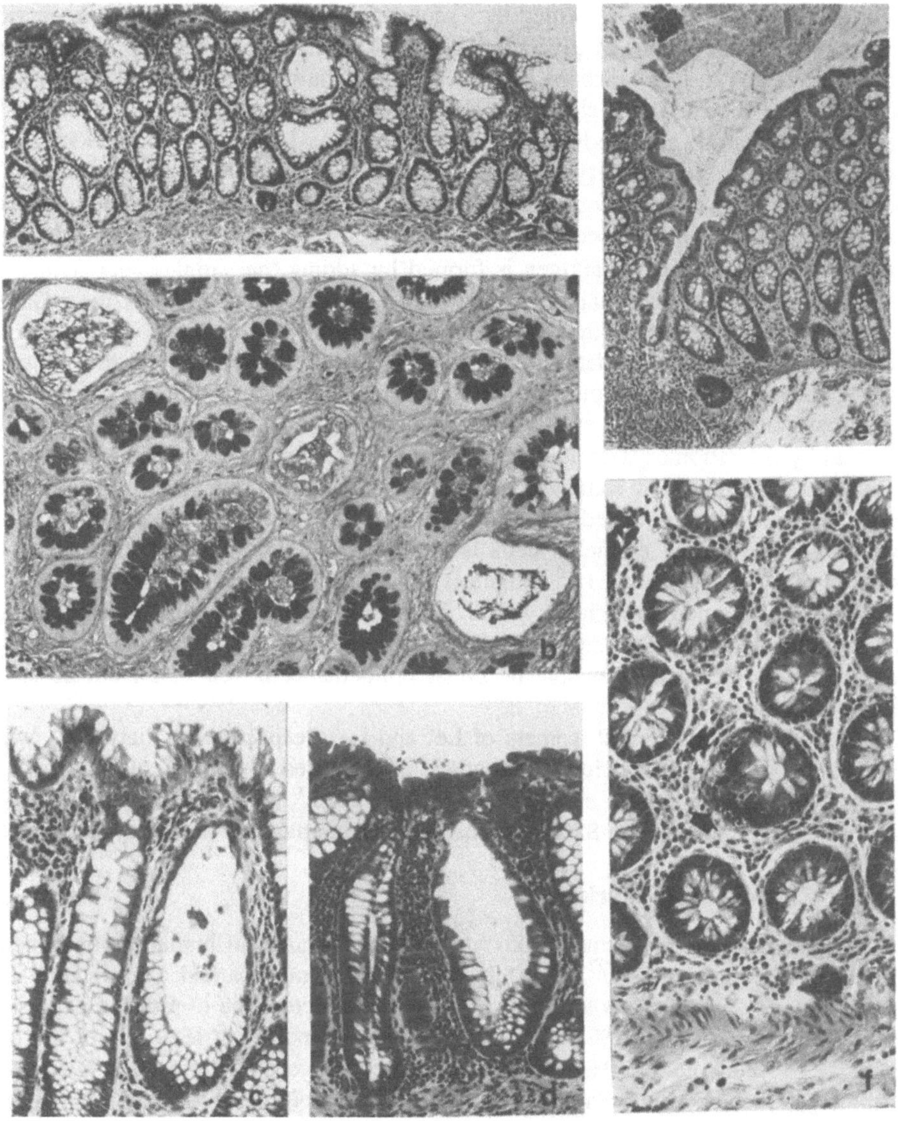

Figure 1. Microscopic appearance of the otherwise normal-appearing (by naked eye) colonic mucosa remote from cancers. (a) Dilated and distorted crypts; (b) special stained section showing an increase in mucus production, as well as altered mucus production in dilated and distorted crypts; (c) and (d) two examples of isolated crypt dilatation; (e) crypt lining cell is similar to the surface epithelium, a lymphoid follicle is underneath the abnormal crypt; and (f) cross section of an atypical crypt showing increased stratification of cells and nuclei and increased basophilia (arrows). (Reproduced from Shamsuddin *et al.*[14])

B. Blood Group-Related Antigen Glycoconjugates

Blood group-related antigens are oligosaccharides ubiquitously present throughout the entire body[22]; they are found in endothelial cells, epithelial cells, red blood cells, and neutrophils to name a few. Human large intestinal epithelial cells are no exception, although they exhibit variation during embryonic developments, birth, and neoplastic transformation.[15] The antigenic determinants are formed by a sequential addition of individual saccharides to the carbohydrate side chains of glycoconjugates. The major blood group specificities are determined by two types of backbone structures, type-1 chains D-galactose-β[1→3]-*N*-acetyl-D-galactosamine-R and type-2 chains D-galactose-β[1→4]-*N*-acetyl-D-galactosamine-R (-R representing -galactose-glucose-lipid/protein).[15] The type-1 chain group includes H(O), A, B, and Le^a and Le^b, while the type-2 chain group has Le^x, and Le^y, and their derivatives. Henceforth, the sugars D-galactose and *N*-acetyl-D-galactosamine may appear in abbreviated forms as Gal and GalNAc respectively.

The determinant of H(O) antigen is formed by adding the simple sugar fucose to the terminal galactose of type-1 chain structure. Subsequent addition of GalNAc or Gal to the terminal galactose residue of the H(O) antigen confers A or B blood group specificity, respectively. Le^a antigenic determinant is formed by fucosylation of the subterminal *N*-acetylgalactosamine residue on the type-1 backbone structure. Additional fucosylation of the terminal galactose on the Le^a confers Le^b specificity. All these antigenic determinants are formed by the action of specific glycosyltransferases. Thus glycosyltransferases (including fucosyltransferase) and sialosyltransferases play an important role in synthesizing a variety of blood group-related antigens. Since the enzymatic expression of H(O) fucosyltransferase on the epithelial cells is controlled by the so-called secretor gene, the expression of H(O), A, B, and Le^b antigen is limited to "secretor" individuals.

Sialyl-Le^a is formed through the action of sialosyltransferase, which transfers sialic acid to the terminal galactose residue of Le^a. The expression of this neosynthesized antigen is increased in fetal as well as neoplastic colon epithelial cells (60-90%), but absent from normal adult colon.[23-25]

Le^x and Le^y are positional isomers of Le^a and Le^b, respectively. Sialyl-Le^x, which is recognized by different monoclonal antibodies, is expressed in human colon adenocarcinomas. According to Itzkowitz *et al.*,[26] sialyldifucosyl-Le^x, recognized by the monoclonal antibody FH6, is expressed in 82% of colorectal carcinomas, but not by the normal colon.

C. Other Carbohydrate Markers

Altered expression of human leukocyte antigens (HLAs) in human colon cancer and precancer has been described.[27-29] While normal colon mucosa had a completely normal HLA phenotype, 96% of the colon tumors had either undetectable or reduced expression of HLA class I antigens, and 92% demonstrated disappearance of at least one of the class II antigens.[30] Since cytotoxic T lymphocytes recognize antigens on tumor cells in conjunction with major histocompatibility complex (MHC) class I "self" antigens, loss or reduction of MHC class I antigens is likely to allow the tumor cells to escape from immune surveillance. Studies in animals and in humans demonstrated that suppression of HLA gene expression increased the tumorigenic potential, whereas enhancement of class I and class II antigens expression by gamma-interferon retarded or inhibited the tumor growth.[31,32] However, because of the diversity in the HLA antigens, it is unlikely to have a real potential in being a specific marker for neoplastic transformation.

D. D-Galactose-β[1→3]-*N*-Acetyl-D-Galactosamine

The disaccharide D-Gal-β[1→3]-D-GalNAc, also known as T-Ag (for Thomsen-Friedenreich antigen) is a precursor substance of the M and N blood group antigen determinants. Transfer of sialic acid (NANA or *N*-acetyl-neuraminic acid) residues to T-Ag confers blood group M and N specificity. The T-Ag determinant D-Gal-β[1→3]-D-GalNAc is recognized by the lectin peanut agglutinin (PNA), which is purified from *Arachis hypogaea*.[33] T-Ag is also detected by polyclonal or monoclonal anti-T antibodies.

During the early stages of the synthesis of the mucin glycoprotein, *N*-acetyl-D-galactosamine (GalNAc, also known as Tn antigen) is first conjugated to serine or threonine residue, followed by an additional conjugation of sialic acid or galactose, resulting in sialosyl Tn antigen or T-Ag, respectively. Each conjugation step is controlled by the specific saliosyltransferase or galactosyltransferase. Net T-Ag expression may be increased when the step of β[1→3] galactosylation of Tn antigen is augmented or when the step of sialosyltransferase reaction from T-Ag to M or N blood group antigens is retarded or blocked.

Differentiation of Tn, T, M, and N Antigens

Tn antigen	:	Gal-NAc-Ser/Thr
T antigen	:	Gal-β[1→3]-GalNAc-Serr/Thr
Sialosyl Tn	:	NANA-GalNAc-Ser/Thr
M antigen	:	Gal-β[1→3]-GalNAc-Ser/Thr I NANA
N antigen	:	Gal-β[1→3]-GalNAc-GalNAc-Ser/Thr I I NANA NANA

T-Ag is not expressed by the normal colonic mucosa, but is extensively expressed by the fetal colon as well as by the colon cancer cell. The latter is detected by PNA, anti-T-Ag antibodies, or enzymatic oxidation.[3,34-37] Not only is the T-Ag expressed by cancer, it is expressed by precancerous lesions as well as by the normal-appearing mucosa remote from cancer.[3,37] Since the enzymatic detection is simple and is the basis of the simple screening assay, the subject matter of this chapter, a description of the method follows.

The enzyme D-galactose oxidase specifically oxidizes C-6 hydroxyl groups of D-galactopyranose and *N*-acetylgalactosamine residues of D-Gal-β[1→3]-D-GalNAc, generating two vicinal aldehyde groups that react with basic fuchsin to give magenta/purple coloration (Figure 2).

Schulte and Spicer[38] first demonstrated the use of galactose oxidase-Schiff procedure (GO-Schiff) to study the T-Ag in rat tracheal gland secretory glycoproteins. Shamsuddin and his co-workers then applied this technique to detect the marker Gal-β[1→3]-GalNAc in precancer and cancer of the colon.[3,39-42] While D-galactose oxidase reacts with both D-Gal-β[1→3]-D-GalNAc and terminal monosaccharide galactose, D-galacto-hexoaldose converted from the latter may not be able to generate magenta coloration with basic fuchsin because of an atypical distance among the participating molecules. In contrast, PNA that binds to either D-Gal-β[1→3]-D-GalNAc (or related structures), or terminal galactose may be equally visualized by the second antibody (or conjugate) that is specific to PNA. This assumption is supported by the fact that not all tumors, tissues, or cells showing PNA reactively may be stained positive with GO-Schiff sequence and vice-versa.[3,42] Shamsuddin postulated that the abnormal mucin in the crypts or in the lumen of the carcinoma as well as normal-appearing mucosa away from the carcinoma site could be exploited as one of the tumor markers.[3,40]

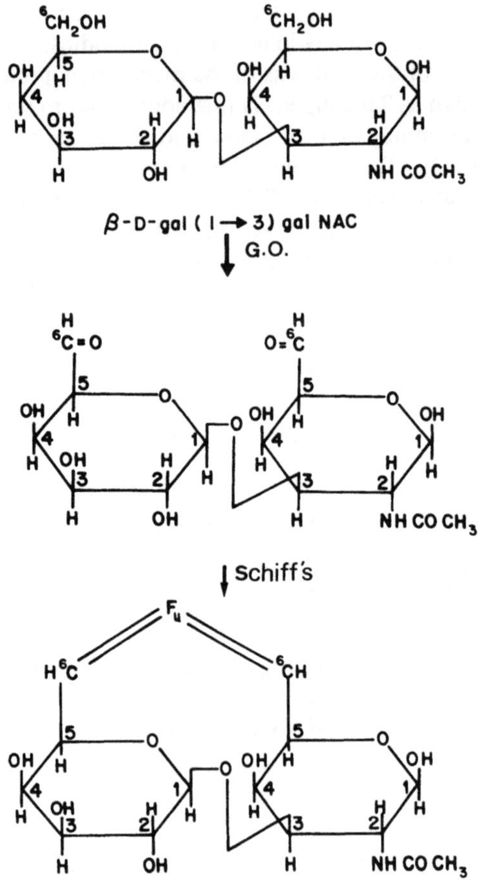

Figure 2. Principle of the GO-Schiff assay. D-Gal-β[1→3]-D-GalNAc is oxidized by the enzyme, GO, generating two vicinal aldehyde groups (middle panel) that react with Schiff's basic fuchsin to give magenta coloration.

E. Field Effect Theory of Colon Carcinogenesis

The concept of field effect in colorectal carcinogenesis is vital in understanding the rationale of the rectal mucin test that predicts, by detecting tumor marker D-Gal-β[1→3]-D-GalNAc in the rectal mucin, the presence of cancer or precancerous lesions somewhere in the large intestine.[3,40] The normal-appearing colonic mucosa that is far distant from the carcinoma site sporadically harbors wide variety of progressive changes. These multifocal changes are commonly observed in the entire colon not only from the experimental animals treated with the carcinogens, but also from the human specimen resected at surgery.[14] Based on morphological and histochemical observations, Shamsuddin hypothesized a field effect carcinogenesis that the alterations in the normal-appearing mucosa are perhaps multifocal areas of initiated (but not fully promoted) foci and that these may be predictors of the cancer away from their site of sampling.[3,14] In other words, as a result of the generalized effect of the carcinogen throughout the entire field of the target tissue (viz., the colonic epithelium), it is most likely that the mucosa away from an obvious cancer would be abnormal. This supports the rationale for testing the mucin of the rectum, particularly since the rectum is a convenient sampling site.

F. Rectal Mucin Test for Large Intestinal Cancer (LIC)

The rectal mucin test exploits the mucus samples in the rectum obtained at the occasion of finger examination, and detects the presence of the marker D-Gal-β[1→3]-D-GalNAc in the mucus. The presence of D-Gal-β[1→3]-D-GalNAc in the rectal mucin would imply the existence of an abnormal mucosa somewhere in the colorectum. The abnormality may be either cancerous or precancerous lesions or the clinical state of precancerous condition, since the mucus samples from a normal subject do not express the marker. The term "precancerous lesion" indicates pathological lesions that carry a high risk of progressing to cancer, whereas "precancerous conditions" are clinical diseases or conditions that increase the probability of cancer in the patient. Although several assays using the rectal mucin have been developed, we will describe the galactose oxidase test here.[3]

The principle of the galactose oxidase test has already been described previously. Briefly, D-Gal-β[1→3]-D-GalNAc is oxidized by the enzyme galactose oxidase, generating two vicinal aldehyde groups that react with basic fuchsin to give magenta coloration (Figure 2). The test procedure is as follows:

1. Examine the rectum with finger (digital rectal examination (DRE)).
2. Smear mucus sample onto test strips (nitrocellulose membrane filter).
3. React with D-galactose oxidase (100 U/ml, pH 7.2, 10 min, room temperature).
4. Wash briefly with distilled water.
5. React with Schiff's reagent (1% basic fuchsin, 1 min).
6. Rinse in running tap water, dry, evaluate for color reaction.

It should be kept in mind that a false-negative result could be due to sampling error. An additional step of reaction with PAS sequence will ensure against that possibility. For further details on this assay, please consult reference 3.

G. Performance of the Galactose Oxidase (GO) Test in Colorectal Cancer Detection

Since the publication of the pilot study by Shamsuddin and Elsayed,[40] various investigators throughout the world have evaluated the sensitivity and specificity of this test for detecting colorectal cancer.[43-54] Table 2 summarizes the results of these and other investigators. Most of these studies varied markedly in their design, thus accounting for the variation in specificity; note that the sensitivity of the assay is rather consistently high. Mackett *et al.*[45] first applied the GO mucin test in the detection of colorectal cancer of primary-care setting. They recruited a total of 676 patients, of which 411 subjects were worked up with flexible sigmoidoscopy. They then compared the result of the GO mucin test to that of the Hemoccult® test. Two cases with carcinoma were detected by the GO mucin test, but not by the Hemoccult® test. Since this was a somewhat skewed population because of their median age, not unexpectedly the false-positive results with the former were 113 (27%) out of 411. Sakamoto *et al.*[48] first used this test to screen asymptomatic populations (n = 330) and detected one case with focal cancer in adenoma. Although their evaluation of the subjects with colonoscopy and/or barium enema fluoroscopy was inadequate, they reported 92.2% specificity. For the accurate evaluation of the test, complete work-up with total colonoscopy in every subject is indispensable.

Table 2. Summary of Performance of the Rectal Mucus Test by GO-Schiff

Institution	Year	N	Specificity %	Sensitivity %
U Maryland (Pilot) Baltimore, MD[40]	1988	73	91.5	100
U Maryland Baltimore, MD[43,44]	1988	240	[a]NA	80
RPMI Buffalo, NY[c]	1988	54	[a]NA	91
Gunma University Maebashi, Japan[46]	1989	85	[a]NA	80
USUHS Washington, DC[45]	1989	411/676	71	100
Red Cross Hospital Kyoto, Japan[47]	1990	94	81.6	82
Gunma University Maebashi, Japan[48]	1991	330	92.2	[b]NA
General Hospital Osijek, Croatia[49]	1991	76	100	100
1st Medical Univ PLA Guangzhou, P.R.C.[50]	1991	77	90	89.6
Fukuoka University Fukuoka, Japan[51]	1992	43	[a]NA	77
Gunma University Maebashi, Japan	1992	166	...study in progress...	
FHCRC Seattle, WA[52]	1992	608+52	[a]NA	92
Inst Rad Medicine Tianjian, P.R.C.[53]	1992	206	82.6	84.5
1st Medical Univ PLA[54] Guangzhou, P.R.C.	1992	6,480	93.8	85.7
Total as of October 1992:		8,995	87.8±9	88.5±8

Due to the nature of samples, the specificity[a] or sensitivity[b] cannot be accurately determined; [c]unpublished.

H. Detection of Extra-Colonic Malignancies by GO Test

It is the current dogma that, by and large, most of our cancers are due to environmental carcinogens; *environment* meaning both the internal and the external environment, our diet included. We are exposed to environmental carcinogenic agents not just at the colonic epithelium but also at other sites, and the metabolites (active or inactive) are excreted via the lungs, kidneys, large intestine, and skin. Thus, the hallmark of carcinogenic exposure—the phenotypic alterations—may be observed in these organs as well. It is therefore not inconceivable that the rectal mucin test may have a potential role in predicting malignancies of extracolonic organs.

According to a pilot study in Japan, the GO mucin test appears to have the potential to detect not only colorectal cancers, but also extracolonic malignancies (Table 3). Further studies are needed to establish this as a fact.

Table 3. Detection of Extracolonic Malignancies by GO Mucin Test[a]

		Positive rectal mucin test	
Organ	N	Number	(%)
Stomach	32	21	65.5
Liver	4	3	75.0
Choledochus	1	1	100.0
Gall bladder	2	2	100.0
Pancreas	1	1	100.0
Ovary	2	2	100.0
Uterus	11	7	63.6
Breast	2	2	100.0
Total	55	39	70.9

[a]Compiled from three unpublished studies in Japan.

III. Intermediate Marker Modulation in Cancer Prevention

Does every person with a positive test for Gal-GalNAc have colorectal cancer? Obviously not. Remember that the marker is not only for cancer, but also for precancerous lesions and conditions. Inasmuch as all persons positive for HIV (human immunodeficiency virus) do not have AIDS, but most of them eventually acquire the disease, individuals with a positive rectal mucin assay are very likely to develop LIC in the course of time. In contrast to AIDS, we fortunately have chemopreventive agents that may change the dreadful outcome.

For instance, let's take the case of an individual who has repeatedly tested positive with the rectal mucin assay for Gal-GalNAc, but careful diagnostic examinations (such as barium enema, complete colonoscopy) reveal no obvious mass lesion. The presence of the marker indicates that the cells are abnormal, but not necessarily cancerous. If we consider this individual to be at a high risk and give a prophylactic chemopreventive agent, we could reduce or perhaps reverse the risk of cancer. In this instance, the marker Gal-GalNAc would not be expressed anymore, indicating that the individual is no longer at high risk.

Evidence for such an optimistic scenario is this: The human colon cancer cell line HT-29 does express Gal-GalNAc. Inositol hexaphosphate ($InsP_6$), a naturally occurring carbohydrate, is a potent anti-cancer agent with chemopreventive and chemotherapeutic properties.[55] Treatment of HT-29 cells with $InsP_6$ results not only in the reduction of cell number, but also a near-total suppression of Gal-GalNAc expression.[56] Thus, Gal-GalNAc detected in the rectal mucus has great potential as an intermediate marker not only for screening, but also for monitoring people at high risk for LIC.

IV. Conclusion

Rectal examination by a physician is a routine procedure in clinical practice. In the event it is not, it must be made one. Besides the diseases of the rectum, that of the prostate in men (prostate cancer being another major cancer in men) and diseases of the uterus and adnexa in women can be detected by this routine and simple examination. If the mucus samples caught on the glove were smeared or collected for the GO-Schiff assay, the information obtained through this simple procedure will be greatly extended. Identifying an individual with high-risk colorectal cancer development is a first step towards an improvement of mortality of LIC. Intensive examination of a preselected population that is screened by a reliable and predictive marker such as D-Gal-β[1→3]-D-GalNAc will increase the efficacy of detecting early stage LIC or precancerous lesions, thus contributing to a saving of national medical expenses. Quantitative assays measuring the concentrations of D-Gal-β[1→3]-D-GalNAc in the mucin might be of use in predicting the malignant potential (prognosis) of precancerous lesions.

References

1. C.C. Boring, T.S. Squires, and T. Tong, Cancer statistics, 1993, *CA* 43:7 (1993).
2. D.M. Eddy, Screening for colorectal cancer, *Ann Intern Med* 113:373 (1990).
3. A.M. Shamsuddin, "*Diagnostic Assays for Colon Cancer*," CRC Press, Boca Raton, FL (1991).
4. J.L. Wagner, B. Duffy, S. Wadhwa, and L. Jakubowski, Costs and effectiveness of colorectal cancer screening in the elderly, *Background paper 5*, Office of Technology Assessment, U.S. Congress, Washington, D.C. (1990).
5. J.E. Allison and F. Feldman, Cost benefits of Hemoccult screening for colorectal carcinoma, *Dig Dis Sci* 30:860 (1985).
6. M.J. Barry, A.G. Mulley, and J.M. Richter, The effect of workup strategy on the cost-effectiveness of fecal occult blood screening for colorectal cancer, *Gastroenterology* 93:301 (1987).
7. W.L. England, J.J. Halls, and V.B. Hunt, Strategies for screening colorectal carcinoma, *Med Decis Making* 9:3 (1989).
8. D.A. Ahlquist, Occult blood screening: Obstacle to effectiveness, *Cancer* 70:1259 (1992).
9. E.J. Stanbridge, Identifying tumor suppressor genes in human colorectal cancer, *Science* 247:12 (1990).
10. D. Sidransky, T. Tokino, S.R. Hamilton, K.W. Kinzler, B. Levine, P. Frost, and B. Vogelstein, Identification of *ras* oncogene mutations in the stool of patients with curable colorectal tumors, *Science* 256:102 (1992).
11. M.I. Filipe and A.C. Branfoot, Abnormal patterns of mucus secretion in apparently normal mucosa of large intestine with carcinoma, *Cancer* 34:282 (1974).
12. A.K.M. Shamsuddin and B.F. Trump, Colonic epithelium. II. *In vivo* studies of colon carcinogenesis. Light microscopic, histochemical, and ultrastructural studies of histogenesis of azoxymethane-induced colon carcinomas in Fischer 344 rats, *JNCI* 66:389 (1981).
13. A.K.M. Shamsuddin and B.F. Trump, Colon epithelium. III. *In vitro* studies of colon carcinogenesis in Fischer 344 rats: *N*-methyl-*N'*-nitro-*N*-nitrosoguanidine-induced changes in colon epithelium in explant culture, *JNCI* 66:403 (1981).
14. A.K.M. Shamsuddin, L. Weiss, P.C. Phelps, and B.F. Trump, Colon epithelium. IV. Changes in human colon mucosa adjacent to and remote from carcinomas of the colon, *JNCI* 66:413 (1981).
15. E.J. Bloom, S.H. Itzkowitz, and Y.S. Kim, Carbohydrate tumor markers in colon cancer and polyps, *in*: *Colon Cancer Cells*, M.P. Moyer and G.H. Poste, eds., Academic Press, New York (1990).

16. A.K.M. Shamsuddin, P.C. Phelps, and B.F. Trump, Human large intestinal epithelium: Light microscopy, histochemistry and ultrastructure, *Hum Pathol* 13:790 (1982).
17. A.P. Corfield, J.B. Rainey, J.R. Clamp, and S.A Wagner, Rat colonic mucosal cell sialic acid metabolism in azoxymethane induced tumors, *Biochem Biophys Acta* 840:264 (1985).
18. P.E. Reid, C.F.A. Culling, W.L. Dunn, and M.G. Clay, Chemical and histochemical studies of normal and diseased human gastrointestinal tract. I. A comparison between histologically normal colon, colonic tumours, ulcerative colitis and diverticular disease of the colon, *Histochem J* 16:235 (1984).
19. C.F.A. Culling, P.E. Reid, and W.L. Dunn, A histochemical comparison of the *O*-acylated sialic acids of the epithelial mucins in ulcerative colitis, Crohn's disease and normal controls, *J Clin Pathol* 32:1272 (1979).
20. J.T. James, A.M. Shamsuddin, and B.F. Trump, A comparative study of the morphological and histochemical changes induced in the large intestine of ICR/Ha and C57Bl/Ha mice by 1,2-dimethylhydrazine, *JNCI* 71:955 (1983).
21. P.A. Dawson and M.I. Filipe, An ultrastructural and histochemical study of the mucous membrane adjacent to and remote from carcinoma of the colon, *Cancer* 37:2388 (1976).
22. D.M. Marcus, The ABO and Lewis blood group system, *N Engl J Med* 280:994 (1969).
23. B.F. Atkinson, C.S. Ernst, M. Herlyn, Z. Steplewski, H.F. Sears, and H. Koprowski, Gastrointestinal cancer-associated antigen in immunoperoxidase assay, *Cancer Res* 42:4820 (1982).
24. J.W. Arends, C. Verstynen, F.T. Bosman, J. Hilgers, and Z. Steplewski, Distribution of monoclonal antibody-defined monosialoganglioside in normal and cancerous human tissues: An immunoperoxidase study, *Hybridoma* 2:219 (1983).
25. E. Gong, S. Hirohashi, Y. Shimosato, M. Watanabe, Y. Ito, S. Teshima, and S. Kodaira, Expression of carbohydrate antigen 19-9 and stage specific embryonic antigen 1 in nontumorous and tumorous epithelia of the human colon and rectum, *JNCI* 75:447 (1985).
26. S.H. Itzkowitz, M. Yuan, Y. Fukushi, A. Palekar, P.C. Phelps, A.M. Shamsuddin, B.F. Trump, S. Hakomori, and Y.S. Kim, Lewis^x- and sialylated Lewis^x-related antigen expression in human malignant and nonmalignant colonic tissues, *Cancer Res* 46:2627 (1986).
27. H.F. van den Ingh, D.J. Ruiter, J. Griffioen, G.N.P. van Muijen, and S. Ferrone, HLA antigens in colorectal tumors - low expression of HLA class I antigens in mucinous colorectal carcinomas, *Br J Cancer* 55:125 (1987).
28. M.E.F. Smith, S.G.E. Marsh, J.G. Bodmer, K. Gelsthorpe, and W.F. Bodmer, Loss of HLA-A,B,C allele products and lymphocyte function-associated antigen 3 in colorectal neoplasia, *Proc Natl Acad Sci (USA)* 86:5557 (1989).
29. G. Tsioulias, T.A. Godwin, M.F. Goldstein, C.J. McDougall, S.S. Ngoi, J.J. DeCosse, and B. Rigas, Loss of colonic HLA antigens in familial adenomatous polyposis, *Cancer Res* 52:3449 (1992).
30. C.J. McDougall, S.S. Ngoi, I.S. Goldman, T. Godwin, J. Felix, J.J. DeCosse, and B. Rigas, Reduced expression of HLA Class I and II antigens in colon cancer, *Cancer Res* 50:8023 (1990).
31. P. Demant, Histocompatibility and the genetics of tumour resistance, *J Immunogenet* 13:61 (1986).
32. W.J. Bateman, R. Fiera, N. Matthews, and A.G. Morris, Inducibility of class II major histocompatibility complex antigens by interferon γ is associated with reduced tumorigenicity in C3H mouse fibroblasts transformed by v-Ki-*ras*, *J Exp Med* 173:193 (1991).
33. R. Lotan, E. Skutelsky, D. Danon, and N. Sharon, The purification, composition and specificity of anti-T lectin from peanut (*Arachis hypogaea*), *J Biol Chem* 250:8518 (1975).
34. C.R. Boland, C.K. Montgomery, and Y.S. Kim, Alterations in human colonic mucin occurring with cellular differentiation and malignant transformation, *Proc Natl Acad Sci USA* 79:2051 (1982).

35. H.S. Cooper, Peanut lectin-binding sites in large bowel carcinoma, *Lab Invest* 47:383 (1982).
36. T.F. Orntoft, N.P. Ole Mors, G. Eriksen, N.O. Jacobsen, and H.S. Poulsen, Comparative immunoperoxidase demonstration of T-antigens in human colorectal carcinomas and morphologically abnormal mucosa, *Cancer Res* 45:447 (1985).
37. K.L. McKenzie, D.M. Purnell, and A.M. Shamsuddin, Expression of carcinoembryonic antigen, T-antigen and oncogene products as markers of neoplastic and preneoplastic colonic mucosa, *Human Pathol* 18:1282 (1987).
38. B.A. Schulte and S.S. Spicer, Light microscopic histochemical detection of sugar residues in secretory glycoproteins of rodent and human tracheal glands with lectin-horseradish peroxidase conjugates and the galactose oxidase-Schiff sequence, *J Histochem Cytochem* 31:391 (1983).
39. A.M. Elsayed and A.M. Shamsuddin, Detection of altered glycoconjugate in preneoplastic and neoplastic human large intestinal epithelium by galactose oxidase-Schiff sequence, *Lab Invest* 56:22A (1987).
40. A.M. Shamsuddin and A.M. Elsayed, A test for detection of colorectal cancer, *Hum Pathol* 19:7 (1988).
41. A.M. Shamsuddin, A.M. Elsayed, and G.A. Jockle, Screening test for large intestinal cancer, *U.S. Patent No. 4,857,457*. (1989).
42. H.M. Xu, K. Sakamoto, and A.M. Shamsuddin, Detection of the tumor marker D-galactose-β[1→3]*N*-acetyl-D-galactosamine in colonic cancer and precancer, *Arch Pathol Lab Med* 116:1234 (1993).
43. J.B. Schreiber, R.G. Knodell, and A.K.M. Shamsuddin, Utilization of alterations in colonic mucus as a screening test for detecting colonic neoplasia, *Am J Gastroenterol* 83:1058(abs) (1988).
44. R.G. Knodell, E. Hale, M. Antos, H. Matossian, and A.K.M. Shamsuddin, A new screening test for detection of colon cancer utilizing neoplasia associated alterations in rectal mucus, *Gastroenterology* 94:A230(abs) (1988).
45. C.W. Mackett III, C. Wilson, T. Brna, and R.D. Kirkham, Comparison of a galactose oxidase test to Hemoccult in colorectal cancer screening, *Am J Gastroenterol* 84:1190(abs) (1989).
46. K. Sakamoto, G. Nakano, and Y. Nagamachi, A pilot study on the usefulness of a new test for mass screening of colorectal cancer in Japan, *Gastroenterol Jpn* 25:432 (1990).
47. M. Tada, H. Ohtsuka, A. Iso, S. Shimizu, M. Okamura, K. Aoki, and K. Kawai, Screening of colorectal cancer by testing rectal mucus for β-D-Gal[1→3]-D-GalNAc (T-antigen), *J Kyoto Pref Univ Med* 99:681 (1990).
48. K. Sakamoto, M. Muratani, T. Ogawa, and Y. Nagamachi, Colon cancer screening by a new test: A prospective study of asymptomatic population, *Proc Am Assoc Cancer Res* 32:168(abs) (1991).
49. A. Vcev, B. Pospihalj, D. Bozic, I. Candrlic, A. Vegar, L. Pinotic, and Z. Krucaj, The galactose oxidase test for detection of colorectal cancer, *Exp Clin Gastroenterol* 1:217 (1991).
50. D.-Y. Zhou, Y. Zhang, Z. Nai, L. Pam, B. Zhang, and W. Zhang, Screening of colorectal cancer by testing rectal mucus for T-antigen, *Chinese J Digestion* 11:261 (1991).
51. C. Sano, R. Kamishiro, H. Uogaeshi, S. Yamazaki, K. Tanaka, S. Mandokoro, R. Beppu, H. Yoshitake, T. Hideshima, T. Maekawa, and S. Inuzuka, A new screening test for colorectal cancer detection by T antigen, *Jpn J Gastroenterol Surg* 25:1568 (1992).
52. A.R. Kristal, M.S. Baker, A.D. Feld, J.T. Ylvisaker, and K.A. Morse, Evaluation of a test for abnormal rectal mucus for early detection of colon cancer, *Cancer Epidemiol Biomarkers Prev* 1:303 (1992).
53. C. Mu, S. Li, B. Liu, and Z. Tian, The screening of colorectal cancer by enzymatic detection of β-D-Gal[1→3]-D-GalNAc in colorectal mucus, *Proc of Chinese Acad Medical Sciences* (1993) in press.

54. D.-Y. Zhou, F. Feng, Y. Zhang, Z. Lai, W. Zhang, L. Li, G. Xu T. Wan, D. Pan, D. Zhou, Y. Zhang, and S. Li, Comparison of Shams' test for rectal mucus to an immunological test for fecal occult blood in large intestinal cancer screening: Analysis of 6480 asymptomatic subjects, submitted for publication (1992).
55. A.M. Shamsuddin and A. Ullah, Inositol hexaphosphate inhibits large intestinal cancer in F344 rats 5 months following induction by azoxymethane, *Carcinogenesis* 10:625 (1989).
56. K. Sakamoto, G. Venkatraman, and A.M. Shamsuddin, Growth inhibition and differentiation of HT-29 cells by inositol hexaphosphate ($InsP_6$), *Proc Am Assoc Cancer Res* 33:42(abs) (1992).

Chapter 7

Chemical and Mutagenic Specificities of Polycyclic Aromatic Hydrocarbon Carcinogens

ANTHONY DIPPLE, KIMMO PELTONEN, S. CHAU CHENG, HELEN ROSS and C. ANITA H. BIGGER

I. Introduction

Polycyclic aromatic hydrocarbons are found in some broiled meats, as a product of the cooking process,[1] and also in vegetables as a contaminant, thus having a widespread distribution through the environment.[2] The hydrocarbons are generated through the process of combustion and it seems unlikely, therefore, that exposure to them will ever be completely eliminated. For this reason, a number of researchers, including ourselves, have focused attention on the mechanism of action of these carcinogens with the idea that an understanding of the details of the process of carcinogenesis might facilitate the development of strategies for intervention, such as chemoprevention.

Like many other organic chemical carcinogens, the hydrocarbons require metabolic activation in order to express their carcinogenic potential. For benzo[*a*]pyrene, Sims *et al.*[3] proposed a three-step activation process leading to the formation of the 7,8-dihydrodiol 9,10-epoxide, shown in Figure 1. Subsequent work by many researchers has confirmed this mechanism and extended it to other polycyclic aromatic hydrocarbons. (For a review of this extensive literature see Dipple *et al.*[4])

The reactive benzo[*a*]pyrene metabolite shown in Figure 1 is responsible for most of the benzo[*a*]pyrene bound to DNA in cellular systems. The predominant adduct formed in DNA results from the *trans* opening of the epoxide ring at carbon-10 by the amino group of deoxyguanosine.[5,6] No other hydrocarbon has been studied as extensively as benzo[*a*]pyrene, but our studies with 7,12-dimethylbenz[*a*]anthracene, which is not found in the environment but which is some 30-fold more potent as a carcinogen than benzo[*a*]pyrene,[7] indicated that

Anthony Dipple and Helen Ross • Chemistry of Carcinogenesis Laboratory, ABL-Basic Research Program, NCI-Frederick Cancer Research and Development Center, Frederick, MD 21702; Kimmo Peltonen • Institute of Occupational Health, Department of Industrial Hygiene and Toxicology, Topeliuksenkatu 41aA, SF-00250, Helsinki, Finland; S. Chau Cheng • Clinical Research Associates, 50 Madison Avenue, New York, NY 10010; C. Anita H. Bigger • Microbiological Associates, 9900 Blackwell Road, Rockville, MD 20850

Diet and Cancer: Markers, Prevention, and Treatment
Edited by M.M Jacobs, Plenum Press, New York, 1994

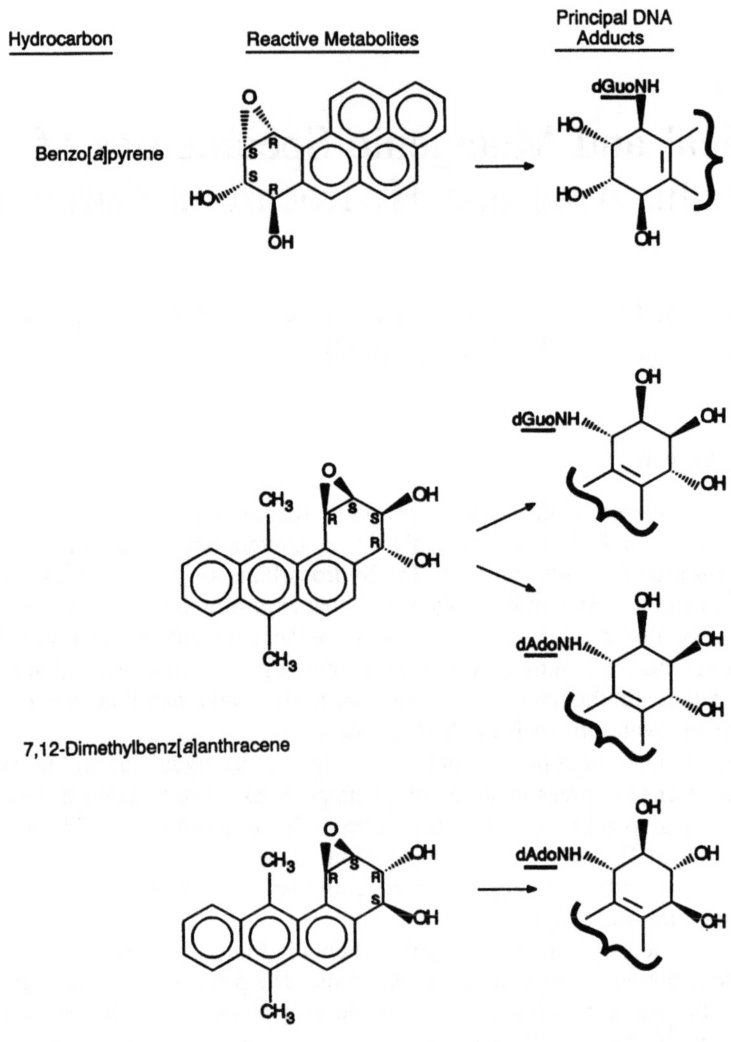

Figure 1. Reactive metabolites of benzo[*a*]pyrene and 7,12-dimethylbenz[*a*]anthracene and the principal adducts that they form with DNA.

two diastereomeric dihydrodiol epoxides contributed to the DNA binding of this potent carcinogen (Figure 1).[8] Moreover, the dihydrodiol epoxide with the same stereochemistry as the benzo[*a*]pyrene active metabolite (i.e., the R,S-dihydrodiol S,R-epoxide) reacted to roughly similar extents with both deoxyadenosine and deoxyguanosine in DNA[9-12] whereas the *syn* diastereomer reacted almost exclusively with deoxyadenosine residues in DNA.[7-9,13] Interestingly, selenium in the form of sodium selenite selectively inhibited binding of 7,12-dimethylbenz[*a*]anthracene to DNA through the R,S-dihydrodiol S,R-epoxide, apparently by inhibiting the induction of a cytochrome P_{450} required for R,S-dihydrodiol S,R-epoxide formation but not required for S,R-dihydrodiol S,R-epoxide formation.[12]

The main point illustrated by Figure 1 is that, though benzo[*a*]pyrene and 7,12-dimethylbenz[*a*]anthracene belong to the same chemical class, i.e., both are hydrocarbons, their mechanisms of action are not identical. Thus, two different dihydrodiol epoxides whose formation is dependent on different P_{450}s are substantially involved in DNA binding for

7,12-dimethylbenz[*a*]anthracene, whereas only one dihydrodiol epoxide mediates the majority of benzo[*a*]pyrene-DNA binding. Moreover, three major DNA adducts are found for 7,12-dimethylbenz[*a*]anthracene, whereas only one major adduct is found for benzo[*a*]pyrene. It has been noted that the large difference in extents of reaction with deoxyadenosine residues in DNA for these two carcinogens might be causally related to their large difference in tumorigenic potencies.[7,11]

II. Chemical Selectivity in Dihydrodiol Epoxide-DNA Reactions

The initial findings described above have prompted investigations of the factors determining adduct formation in DNA for hydrocarbon dihydrodiol epoxides. From studies with reactive metabolites from several hydrocarbons, notably benzo[*a*]pyrene, 7,12-dimethylbenz[*a*]anthracene, 7-methylbenz[*a*]anthracene, 5-methylchrysene, and benzo[*c*]phenanthrene, we have found that when the stereochemistry of the dihydrodiol epoxide is kept constant but the hydrocarbon residue is varied, the percentage of dihydrodiol epoxide trapped by DNA, as opposed to water, upon reaction with an aqueous solution of DNA, varies widely. This is illustrated for four R,S-dihydrodiol S,R-epoxides[6,14-16] in Figure 2, where it can be seen that the percentage of dihydrodiol epoxide trapped by DNA ranges from a few percent for benzo[*a*]pyrene dihydrodiol epoxide to as much as 75% for benzo[*c*]phenanthrene dihydrodiol epoxide. The reasons for these profound differences lie within the structures of the reactive metabolites. One possible factor would be the relative ease with which the various dihydrodiol epoxides ionize to a carbocation, presumably after protonation of the epoxide oxygen. An ion should react readily with water, whereas, for a somewhat less ionic intermediate, the greater nucleophilicity of DNA over water would lead to the capture of more reactive metabolite by DNA. Certainly, one would expect benzo[*a*]pyrene dihydrodiol epoxide to be the most, and benzo[*c*]phenanthrene dihydrodiol epoxide to be the least, readily ionized epoxides in this series, in concert with the extents to which they are trapped by DNA.[17]

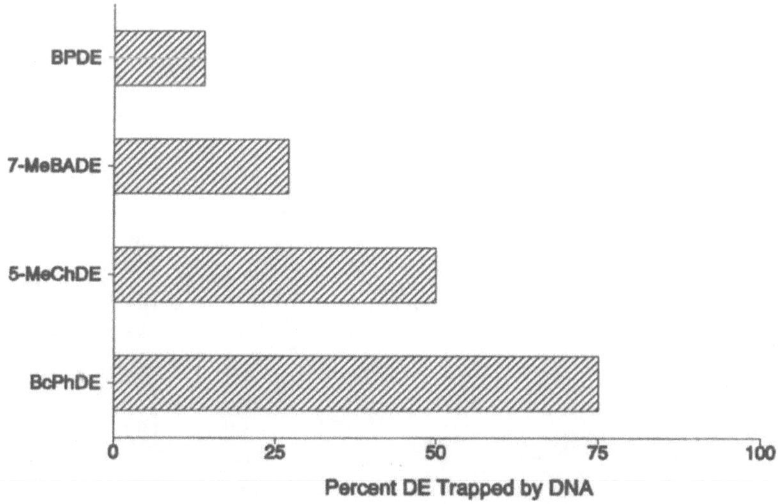

Figure 2. Extents of reaction with DNA for various R,S-dihydrodiol S,R-epoxides. BPDE, 5-MeChDE, 7-MeBADE, and BcPhDE indicate the bay region dihydrodiol epoxides of benzo[*a*]pyrene, 5-methylchrysene, 7-methylbenz[*a*]anthracene, and benzo[*c*]phenanthrene, respectively.

Extents of reaction are not the only variable in the reactions of dihydrodiol epoxides with DNA. As shown in Figure 1, products vary substantially also with the nature of the hydrocarbon and this can be seen for a range of hydrocarbons in Figure 3. In this figure, the percentage of total products represented by the *trans* opening of the epoxide by either the amino group of deoxyadenosine or deoxyguanosine is shown.[6,14-16,18] The dihydrodiol epoxide that is trapped the least efficiently by DNA, benzo[*a*]pyrene dihydrodiol epoxide, is also the one that reacts almost exclusively with deoxyguanosine residues. In contrast, a dihydrodiol epoxide that is efficiently trapped by DNA, benzo[*c*]phenanthrene dihydrodiol epoxide, reacts more extensively with deoxyadenosine residues than with deoxyguanosine residues. It is also true that benzo[*c*]phenanthrene dihydrodiol epoxide is a more potent tumorigen than is benzo[*a*]pyrene dihydrodiol epoxide.[19] One obvious structural difference between these derivatives that exhibit very different chemical behaviors is that the bay region of benzo[*c*]phenanthrene is a sterically hindered fjord region that causes the parent molecule to be nonplanar, whereas benzo[*a*]pyrene is a relatively planar structure.

In addition to the chemical differences described above for different hydrocarbons, the different configurational isomers of the dihydrodiol epoxide of the same hydrocarbon also exhibit profound chemical differences. For example, in collaborative studies with Jerina and his colleagues,[14,18] we have found that the four configurational isomers of benzo[*c*]phenanthrene dihydrodiol epoxide each react quite differently with DNA (Figure 4) and their chemistry closely parallels that for the dihydrodiol epoxides of 7,12-dimethylbenz[*a*]anthracene, which also has a sterically crowded bay region.[13,16,20-23] For example, the S,R-dihydrodiol S,R-epoxide of benzo[*c*]phenanthrene reacts almost exclusively (~ 90%) with deoxyadenosine residues in DNA, as does this configurational isomer from 7,12-dimethylbenz[*a*]anthracene[13] (Figure 1). Secondly, the R,S-dihydrodiol S,R-epoxide of benzo[*c*]phenanthrene, like its 7,12-dimethylbenz[*a*]anthracene equivalent in Figure 1, reacts extensively with both deoxyguanosine and deoxyadenosine residues in DNA. Although the S,R-dihydrodiol R,S-epoxide of benzo[*c*]phenanthrene also reacts extensively with deoxyadenosine and deoxyguanosine in DNA, the deoxyadenosine adducts arise in a *cis*/*trans* ratio of 0.48 rather

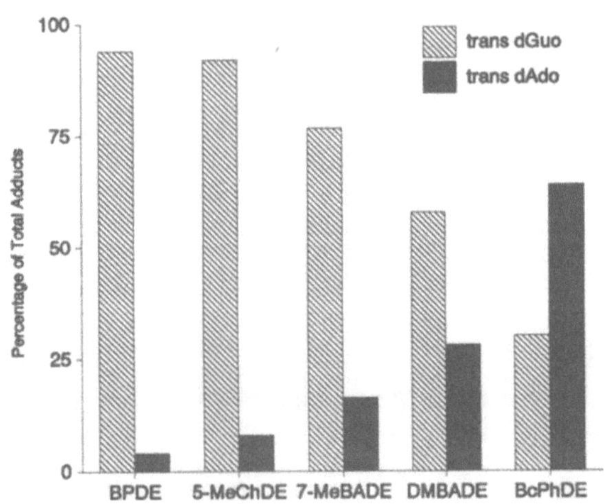

Figure 3. Distribution of R,S-dihydrodiol S,R-epoxide adducts over deoxyadenosine (dAdo) and deoxyguanosine (dGuo) residues in DNA. DMBADE indicates the bay region dihydrodiol epoxide of 7,12-dimethylbenz[*a*]anthracene.

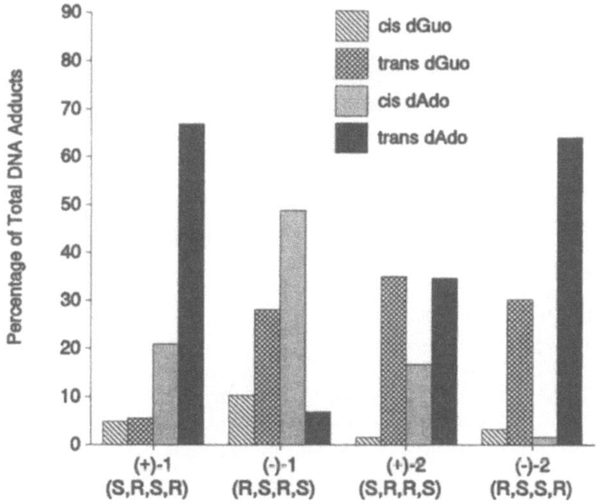

Figure 4. Different configurational isomers of benzo[*c*]phenanthrene dihydrodiol epoxide react differently with DNA. Dihydrodiol epoxides where the epoxide oxygen and benzylic OH are *cis* or *trans* are labeled 1 or 2, respectively. The absolute stereochemistry of each is indicated by the letters R and S in the sequence of the carbons of the dihydrodiol followed by the carbons of the epoxide ring.

than the ratio of 0.025 found for its R,S-dihydrodiol S,R-epoxide enantiomer. The R,S-dihydrodiol R,S-epoxide also reacts with both deoxyadenosine and deoxyguanosine residues in DNA, but this configurational isomer is unusual in that the major adduct it forms is a *cis* adduct. Thus, not only is the chemistry of DNA adduct formation dependent upon the particular hydrocarbon residue concerned, but it is also dependent upon the absolute stereochemistry in the dihydrodiol epoxide ring.

III. Mutagenic Specificity

In order to determine whether the different chemistries of DNA adduct formation for different dihydrodiol epoxides lead to different biological consequences, we have been exploring the mutagenic specificities of these agents. The system we have used for these studies is that developed by Seidman and Dixon and their collaborators[24] (Figure 5). In this system, a shuttle vector, pS189,[25] is treated *in vitro* with the dihydrodiol epoxide and transfected into human Ad293 cells to allow adducts to be converted to mutations. The vector is then recovered and used to transform *Escherichia coli* that have an amber mutation in the *lacZ* gene. This mutation is suppressed by the *supF* gene in the original vector, so that β-galactosidase is made and, on X-Gal plates, the colonies are blue. However, if the *supF* gene has been inactivated by mutation, the amber mutation is not suppressed and white or pale blue colonies are formed. Vector DNA can be isolated by growing up the white or pale blue colonies and this DNA is sequenced to define the specific site and type of mutation generated in the *supF* gene.

Some of the results obtained with this assay are summarized and compared with chemical data in Figure 6. This figure compares the mutagenic selectivity and chemical selectivity for adenine and guanine residues in DNA.[26] Mutagenic selectivity has been defined as the percentage of total mutations at A-T or G-C pairs divided by the percentage of A-T or G-C in the 85-base pair (bp) sequence coding for the mature tRNA in the *supF*

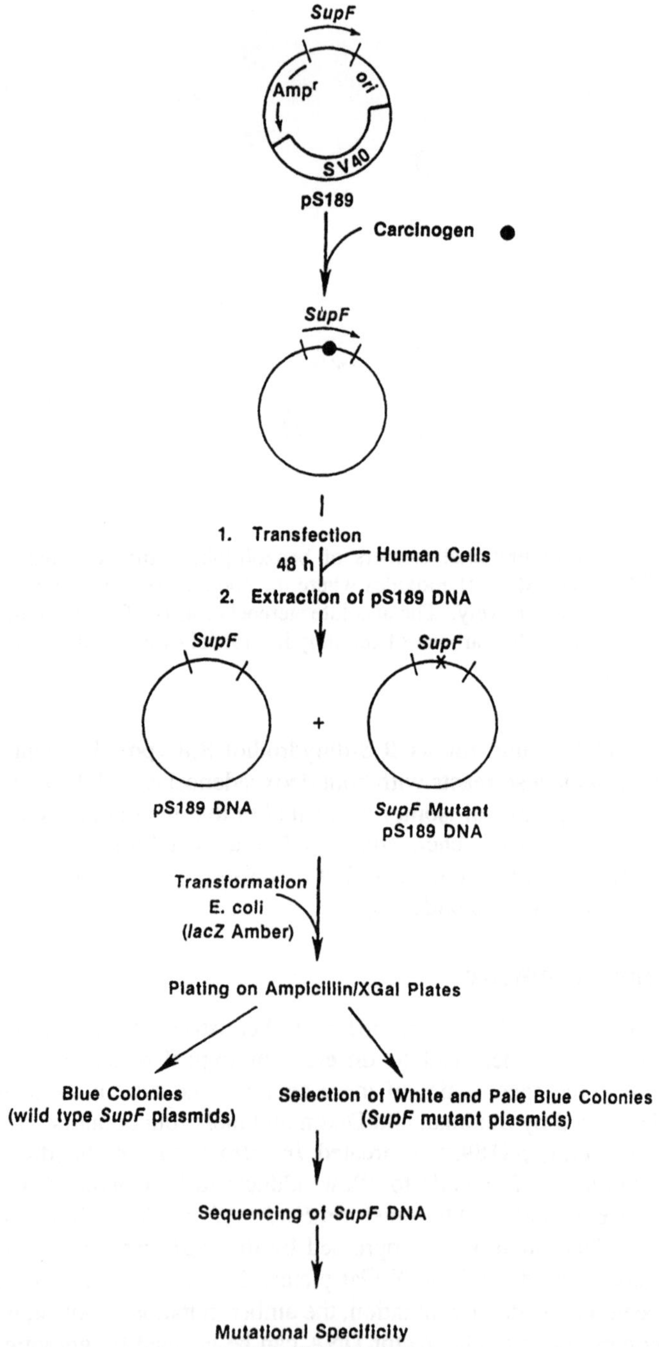

Figure 5. Description of the mutational assay using the *supF* gene of shuttle vector pS189.

gene. Chemical selectivity was determined from the data in Figure 4 by dividing the percentage of reaction with A-T or G-C pairs in calf thymus DNA by the percentage occurrence of A-T or G-C pairs in that DNA.

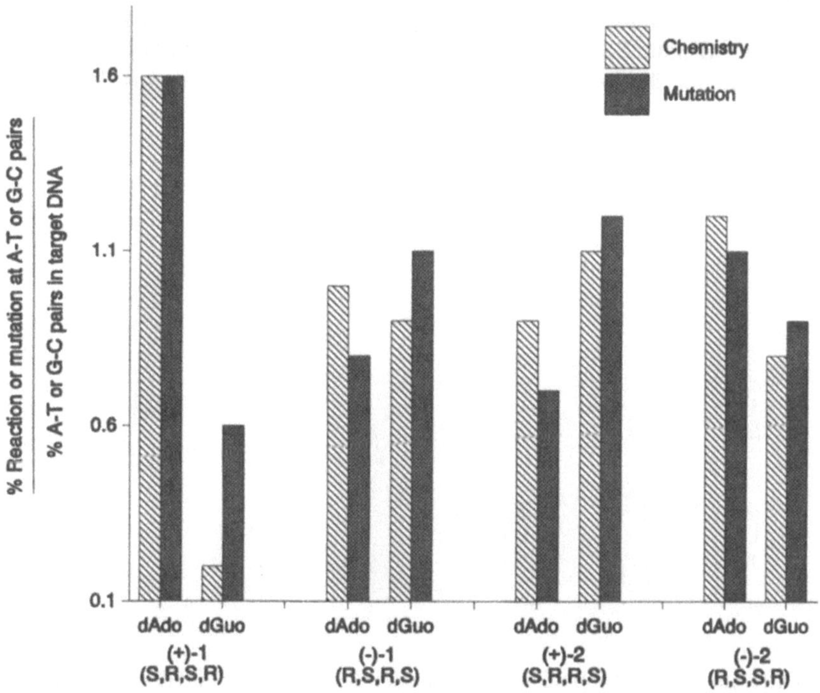

Figure 6. Comparison of chemical and mutagenic selectivities of configurationally isomeric benzo[*c*]phenanthrene dihydrodiol epoxides. Abbreviations are defined in legends to Figures 3 and 4.

It can be seen that there is a reasonable correspondence between the chemical data and the mutation data, although they do not follow one another exactly. One discrepancy is that whereas the mutational preference for A-T pairs is either the same or less than the chemical preference, the mutational preference at G-C pairs is always somewhat greater than the chemical preference. This cannot be attributed to silent mutations being more prevalent at A-T than G-C pairs because, of four silent single-base substitution mutations known for this gene, three are at G-C pairs.[27]

Since the chemistry of each of the configurationally isomeric benzo[*c*]phenanthrene dihydrodiol epoxides is distinctly different, it was not surprising that their mutagenic specificities for the *supF* gene were also different.[26,28] The distribution of the base substitution mutations found through the four possible transversion and two possible transition mutations is summarized in Figure 7. These data were derived from an analysis of 148, 90, 141, and 111 mutations for the (–)-2, (+)-2, (–)-1, and (+)-1 isomers respectively (see legend to Figure 4 for definition of abbreviations). If larger numbers of mutants were examined, some changes in these distributions might be seen, but it is unlikely that the changes would be dramatic.

Analysis of mutations induced by racemic 7-methylbenz[*a*]anthracene dihydrodiol epoxide-2,[29] by racemic 5-methylchrysene dihydrodiol epoxide-2[30] and by racemic benzo[*a*]pyrene dihydrodiol epoxide-2[31] indicates that these reactive carcinogen derivatives generate mutations principally at G-C pairs in concert with their chemical preference for reaction at guanines (Figure 3). However, significant differences, as well as some similarities in the distribution of these mutations through the target *supF* gene were noted.[29]

The distribution of benzo[*c*]phenanthrene dihydrodiol epoxide-induced base substitution mutations through the *supF* gene is also unique for each benzo[*c*]phenanthrene dihydrodiol epoxide isomer[26] (Figure 8). It seemed conceivable that this distribution of mutations might simply follow the distribution of benzo[*c*]phenanthrene dihydrodiol epoxide adducts through

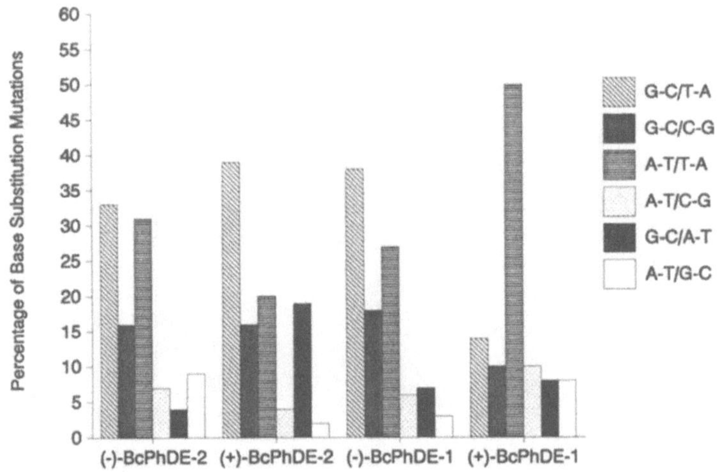

Figure 7. Different configurational isomers of benzo[*c*]phenanthrene dihydrodiol epoxide (BcPhDE) generate different distributions of base substitution mutations.

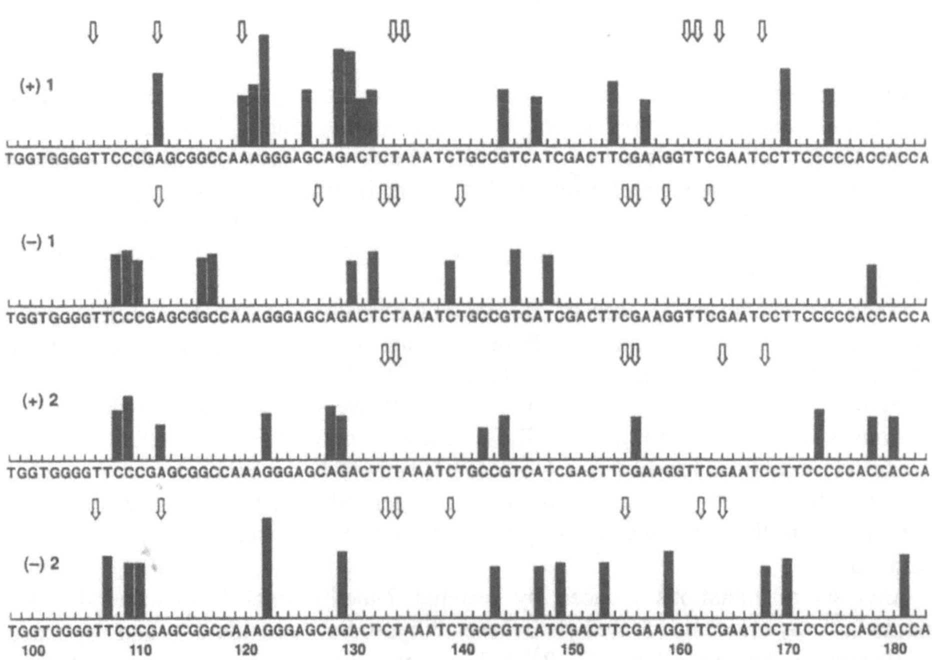

Figure 8. Sites of DNA polymerase arrest in the *supF* gene of pS189 treated with the configurational isomers of benzo[*c*]phenanthrene dihydrodiol epoxide. Vertical bars represent the site and intensity of polymerase arrest bands; downward arrows represent the site of mutation hotspots.

this target gene. In order to monitor the latter, experiments have been undertaken in which the sites of adduct formation have been monitored by examining the sites at which DNA polymerase is arrested when it is asked to replicate the carcinogen-modified *supF* gene.[32] Figure 8 shows the sites of mutation hotspots through the *supF* tRNA structural gene (arrows). It also shows the relative intensity of the electrophoretic bands corresponding to the most intense sites of polymerase arrest. Although there are minor arrest bands in the vicinity of mutation hotspots[32] that are not shown in this figure, it is clear that the most prominent sites of polymerase arrest are clearly not the most prominent sites of mutation induction.

IV. Conclusions

The data discussed above are relevant to the interaction of genotoxic carcinogens with DNA and the expression of the DNA damage in the form of mutation. The initiation stage of the carcinogenic process is somewhat more complex than mutation, but the events discussed above are considered to be key events in initiation. Although a wide array of chemical structures are capable of initiating the carcinogenic process (reviewed in [33,34]), we have focused in this discussion on a few members of one family of carcinogens, the polycyclic aromatic hydrocarbons.

All of the compounds discussed are activated through the same general dihydrodiol epoxide mechanism as a consequence of our very narrow selection. Nevertheless, substantial differences in mechanisms of action have been documented. For example, two dihydrodiol epoxides generated by at least two different P_{450}s are substantially involved in 7,12-dimethylbenz[*a*]anthracene-DNA binding vs. one dihydrodiol epoxide for benzo[*a*]pyrene-DNA binding. Once the reactive dihydrodiol epoxides are formed, even those with the same stereochemistry but derived from different hydrocarbons, react quite differently with DNA in terms of both extents of reaction (Figure 2) and products of reaction (Figure 3). Moreover, different isomeric dihydrodiol epoxides from the same hydrocarbon display quite varied reaction preferences with DNA (Figure 4). Thus, even for closely related carcinogens, qualitative and quantitative differences in activation and reactions with DNA make each carcinogen unique.

This range of variation is not limited to the chemical properties of these agents, but applies also to their biological activities, monitored here by their mutational specificities (Figures 7 and 8). Although the mutation data follows the chemical findings to some extent, there is not a direct relationship, as shown by comparison of the sites of hotspots for mutation and sites of the most extensive polymerase arrest (Figure 8).

In the longer term, we would hope that these kinds of investigations might lead to the identification of specific adducts, specific mutations, or specific mutation targets that are instrumental in the initiation of the carcinogenic process. At present, however, it leaves us with a range of mechanistic choices for events involved in initiation even for the narrow range of carcinogens examined here. When one considers the wider range of possibilities that arise when the whole spectrum of genotoxic carcinogens is considered, the options of chemopreventive approaches are similarly very wide and varied. However, the challenge of defining chemopreventive agents that might be effective against initiation by the wide range of genotoxic carcinogens known is clearly substantial.

Acknowledgment

We wish to acknowledge the contributions of our collaborators, Dr. Bruce Hilton, Chemical Synthesis and Analysis Laboratory, PRI/DynCorp, Inc., Frederick, MD; Ronald G.

Harvey, Ben May Institute, Chicago, IL; Dr. Donald M. Jerina, National Institute of Diabetes and Digestive and Kidney Diseases, National Institutes of Health, Bethesda, MD; and their colleagues. This research was sponsored by the National Cancer Institute, DHHS, under contract No. NO1-CO-74101 with ABL. The contents of this publication do not necessarily reflect the views or policies of the Department of Health and Human Services, nor does mention of trade names, commercial names, or organizations imply endorsement by the U.S. Government.

References

1. W. Lijinsky, The formation and occurrence of polynuclear aromatic hydrocarbons associated with food, *Mutat Res* 259:251 (1991).
2. A. Dipple, Formation, metabolism, and mechanism of action of polycyclic aromatic hydrocarbons, *Cancer Res* 43:2422s (1983).
3. P. Sims, P.L. Grover, A. Swaisland, K. Pal, A. Hewer, Metabolic activation of benzo[*a*]pyrene proceeds by a diol-epoxide, *Nature* 252:326 (1974).
4. A. Dipple, R.C. Moschel, C.A.H. Bigger, Polynuclear aromatic carcinogens, *in*: *Chemical Carcinogens*, C.E. Searle, ed., American Chemical Society, Washington, D.C., pp. 41-163 (1984).
5. A.M. Jeffrey, K.W. Jennette, S.H. Blobstein, I.B. Weinstein, F.A. Beland, R.G. Harvey, H. Kasai, I. Miura, K. Nakanishi, Benzo[*a*]pyrene-nucleic acid derivative found in vivo: Structure of a benzo[*a*]pyrene dihydrodiol epoxide-guanosine adduct, *J Am Chem Soc* 98:5714 (1976).
6. S.C. Cheng, B.D. Hilton, J.M. Roman, A. Dipple, DNA adducts from carcinogenic and noncarcinogenic enantiomers of benzo[*a*]pyrene dihydrodiol epoxide, *Chem Res Toxicol* 2:334 (1989).
7. A. Dipple, J.T. Sawicki, R.C. Moschel, C.A.H. Bigger, 7,12-Dimethylbenz[*a*]anthracene-DNA interactions in mouse embryo cell cultures and mouse skin, *in*: *Extrahepatic Drug Metabolism and Chemical Carcinogenesis*, J. Rydstrom, J. Montelius, M. Bengtsson, eds., Elsevier, Amsterdam, pp. 439-448 (1983).
8. J.T. Sawicki, R.C. Moschel, A. Dipple, Involvement of both syn- and anti-dihydrodiol-epoxides in the binding of 7,12-dimethylbenz[*a*]anthracene to DNA in mouse embryo cell cultures, *Cancer Res* 43:3212 (1983).
9. A. Dipple, M. Pigott, R.C. Moschel, N. Costantino, Evidence that binding of the 7,12-Dimethylbenz[*a*]anthracene to DNA in mouse embryo cell cultures results in extensive substitution of both adenine and guanine residues, *Cancer Res* 43:4132 (1983).
10. A. Dipple, M.A. Pigott, C.A.H. Bigger, D.M. Blake, 7,12-Dimethylbenz[*a*]anthracene-DNA binding in mouse skin: Response of different mouse strains and effects of various modifiers of carcinogenesis, *Carcinogenesis* 5:1087 (1984).
11. C.A.H. Bigger, J.T. Sawicki, D.M. Blake, L.G. Raymond, A. Dipple, Products of binding of 7,12-dimethylbenz[*a*]anthracene to DNA in mouse skin, *Cancer Res* 43:5647 (1983).
12. J.A. Milner, M.A. Pigott, A. Dipple, Selective effects of selenium on 7,12-dimethylbenz[*a*]-anthracene-DNA binding in fetal mouse cell cultures, *Cancer Res* 45:6347 (1985).
13. S.C. Cheng, A.S. Prakash, M.A. Pigott, B.D. Hilton, H. Lee, R.G. Harvey, A. Dipple, A metabolite of the carcinogen 7,12-dimethylbenz[*a*]anthracene that reacts predominantly with adenine residues in DNA, *Carcinogenesis* 9:1721 (1988).
14. S.K. Agarwal, J.M. Sayer, H.J.C. Yeh, L.K. Pannell, B.D. Hilton, M.A. Pigott, A. Dipple, H. Yagi, D.M. Jerina, Chemical characterization of DNA adducts derived from the configurationally isomeric benzo[*c*]phenanthrene-3,4-diol 1,2-epoxides, *J Am Chem Soc* 109:2497 (1987).

15. D.B. Reardon, A.S. Prakash, B.D. Hilton, J.M. Roman, J. Pataki, R.G. Harvey, A. Dipple, Characterization of 5-methylchrysene-1,2-dihydrodiol-3,4-epoxide-DNA adducts, *Carcinogenesis* 8:1317 (1987).
16. K. Peltonen, S.C. Cheng, B.D. Hilton, H. Lee, C. Cortez, R.G. Harvey, A. Dipple, Effect of bay region methyl group on reactions of *anti*-benz[*a*]anthracene 3,4-dihydrodiol 1,2-epoxides with DNA, *J Org Chem* 56:4181 (1991).
17. A. Dipple, K. Peltonen, S.C. Cheng, B.D. Hilton, Chemistry of DNA adduct formation by dihydrodiol epoxides of polycyclic aromatic hydrocarbons, *in*: *Polycyclic Aromatic Compounds: Synthesis, Properties, Analytical Measurements, Occurence and Biological Effects*, P. Garrigues, M. Lamotte, eds., Gordon and Breach, Philadelphia, pp. 811-820 (1993).
18. A. Dipple, M.A. Pigott, S.K. Agarwal, H. Yagi, J.M. Sayer, D.M. Jerina, Optically active benzo[*c*]phenanthrene diol epoxides bind extensively to adenine in DNA, *Nature* 327:535 (1987).
19. W. Levin, R.L. Chang, A. Wood, D.R. Thakker, H. Yagi, D.M. Jerina, A.H. Conney, Tumorigenicity of optical isomers of the diastereomeric bay region 3,4-diol epoxides of benzo[*c*]phenanthrene 3,4-diol-1,2-epoxides, *Cancer Res* 46:2257 (1986).
20. J.A. Vericat, S.C. Cheng, A. Dipple, Absolute configuration of 7,12-dimethylbenz[*a*]anthracene-DNA adducts in mouse epidermis, *Cancer Lett* 57:237 (1991).
21. H. Schmeiser, A. Dipple, M.E. Schurdak, E. Randerath, K. Randerath, Comparison of ^{32}P postlabeling and high pressure liquid chromatographic analyses for 7,12-dimethylbenz[*a*]-anthracene-DNA adducts, *Carcinogenesis* 9:633 (1988).
22. S.C. Cheng, A.S. Prakash, M.A. Pigott, B.D. Hilton, J.M. Roman, H. Lee, R.G. Harvey, A. Dipple, Characterization of 7,12-dimethylbenz[*a*]anthracene-adenine nucleoside adducts, *Chem Res Toxicol* 1:216 (1988).
23. J.A. Vericat, S.C. Cheng, A. Dipple, Absolute stereochemistry of the major 7,12-dimethylbenz[*a*]anthracene-DNA adducts formed in mouse cells, *Carcinogenesis* 10:567 (1989).
24. M.M. Seidman, K. Dixon, A. Razzaque, R.J. Zagursky, M.L. Berman, A shuttle vector plasmid for studying carcinogen-induced point mutations in mammalian cells, *Gene* 38:233 (1985).
25. M. Seidman, The development of transient SV40 based shuttle vectors for mutagenesis studies: Problems and solutions, *Mutat Res* 220:55 (1989).
26. C.A.H. Bigger, J. St. John, H. Yagi, D.M. Jerina, A. Dipple, Mutagenic specificities of four stereoisomeric benzo[*c*]phenanthrene dihydrodiol epoxides, *Proc Natl Acad Sci USA* 89:368 (1992).
27. K.H. Kraemer, M.M Seidman, Use of *supF*, an *Escherichia coli* tyrosine suppressor tRNA gene, as a mutagenic target in shuttle-vector plasmids, *Mutat Res* 220:61 (1989).
28. C.A.H. Bigger, J. Strandberg, H. Yagi, D.M. Jerina, A. Dipple, Mutagenic specificity of a potent carcinogen, benzo[*c*]phenanthrene (4R,3S)-dihydrodiol (2S,1R)-epoxide, which reacts with adenine and guanine in DNA, *Proc Natl Acad Sci USA* 86:2291 (1989).
29. C.A.H. Bigger, D.H. Flickinger, J. St. John, R.G. Harvey, A. Dipple, Preferential mutagenesis at G.C base pairs by the anti 3,4-dihydrodiol 1,2-epoxide of 7-methylbenz[*a*]anthracene, *Mol Carcinogen* 4:176 (1991).
30. C.A.H. Bigger, D.J. Flickinger, J. Strandberg, J. Pataki, R.G. Harvey, A. Dipple, Mutational specificity of the anti 1,2-dihydrodiol 3,4-epoxide of 5-methylchrysene, *Carcinogenesis* 11:2263 (1990).
31. J.-L. Yang, V.M. Maher, J.J. McCormick, Kinds of mutation formed when a shuttle vector containing adducts of (+/–)-7β,8α-dihydroxy-9α,10α-epoxy-7,8,9,10-tetrahydrobenzo-[*a*]pyrene replicates in human cells, *Proc Natl Acad Sci USA* 84:3787 (1987).
32. H. Ross, C.A.H. Bigger, H. Yagi, D.M. Jerina, A. Dipple, Sequence specificity in the interaction of the four stereoisomeric benzo[*c*]phenanthrene dihydrodiol epoxides with the *supF* gene, *Cancer Res* 53:1273 (1993).
33. A. Dipple, C.J. Michejda, E.K. Weisburger, Metabolism of chemical carcinogens, *Pharmacol Ther* 27:265 (1985).

34. E.C. Miller, J.A. Miller, Mechanisms of chemical carcinogens: Nature of proximate carcinogens and interactions with macromolecules, *Pharmacol Rev* 18:805 (1966).

Chapter 8

Inhibition of Tumorigenesis by Chemicals from Garlic and Tea

CHUNG S. YANG, ZHI-YUAN WANG and JUN-YAN HONG

I. Introduction

Garlic (*Allium sativum*) has been used widely in culinary practice and folk remedies in many cultures. Its biological functions are beginning to be understood by studies with modern research tools. Epidemiological studies in China and Italy indicate that more frequent consumption of garlic and other allium vegetables was associated with a lower incidence of gastric cancer.[1,2] The inhibition of carcinogenesis by garlic constituents has also been demonstrated in animal models. In previous studies, we demonstrated the selective inhibition and induction of cytochrome P_{450} enzymes by a garlic compound, diallyl sulfide (DAS), as a possible mechanism for inhibitory action against carcinogenesis.[3] In the present paper, this mechanism is discussed further along with new results on the inhibition of tumorigenesis as predicted from metabolism studies.

Tea (*Camellia sinensis*) is one of the most ancient and popular beverages consumed worldwide. The effects of tea consumption on carcinogenesis are an important concern. However, epidemiological studies on tea and cancer have yielded inconsistent and inconclusive results. Early studies on the carcinogenicity of tea topically applied or tea extract subcutaneously injected into animals have yielded controversial conclusions. On the other hand, many recent studies have shown an inhibitory action of tea or tea components on tumorigenesis in animals. In the present paper, we describe the inhibitory action of green tea and black tea on nitrosamine-induced tumorigenesis in the lung, forestomach, and esophagus.

Abbreviations used: DAS, diallyl sulfide; $DASO_2$, diallyl sulfone; NDMA, *N*-nitrosodimethylamine; DMH, 1,2-dimethylhydrazine; NNK, 4-(methylnitrosamino)-1-(3-pyridyl)-1-butanone; EGCG, (–)epigallocatechin-3-gallate; EGC, (–)epigallocatechin; ECG, (–)epicatechin gallate; EC, (–)epicatechin; NMBzA, *N*-nitrosomethylbenzylamine; and NDEA, *N*-nitrosodiethylamine.

Chung S. Yang, Zhi-Yuan Wang, and Jun-Yan Hong • Laboratory for Cancer Research, College of Pharmacy, Rutgers University, Piscataway, NJ 08855

Diet and Cancer: Markers, Prevention, and Treatment
Edited by M.M Jacobs, Plenum Press, New York, 1994

II. Effects of DAS on Cytochrome P_{450}, Carcinogen Activation, and Carcinogenesis

Fresh garlic contains a vast number of organosulfur compounds, among which alliin is a major component. Allicin, diallyl disulfide, DAS, methyl allyl sulfide, and other organosulfur compounds are formed through the action of the enzyme allinase, cooking, or metabolism in animals. The structures of some of these compounds are shown in Figure 1. The actions of DAS have been studied extensively and will be discussed herein to illustrate the possible mechanisms of its inhibitory actions against chemically induced tumorigenesis. Some specific effects of DAS on different P_{450} enzymes are demonstrated in the following experiment.[3] After an oral dose of DAS (200 mg/kg) to the rat, the total P_{450} content and P_{450} reductase activity in liver microsomes were affected only slightly. However, a time-dependent decrease in the P_{450} 2E1-dependent *N*-nitrosodimethylamine (NDMA) demethylase activity was observed; the activity was lowest (<20% of the control) at 15 hr and gradually returned to the normal level (1.4 nmol per min per mg protein) after 2 days. On the other hand, the pentoxyresorufin dealkylase activity (mainly due to P_{450} 2B1), which was very low in untreated rat liver microsomes, was greatly enhanced, reaching a plateau of 100-fold increase between 24 to 48 hr. When diallyl sulfone ($DASO_2$) was given to the rat, the NDMA demethylase activity decreased very rapidly. The activity was decreased significantly at 10 min and decreased to 25% of the control at 2 hr, but was partially recovered at 15 hr. Similar results were also observed with mice in which a significant decrease in ethoxyresorufin dealkylase activity (mainly due to P_{450} 1A2) was also observed.

Subsequent studies demonstrated that DAS was metabolized to diallyl sulfoxide and then to $DASO_2$. Diallyl sulfone was further metabolized by P_{450} 2E1 to a reactive intermediate that could attack the heme moiety of the enzyme and thus inactivate P_{450} 2E1 in a mechanism commonly known as suicide inhibition. In addition, DAS, $DASO_2$, and diallyl sulfoxide were all competitive inhibitors of P_{450} 2E1. The structures and the metabolism of these compounds are shown in Figure 2. A weaker activity in the inactivation of P_{450} 2E1 was observed with methyl allyl sulfide, but diallyl disulfide or dipropyl sulfide did not inactivate P_{450} 2E1. Since P_{450} 2E1 is vital in catalyzing the activation of NDMA, 1,2-

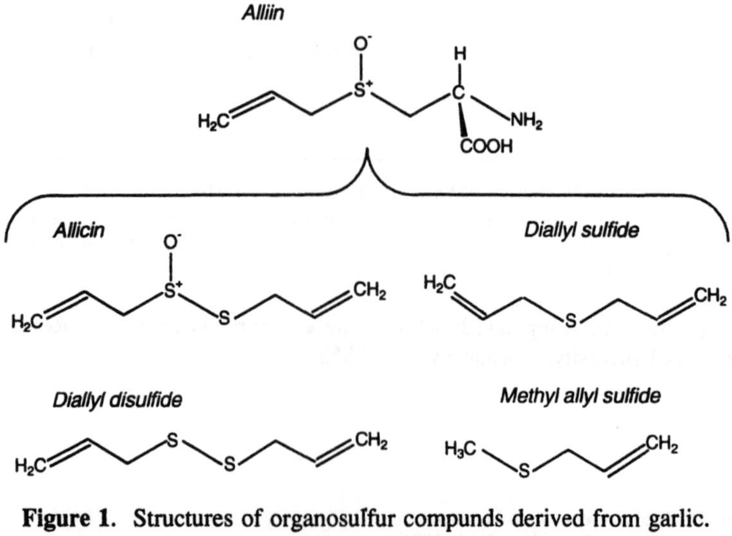

Figure 1. Structures of organosulfur compunds derived from garlic.

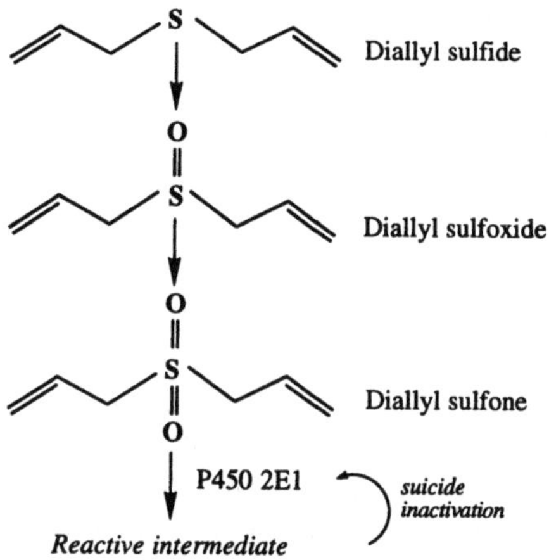

Figure 2. Metabolism of DAS and inactivation of P_{450} 2E1.

dimethylhydrazine (DMH), benzene, alkanes, halogenated hydrocarbons, and many other low-molecular-weight environmental chemicals,[4,5] inhibition of P_{450} 2E1 is expected to block the toxicity and carcinogenicity of these compounds. The inhibition of NDMA–, CCL_4–, and acetaminophen-induced hepatotoxicity by DAS has been demonstrated.[3] The inhibition of P_{450} 2E1-dependent activation of carcinogens is the most likely molecular mechanism for the reported inhibitory action of DAS against DMH-induced hepatotoxicity and colon carcinogenesis in rats.[6,7] In the activation of DMH, P_{450} 2E1 is involved in the oxidation of azoxymethane to methylazoxymethanol, which is further converted to the methylating species, methyldiazonium hydroxide.[8]

In theory, inhibition of the metabolic activation of a toxicant or a carcinogen in the liver could reduce hepatotoxicity and hepatocarcinogenesis. However, it may increase the exposure of nonhepatic tissues to this toxic compound and thus enhance carcinogenesis in nonhepatic organs.[9] When given together with NDMA, ethanol enhanced the tumorigenesis in the nasal cavity[10] and lung.[9] This may be one of the mechanisms by which ethanol consumption increases the incidence of extrahepatic carcinogenesis in humans.

III. Inhibition of 4-(Methylnitrosamino)-1-(3-pyridyl)-1-butanone Bioactivation and Lung Tumorigenesis by DAS and $DASO_2$

4-(Methylnitrosamino)-1-(3-pyridyl)-1-butanone (NNK) is a potent tobacco carcinogen believed to be important in human cancer etiology, especially in the causation of oral cancer in tobacco chewers and lung cancer in cigarette smokers.[11] It was observed previously in our laboratory that oral administration of DAS markedly decreased the rate of NNK bioactivation in mouse lung and rat nasal mucosa microsomes.[12,13] The α-methylene oxidation pathway, which leads to the generation of a methylating agent and the keto aldehyde [4-oxo-1-(3-pyridyl)-1-butanone], was decreased by 80% in mouse lung microsomes. The rate of the bioactivation of NNK was decreased even more extensively in rat nasal mucosa microsomes, but only moderately in liver microsomes. This decrease in metabolic activity appears not to be related to P_{450} 2E1 inactivation, and the mechanisms remain to be elucidated. However, this result suggested that DAS would inhibit NNK-induced lung tumorigenesis. This activity was demonstrated experimentally. When a single dose of NNK was given (2 mg/mouse, i.p.) to female A/J mice (7 weeks of age maintained on an AIN-76 diet), almost all the animals developed lung adenomas after 16 weeks, with an average of 9 ± 1.3 tumors per mouse. When DAS was given to the mice orally at a daily dose of 200 mg/kg for 3 days with the final dose given 2 hr prior to the NNK administration, the tumor incidence was reduced by 60% and tumor multiplicity was decreased by 90%.[13] A similar inhibitory action was also produced with $DASO_2$ at a dose of 100 mg/kg; at a dose of 20 mg/kg the tumor multiplicity was inhibited by 38%, but the tumor incidence was not affected.

IV. Mechanisms of Inhibition of Tumorigenesis by DAS and Related Compounds

The inhibitory activity of DAS against DMH– and NNK–induced tumorigenesis is probably due to inhibition of the bioactivation of these carcinogens. In the case of DMH, inactivation and inhibition of P_{450} 2E1 is the most likely mechanism. Based on the P_{450} 2E1 inactivation mechanism, allyl mecaptan, diallyl disulfide, and dipropyl sufide, which are not effective P_{450} 2E1 inactivators, as expected were ineffective tumorigenesis inhibitors. However, diallyl disulfide and allyl mercaptan were found to be effective, whereas DAS was ineffective, in inhibiting *N*-nitrosodiethylamine (NDEA)-induced tumorigenesis in the forestomach of female A/J mice.[14] Allyl methyl trisulfide, diallyl trisulfide, and DAS, when given 96 and 48 hr prior to benzo(*a*)pyrene, also inhibited forestomach neoplasia in female A/J mice.[15] Diallyl sulfide was also very effective in inhibiting *N*-nitrosomethylbenzylamine (NMBzA)-induced esophageal tumorigenesis in rats.[16] In all these studies, the organosulfur compounds appeared to inhibit the activation of the carcinogen or to trap the activated carcinogenic species. The detailed mechanisms remain to be elucidated. Other garlic-related compounds or products such as 1-propenyl sulfide, ajoene, and garlic oil have been shown to inhibit tumor promotion by phorbol-myristate acetate on mouse skin.[17] Inhibition of lipoxygenase and ornithine decarboxylase are the possible mechanisms of such an inhibition.

V. Effects of Oral Administration of Green Tea on NDEA-Induced Lung and Forestomach Tumorigenesis

Green tea leaves (12.5 g) were placed in 500 ml of freshly boiled deionized water for 15 min and then filtered. The tea leaves were steeped a second time with 500 ml of freshly boiled water and filtered. The combined filtrate is referred to as the 1.25% infusion. A 0.63% infusion was prepared from the 1.25% infusion via a 1 to 2 dilution. The amount of solids present in the 1.25% infusion was 4.69 mg/ml. (–)Epigallocatechin-3-gallate (EGCG), caffeine, (–)epigallocatechin (EGC), (–)epicatechin gallate (ECG), and (–)epicatechin (EC) were the major components, accounting for 15.1%, 8.1%, 6.9%, 3.0%, and 1.8% of the dry solid, respectively. The structures of these compounds are shown in Figure 3.

Treatment of female A/J mice with NDEA caused lung tumors in more than 90% of the animals, with an average of 8.3 ± 1.0 tumors per mouse 16 weeks after the last dose of NDEA. The administration of 0.63% or 1.25% of green tea infusion as the sole source of drinking water during the NDEA treatment period, post-NDEA treatment period, or the entire experimental period significantly decreased the forestomach tumor multiplicity by 31% to 63%, and decreased the lung tumor multiplicity by 36% to 60%. The 1.25% green tea infusion also decreased lung tumor incidence in all three protocols by 36% to 44% and significantly decreased forestomach tumor incidence when given during the initiation period or the entire experimental period. Inhibitory effects on lung and forestomach tumorigenesis by green tea infusion were also observed in a similar experiment using a high dosage of NDEA (20 mg/kg). Histopathological examination showed that almost all lung tumors were pulmonary adenomas. Most of the forestomach lesions were hyperplasia or papillomas. Carcinoma *in situ* and squamous cell carcinomas were also observed. In all the tea treatment

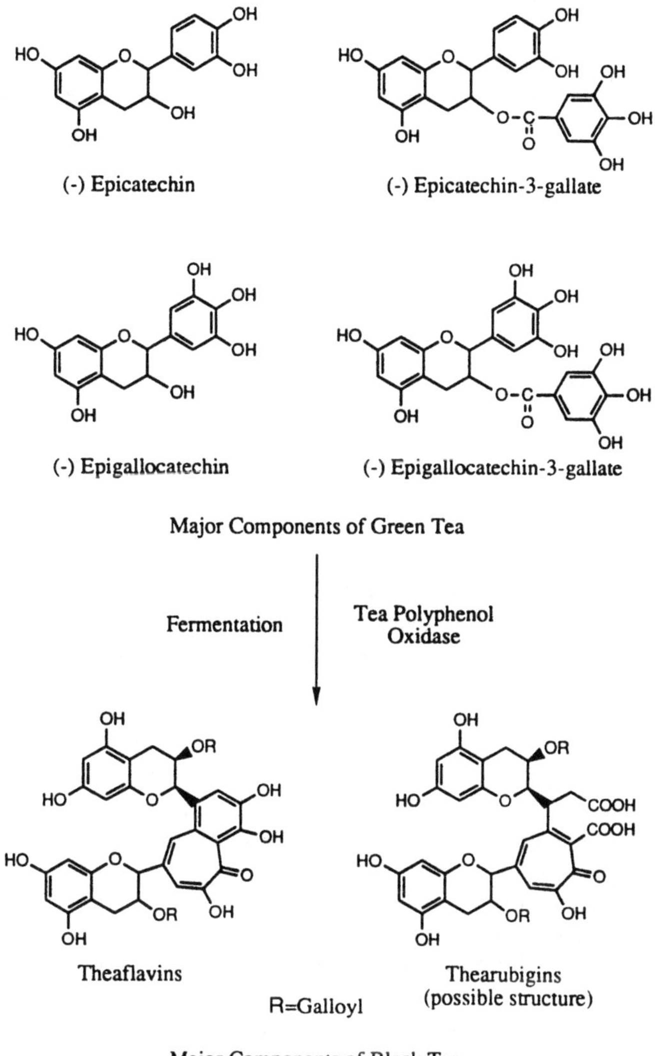

Figure 3. Major components of green tea and black tea.

groups, the incidence of papilloma was lower and the tumor size was smaller than the positive control group.[18]

VI. Effects of Oral Administration of Decaffeinated Green Tea and Black Tea on NNK-Induced Lung Tumorigenesis

Decaffeinated green tea powder and black tea powder, prepared from water extracts of decaffeinated tea leaves, were supplied by the Thomas J. Lipton Company (Englewood Cliffs, NJ). The decaffeinated tea leaves were prepared by extracting tea leaves with supercritical CO_2. EGCG, EGC, ECG and EC were the major catechins in decaffeinated green tea powder, accounting for 11.1%, 11.0%, 2.6%, and 3.1% of the dry weight, respectively, and caffeine was decreased to 0.3% of the dry weight. In the decaffeinated black tea powder, EGCG, EGC, ECG, EC, and caffeine accounted for 2.2%, 0.9%, 1.3%, 0.9%, and 0.4% of the dry weight, respectively. The amounts of theaflavins were 1.7% and thearubigins were about 25% of the dry weight.

Treatment of A/J mice with a single dose of NNK (103 mg/kg) resulted in 96% of mice bearing lung tumors and an average tumor multiplicity of 9.3 ± 1.3 tumors per mouse after 16 weeks. When 0.6% decaffeinated green tea or black tea was given during the NNK treatment period as the sole source of drinking water, tumor multiplicity was reduced by 67% or 65%, respectively. When the tea extract was given after the NNK treatment period until the end of the experiment, 0.3% and 0.6% decaffeinated green tea reduced the tumor multiplicity by 74% and 85%, respectively, and the tumor incidence by 14% and 30%, respectively. Under the same conditions, the decaffeinated black tea extracts decreased tumor multiplicity by about 63%, but did not significantly reduce the tumor incidence.[18] The results suggest that when given during the carcinogen treatment period, green tea and black tea are equally effective. However, when given after the carcinogen treatment period, green tea is more effective than black tea.

VII. Effects of Tea Consumption on Esophageal Cancer

The effects of decaffeinated green tea and black tea on esophageal tumorigenesis were studied in a rat model. In this study, Sprague-Dawley rats were treated by subcutaneous injection with NMBzA (2.5 mg/kg) twice weekly for 5 weeks, and 65% of the rats developed papillomas with an average of 1.4 ± 0.3 tumors per mouse at week 39. Administration of 0.6% decaffeinated green tea or black tea extract as the sole source of drinking water during the NMBzA treatment period decreased the tumor incidence and multiplicity by more than 70%. When green or black tea extracts were given to rats after the carcinogen treatment period, the tumor incidence and multiplicity also appeared to decrease, but the difference was not statistically significant. A more pronounced effect in this protocal were the decreases in esophageal papilloma sizes by 89 and 97% in the green and black tea groups, respectively (manuscript in preparation).

Our results are consistent with those of Han and Xu,[19] who reported that oral administration of infusions made from five different brands of green tea and black tea as the sole source of drinking water during the entire experimental period inhibited NMBzA-induced esophageal tumorigenesis in rats. The positive control group (2.5 mg NMBzA/kg administered p.o. once weekly for 2 weeks followed by administration twice weekly for another 10 weeks) had a tumor incidence of 90%, with an average of 4.7 esophageal tumors per rat. Treatment with 2% tea infusions via the drinking water decreased tumor incidence by 26-53% and reduced tumor multiplicity by 68-75%. The average tumor size was also significantly reduced in all treatment groups. Oral administration of these tea infusions also markedly inhibited esophageal tumorigenesis caused by NMBzA precursors (methylbenzylamine plus sodium nitrite) in rats, reducing tumor incidence by 80% to 95%.[20] In these studies, all the different brands of tea had inhibitory activities, although some brands had higher activities than others.

VIII. Possible Mechanisms of the Inhibitory Action of Tea Components against Tumorigenesis

Our present results demonstrate inhibitory effects of orally administered tea infusions on NDEA- and NNK-induced tumorigenesis in A/J mice and NMBzA-induced esophageal tumorigenesis in rats. In studies with the green tea infusion, the body weights of the mice that received tea during the post-NDEA treatment period were lower than those in the control group (by 5% to 13%). A question may be raised as to whether the observed inhibitory effects were due to the reduced growth rate. The studies on NNK-induced lung tumorigenesis showed that decaffeinated tea extract inhibited tumorigenesis without affecting the body weight. If we assume that, after the carcinogen treatment period, the biological processes in the NDEA- and NNK-induced tumorigenesis models are similar, then the results from the NNK model suggest that it was the tea components rather than the retarded growth rate that inhibited NDEA-induced tumorigenesis.

The components in tea extracts responsible for the presently observed inhibitory action are not known. The studies with decaffeinated tea extracts indicated that the effect was not due to caffeine. In studying the inhibitory effect of tea, many investigators have focused on green tea and green tea polyphenols, especially EGCG. Our results demonstrate that black tea, which contains much lower amounts of catechins than green tea, was just as effective as green tea in inhibiting tumorigenesis when given to the mice during the carcinogen treatment period. Inhibition of the metabolic activation of NNK may be one of the mechanisms involved. Preliminary results from our laboratory indicate that under the same conditions of our NNK-induced lung tumorigenesis experiment, oral administration of green or black decaffeinated tea inhibits NNK-caused lung DNA methylation by 37%. Studies *in vitro* also show that EGCG, ECG, EGC, EC, and the ethyl acetate extractable polyphenol fractions from green tea and black tea inhibit the oxidative metabolic conversion of NNK to its *N*-oxide, keto alcohol, and keto aldehyde metabolites in mouse lung microsomes. EGCG is a more potent inhibitor than the other compounds tested (manuscript in preparation). Previously Wang *et al.*[21] reported that flavan-3-ols, green tea polyphenols, and black tea polyphenols inhibited benzo(*a*)pyrene metabolism in rat liver microsomes. Thus tea preparations may have the ability to inhibit the activation of a variety of carcinogens. Other mechanisms, such as the induction of phase II enzymes[22,23] and the trapping of ultimate carcinogens,[24] have also been suggested for the anticarcinogenic activities of tea. However, the importance of these mechanisms for the inhibition of NDEA-, NNK-, and NMBzA-induced carcinogenesis remain to be investigated.

The antioxidant properties of tea components may be important in the protection against cellular oxidative damage such as lipid peroxidation, DNA single-strand breakage, or the formation of 8-hydroxydeoxyguanosine. These insults may contribute to carcinogenesis at both the initiation and post-initiation stages. (+)-Catechin was reported to inhibit NNK-induced DNA single-strand breaks in rat hepatocytes *in vitro* and *in vivo*.[25] EGCG was shown to inhibit TPA-induced 8-hydroxydeoxyguanosine formation in HeLa cells.[26] Oral administration of green tea was shown to inhibit the formation of 8-hydroxydeoxyguanosine in mice. [27]

IX. Relevance of This Research to the Prevention of Human Cancer

Although the present work has demonstrated interesting biological activities of chemicals derived from garlic and tea in the inhibition of tumorigenesis in animals, the relevance of this work to the prevention of human cancer remains to be determined. In extrapolating results from animal experiments to human situations, the relevance of the animal model and the effective doses of the agents are key factors. Both mouse and human lungs are believed to be target organs and both can metabolically activate NNK to alkylating agents. In the mouse lung, the activation is mainly mediated by cytochrome P_{450} enzymes.[28] However, in human lung microsomes, only a fraction of the activity in the activation of NNK is due to P_{450} enzymes.[29] Thus, if the observed inhibitory action of a certain agent against tumorigenesis in mice is due to the blocking of P_{450}-dependent activation of NNK, then we may not expect a similar inhibition in humans because the bioactivation would be blocked only partially. In terms of dosage, the amounts of tea administered to the mice was much higher than human consumption on a per gram body weight basis. However, based on caloric consumption, the amounts administered to mice correspond to the consumption of 6 cups of tea per day by a human. The amount of DAS administered to the mice, 100 to 200 mg/kg body weight, is much higher than that derived from daily consumption of garlic (estimated to contain 50 µg DAS per g fresh garlic) by humans. If garlic indeed has protective actions against cancer, it is most likely to be due to the combined effects of many garlic components.

Definitive answers concerning the role of garlic and tea on human cancer have to come from studies with human populations. Although an association between the consumption of garlic (and other allium vegetables) and lower gastric cancer rate has been observed,[1,2] additional studies are needed to elucidate the cause-effect relationship. The relationship between tea consumption and human cancer has been studies by many epidemiologists. In a recent review by the International Agency for Research on Cancer,[30] it was concluded that "There is inadequate evidence for the carcinogenicity in humans and experimental animals of tea drinking." Even less well known is the protective role against cancer in human populations. Because the causative factors are different for different cancers or for the same cancer in different populations, it is possible that tea consumption may affect carcinogenesis only in selected situations rather than having a general effect on all cancers. Therefore, in future studies, it is important to consider the etiological factors of the specific cancers and to study the same population repeatedly in order to elucidate the role of tea consumption on human cancer incidence.

Acknowledgments

This work was supported by Grant 88B18R90B from the American Institute for Cancer Research and NIH grant CA 56673. The authors thank Ms. Marie Liethauser for editorial assistance and Ms. Dorothy Wong for excellent secretarial assistance.

References

1. W.-C. You, W.J. Blot, Y.-S. Chang, A. Ershow, Z.T. Yang, Q. An, B.E. Henderson, J.F. Fraumeni, Jr., T.G. Wang, Allium vegetables and reduced risk of stomach cancer, *J Natl Canc Inst* 81:162-164 (1989).
2. E. Buiatti, D. Palli, A. Decarli, D. Amadori, C. Avellini, S. Bianchi, R. Biserni, F. Cipriani, P. Cocco, A. Giacosa, E. Marubini, R. Puntoni, C. Vindigni, J.J. Fraumeni, W. Blot, A case-control study of gastric cancer and diet in Italy, *Int J Canc* 44:611-616 (1989).
3. J.F. Brady, M.-H. Wang, J.-Y. Hong, F. Xiao, Y. Li, J.-S.H. Yoo, S.M. Ning, J.M. Fukuto, J.M. Gapac, C.S. Yang, Modulation of rat hepatic microsomal monooxygenase activities and cytotoxicity by diallyl sulfide, *Toxicol Appl Pharmacol* 108:342-354 (1991).
4. C.S. Yang, J.-S.H. Yoo, H. Ishizaki, J.-Y. Hong, Cytochrome P450IIE1: Roles in nitrosamine metabolism and mechanisms of regulation, *Drug Metab Rev* 22:147-160 (1990).
5. C.S. Yang, J.F. Brady, J.-Y. Hong, Dietary effects on cytochromes P-450, xenobiotics metabolism, and toxicity, *FASEB J* 6:737-744 (1992).
6. M.A. Hayes, T.H. Rushmore, M.T. Goldberg, Inhibition of hepatocarcinogenic responses to 1,2-dimethylhydrazine by diallyl sulfide, a component of garlic oil, *Carcinogenesis* 8:1155-1157 (1987).
7. M.J. Wargovich, Diallyl sulfide, a flavor component of garlic (*Allium sativum*), inhibits dimethylhydrazine-induced colon cancer, *Carcinogenesis* 8:487-489 (1987).
8. O.S. Sohn, H. Ishizaki, C.S. Yang, E.S. Fiala, Metabolism of azoxymethane, methylazoxymethanol and *N*-nitrosodimethylamine by cytochrome P-450IIE1, *Carcinogenesis* 12:127-131 (1991).

9. L.M. Anderson, Increased numbers of *N*-nitrosodimethylamine-initiated lung tumors in mice by chronic co-administration of ethanol, *Carcinogenesis* 9:1717-1719 (1988).
10. L. Griciute, M. Castegnaro, J.-C. Bereziat, Influence of ethyl alcohol on carcinogenesis with *N*-nitrosodimethylamine, *Canc Lett* 13:345-352 (1981).
11. S.S. Hecht, D. Hoffman, Tobacco-specific nitrosamines, an important group of carcinogens in tobacco and tobacco smoke, *Carcinogenesis* 9:875-884 (1988).
12. J.-Y. Hong, T. Smith, M.-J. Lee, W. Li, B.-L. Ma, S.-M. Ning, J.F. Brady, P.E. Thomas, C.S. Yang, Metabolism of carcinogenic nitrosamines by rat nasal mucosa and the effect of diallyl sulfide, *Canc Res* 51:1509-1514 (1991).
13. J.-Y. Hong, Z.-Y. Wang, T. Smith, S. Zhou, S. Shi, C.S. Yang, Inhibitory effects of diallyl sulfide on metabolism and tumorigenicity of the tobacco-specific carcinogen 4-(methylnitrosamino)-1-(3-pyridyl)-1-butanone (NNK) in A/J mouse lung, *Carcinogenesis* 13:901-904 (1992).
14. L.W. Wattenberg, V.L. Sparnins, G. Barany, Inhibition of *N*-nitrosodiethylamine carcinogenesis in mice by naturally occurring organosulfur compounds and monoterpenes, *Canc Res* 49:2689-2692 (1989).
15. V.L. Sparnins, G. Barany, and L.W. Wattenberg, Effects of organosulfur compounds from garlic and onions on benzo[*a*]pyrene-induced neoplasia and glutathione *S*-transferase activity in the mouse, *Carcinogenesis* 9:131-134 (1988).
16. M.J. Wargovich, C. Woods, V.W.S. Eng, L.C. Stephens, K. Gray, Chemoprevention of *N*-nitrosomethylbenzylamine-induced esophageal cancer in rats by the naturally occurring thioether, diallyl sulfide, *Canc Res* 48:6872-6875 (1988).
17. S. Belman, Onion and garlic oils inhibit tumor promotion, *Carcinogenesis* 4:1063-1065 (1983).
18. Z.Y. Wang, J.-Y. Hong, M.-T. Huang, K. Reuhl, A.H. Conney, C.S. Yang, Inhibition of *N*-nitrosodiethylamine and 4-(methylnitrosamino)-1-(3-pyridyl)-1-butanone-induced tumorigenesis in A/J mice by green tea and black tea, *Canc Res* 52:1943-1947 (1992).
19. C. Han, Y. Xu, The effect of Chinese tea on occurrence of esophageal tumor induced by *N*-nitrosomethylbenzylamine in rats, *Biomed Environ Sci* 3:35-42 (1990).
20. Y. Xu, C. Han, The effect of Chinese tea on the occurrence of esophageal tumors induced by *N*-nitrosomethylbenzylamine formed *in vivo*, *Biomed Environ Sci* 3:406-412 (1990).
21. Z.Y. Wang, D.R. Bickers, H. Mukhtar, Interaction of epicatechins derived from green tea with rat hepatic cytochrome P-450, *Drug Metab Dispos* 16:93-103 (1988).
22. V.L. Sparnins, P.L. Venegas, L.W. Wattenberg, Glutathione *S*-transferase activity: Enhancement by compounds inhibiting chemical carcinogenesis and by dietary constituents, *J Natl Canc Inst* 68:493-496 (1982).
23. S.G. Khan, S.K. Katiyar, R. Agarwal, H. Mukhtar, Enhancement of antioxidant and phase II enzymes by oral feeding of green tea polyphenols in drinking water to SKH-1 hairless mice: Possible role in cancer chemoprevention, *Canc Res* 52:4050-4052 (1992).
24. Z.Y. Wang, S.J. Cheng, Z.C. Zhou, M. Athar, W.A. Khan, D.R. Bickers, H. Mukhtar, Antimutagenicity of green tea polyphenols, *Mutat Res* 223:273-285 (1989).
25. L. Liu, A. Castonguay, Inhibition of the metabolism and genotoxicity of 4-(methylnitrosamino)-1-(3-pyridyl)-1-butanone (NNK) in rat hepatocytes by (+)-catechin, *Carcinogenesis* 12:1203-1208 (1991).
26. R. Bhimani, K. Frankel, Suppression of H_2O_2 production and oxidative DNA damage in HeLa cells by (–)-epigallocatechin gallate (EGCG), *Proc Amer Assoc Canc Res* 32:126 (1991).
27. Y. Xu, C.-T. Ho, S.G. Amin, C. Han, F.-L. Chung, Inhibition of tobacco-specific nitrosamine-induced lung tumorigenesis in A/J mice by green tea and its major polyphenol as antioxidants, *Canc Res* 52:3875-3879 (1992).
28. T.J. Smith, Z.-Y. Guo, P.E. Thomas, F.-L. Chung, M.A. Morse, K. Eklind, C.S. Yang, Metabolism of 4-(methylnitrosamino)-1-(3-pyridyl)-1-butanone in mouse lung microsomes and its inhibition by isothiocyanates, *Canc Res* 50:6817-6822 (1990).

29. T.J. Smith, Z. Guo, F.J. Gonzalez, F.P. Guengerich, G.D. Stoner, C.S. Yang, Metabolism of 4-(methylnitrosamino)-1-(3-pyridyl)-1-butanone (NNK) in human lung and liver microsomes and cytochromes P_{450} expressed hepatoma cell, *Canc Res* 52:1757-1763 (1992).
30. World Health Organization International Agency for Research on Cancer, Coffee, tea, mate, methylxanthines and methylglyoxal, *IARC Monographs on the Evaluation of Carcinogenic Risks to Humans* 51:207-271 (1991).

Chapter 9

Cancer Chemoprevention by Green Tea Components

HASAN MUKHTAR, SANTOSH K. KATIYAR and RAJESH AGARWAL

I. Introduction

Since the treatment of cancer is often ineffective once the tumor is diagnosed, considerable efforts have been made in recent years to develop chemopreventive agents, i.e., agents that would prevent cancers from developing in the first place.[1] Cancer chemoprevention therefore has become an important emerging area of research that, in addition to providing a practical approach to identifying potentially useful inhibitors of cancer development, also affords opportunities to study the mechanisms of carcinogenesis.[2,3] It is appreciated that extrinsic factors such as environmental pollutants and lifestyle play major roles in the development of some forms of human malignancies.[4] It is also becoming clear that diet can influence the risk of developing cancer.[4] Our food contains not only mutagens and carcinogens, but also a variety of chemicals that are antimutagenic in bacterial test systems and that block carcinogenesis in animal tumor bioassay systems.[4] For these reasons, changes in dietary habits with the routine intake of more cancer chemopreventive agents, specifically those which have shown preventive effects in a wide range of animal tumor bioassay systems, are an attractive approach for cancer prevention. Experimental studies in animal tumor models have shown that cancer can be prevented by administration of a variety of chemical compounds, some of which are naturally occurring whereas others are synthetic.[2]

At present, several classes of chemicals with anticarcinogenic effects in animal tumor bioassay systems are known.[2] Among these polyphenols present in fruits, vegetables, and beverages are receiving increasing attention. In the present discussion, we have summarized the work done by our group as well as in other laboratories showing that green tea contains potentially useful antimutagenic and anticarcinogenic agents. In doing this, we have given emphasis on the effects of a polyphenolic fraction isolated from green tea (hereafter referred as GTP), water extract of green tea (WEGT), and individual epicatechin derivatives present

Hasan Mukhtar, Santosh K. Katiyar, and Rajesh Agarwal • Department of Dermatology, University Hospitals of Cleveland, Case Western Reserve University, Cleveland, OH 44106

Diet and Cancer: Markers, Prevention, and Treatment
Edited by M.M Jacobs, Plenum Press, New York, 1994

in green tea against various genotoxicity test systems and against animal tumor bioassay protocols. Emphasis has also been placed on defining the possible mechanism of such effects.

II. Inhibition of Mutagenicity by Green Tea

We examined in detail the antimutagenic activity of WEGT and GTP.[5] GTP significantly inhibited the mutagenicity induced by benzo(α)pyrene (BP), aflatoxin B_1 (AFB_1), 2-aminofluorene, and methanol extract of coal tar pitch in *Salmonella typhimurium* TA 100 and/or TA 98 strains in the presence of a microsomal activation system.[5] In additional studies, it was shown that GTP inhibited gene forward mutation in V79 cells treated with AFB_1 and BP, and also decreased the frequency of sister-chromatid exchange and chromosomal aberrations in V79 cells treated with AFB_1.[5] Jain *et al.*[6] have shown the effectiveness of tea extracts in inhibiting *N*-methyl-*N'*-nitro-*N*-nitrosoguanidine (MNNG)-induced mutagenicity, and suggested that the habitual drinking of tea may reduce the tumor-initiating potency of MNNG-type of nitrosoureido compounds. In another study, Ruch *et al.*[7] demonstrated that an antioxidant fraction of Chinese green tea prevents oxygen radical-induced cytotoxicity and inhibits intercellular communication, two possible mechanisms whereby tumor promoters may exert their biological effects.[7] This was shown both in cultured B6C3F1 mouse hepatocytes and in human keratinocytes.

III. Chemoprevention of Skin Cancer by Green Tea

It is recognized that carcinogenesis in murine skin and possibly in human skin and other tissues is a stepwise process comprising three stages: initiation, promotion, and progression. For this discussion, chemoprevention of skin cancer refers to the administration of chemical agent(s) to prevent the initiation (mutation) and/or promotion and/or progression events that occur during the multi-stage process of neoplastic development. In Table 1 we have summarized the skin cancer chemopreventive effects of green tea in relation to the first two events, *viz.*, initiation and promotion along with complete carcinogenesis induced by chemical carcinogens and ultraviolet B (UVB) radiation. Experiments are under way in our laboratory to study the possible protective effects of green tea against the tumor progression stage of skin carcinogenesis.

A. Protection of Mouse Skin Tumorigenicity by GTP

We showed that in a complete skin tumorigenesis protocol in BALB/c mice using 3-methylcholanthrene (3-MC) as the tumorigen, topical application of GTP (1.2 mg in 0.2 ml acetone) for 7 days prior to the beginning of 3-MC application afforded significant protection against the onset and subsequent development of skin tumors.[8] This was the first study using a murine skin tumor bioassay system that showed GTP possesses substantial anticarcinogenic activity against a known potent skin carcinogen.

B. Protection of Mouse Skin Tumor Initiation by GTP and (-)Epigallocatechin-3-gallate (EGCG)

We further assessed the anti-tumor-initiating effect of GTP in SENCAR mice using a two-stage carcinogenesis protocol. Topical application of GTP at the dose of 10 mg in 0.2 ml acetone/day per mouse for 7 days prior to initiation by 7,12-dimethylbenz(a)-anthracene (DMBA), followed by twice-weekly applications of 12-O-tetradecanoylphorbol-13-acetate (TPA) as the promoting agent resulted in significant protection against skin tumorigenicity.[8] In this study, considerable delay in the time of first tumor appearance, tumor growth, and

Table 1. Protective Effects of Oral Feeding in Drinking Water or Topical Application of GTP, WEGT, and EGCG against Chemical Carcinogen- and Ultraviolet B Radiation-Induced Tumorgenesis in Murine Skin[a]

Mouse strain	Carcinogen used	Tumorigenesis protocol	Dose and mode of GTP/WEGT/ECGC treatment	Observed protection (%)[b]	Reference
BALB/c	3-MC	Complete carcinogenesis	1.2 mg GTP, topically	50	8
SENCAR	DMBA/TPA	Anti-tumor-initiating	10 mg GPT, topically	42	8
SENCAR	DMBA/TPA	Anti-tumor-initiating	0.05% GPT, drinking water	44	8
SENCAR	BPDE-2/TPA	Anti-tumor-initiating	24 mg GTP, topically	38	9
SENCAR	DMBA/TPA	Anti-tumor-initiating	5 μmol EGCG, topically	51	10
SENCAR	DMBA/TPA	Anti-tumor-promoting	1-24 mg GTP, topically	31-84	11
CD-1	DMBA/TPA	Anti-tumor-promoting	3.6 mg GTP, topically	95	12
SKH-1 hairless	DMBA/TPA	Anti-tumor-promoting	1.25% WEGT, drinking water	55-84	13
CD-1	DMBA/teleocidin	Anti-tumor-promoting	5 mg EGCG, topically	95	14
SKH-1 hairless	UVB	Complete carcinogenesis	10 mg GTP, topically	20	15
SKH-1 hairless	UVB	Complete carcinogenesis	0.1% GTP, drinking water	41	15
SKH-1 hairless	UVB/TPA	Anti-tumor-initiating	1.25% WEGT, drinking water	64-82	13
SKH-1 hairless	DMBA/UVB	Anti-tumor-promoting	1.25% WEGT, drinking water	38-87	13

[a] Delay in latency period of tumors was found in each tumorigenesis protocol.
[b] Percent protection in number of tumors per animal at the termination of the experiment.

tumor multiplicity occurred in the GTP-pretreated group as compared to non-GTP-treated group of animals.[8] Similarly, oral feeding of GTP (0.5 g/liter, w/v) in drinking water for 50 days prior to the start of DMBA-TPA treatment or its continuous feeding during the entire initiation-promotion protocol also resulted in significantly fewer skin tumors per mouse (greater than 40% protection) as compared to non-GTP fed animals.[8] In additional experiments, we used *anti-trans*-BP-7, 8-diol-9, 10-epoxide (*anti-trans*-BPDE or BPDE-2) as the tumor-initiating agent and TPA as the tumor-promoting agent to study the protective effect of GTP on skin tumor initiation in SENCAR mice. In this study, topical application of GTP for 7 days prior to the application of BPDE-2 (at the initiating dose) to the skin of SENCAR mice showed significant protection against skin tumor induction.[9] As in other studies, considerable delay in the appearance of the tumor (latency period), as well as reduction in the total number of tumors per mouse, was observed in the GTP-pretreated group compared to the group of mice treated with only BPDE-2 and TPA.[9]

In a recent study it was shown that topical application of EGCG (5 μmol in 0.2 ml acetone/animal per day up to 7 days) prior to challenge with DMBA on the skin of SENCAR mice resulted in significant protection against tumor initiation.[10] The protective effects of EGCG were evident when tumor data were considered as the percentage of mice with tumors, the cumulative number of tumors, and the tumors per mouse.[10] These results suggested that EGCG may be the most effective cancer chemopreventive polyphenol present in GTP, more specifically in green tea.

C. Protection of Mouse Skin Tumor Promotion by GTP, WEGT, and EGCG

We also assessed the anti-skin tumor-promoting effects of GTP in SENCAR mice utilizing a two-stage initiation-promotion protocol. Topical application of varying doses of GTP (1-24 mg in 0.2 ml acetone/animal) 30 min prior to that of each TPA application in DMBA-initiated mice resulted in highly significant protection against skin tumor promotion in a dose-dependent manner.[11] The animals pretreated with GTP showed fewer tumors per mouse, a decrease in tumor volume per mouse, and a decrease in the average tumor size per tumor as compared to non-GTP-fed animals.[11] In CD-1 mice, Huang *et al.*[12] have shown similar protective effects of topical application of GTP on TPA-caused tumor promotion in DMBA-initiated skin. In SKH-1 hairless mice, Wang *et al.*[13] have shown the protective effects of oral feeding of WEGT in drinking water on TPA-caused tumor promotion in DMBA-initiated skin. In another study, anti-skin tumor-promoting effects of EGCG, the major epicatechin derivative present in GTP and WEGT, have been shown against teleocidin-caused tumor promotion in DMBA-initiated mouse skin.[14]

D. Protection of UVB Radiation-Induced Skin Tumorigenicity by GTP and WEGT

In a further study, we assessed the effect of feeding of GTP in drinking water or its topical application to SKH-1 hairless mice on UVB radiation-induced photocarcinogenesis. Chronic oral feeding of GTP (0.1%, w/v) in drinking water to SKH-1 hairless mice was shown to result in significantly lower tumor yield as compared to the animals that did not receive GTP. The effect of GTP on photocarcinogenesis was evident when tumor data were considered as the percent of mice with tumors and number of tumors per mouse.[15] Topical application of GTP before UVB irradiation also afforded some protection against photocarcinogenesis, however, the protective response was lower than that observed after oral feeding of GTP in drinking water.[15] In a recent study from the laboratory of Conney[13] it was shown that feeding WEGT (1.25%, w/v) as a sole source of drinking water to SKH-1 hairless mice afforded protection in a dose-dependent manner against UVB radiation-induced intensity of red color and area of skin lesions, as well as UVB radiation-induced skin tumor initiation (TPA was used as tumor promoter) and skin tumor promotion (DMBA was used as tumor initiator). These results, in conjunction with our studies, suggested that green tea may reduce the risk of skin cancer induction in humans by solar UV radiation.

IV. Chemoprevention of Lung and Forestomach Cancer by Green Tea

Since tea is consumed as a beverage, several studies have been conducted to assess whether oral consumption of tea components produces anticarcinogenic effects against carcinogenesis induced in internal body organs. As summarized in Table 2, in a series of

Table 2. Protective Effects of Oral Feeding in Drinking Water of WEGT and GTP against Chemical Carcinogen-Induced Forestomach and Lung Tumorigenesis in A/J Mice

Tumorigenesis protocol	Carcinogen used	Dose of GTP/WEGT	Target organ	% Protection (tumors/mouse)	Reference
Complete carcinogenesis	DEN	1.25% WEGT	Forestomach and lung	80 55	16
Complete carcinogenesis	BP	1.25% WEGT	Forestomach and lung	71 56	16
Complete carcinogenesis	DEN	0.63 & 1.25% WEGT	Forestomach and lung	59-63 36-60	17
Anti-tumor-initiating, -promoting, and complete carcinogenesis	DEN	2.5% WEGT	Forestomach and lung	80-85 43-62	a
Anti-tumor-initiating, -promoting, and complete carcinogenesis	BP	2.5% WEGT	Forestomach and lung	61-71 25-51	a
Anti-tumor-initiating, -promoting, and complete carcinogenesis	DEN	0.2% GTP	Forestomach and lung	68-82 38-43	a
Anti-tumor-initiating, -promoting, and complete carcinogenesis	BP	0.2% GTP	Forestomach and lung	39-66 25-46	a

[a] S.K. Katiyar, R. Agarwal, and H. Mukhtar, unpublished data.

studies from our laboratory as well as by other investigators, cancer chemopreventive effects of WEGT and GTP have been reported against chemical carcinogen-induced tumorigenicity in the lung and forestomach of A/J mice.

A. Protection of Lung and Forestomach Tumorigenicity by WEGT

In a complete carcinogenesis protocol, oral feeding of WEGT (1.2%, w/v) as the sole source of drinking water to A/J mice during the entire period of the experiment resulted in significant protection against BP- and *N*-nitrosodiethylamine (DEN)-induced lung and forestomach neoplasia formation.[16] At the termination of the experiment (33 weeks), WEGT-fed animals developed 55% fewer tumors in lungs and 70% fewer tumors in forestomach than did the non-WEGT-fed control group of animals.[16] Similar protective effects of WEGT have also been shown from the laboratory of Yang,[17] where oral feeding of 0.63 and 1.25% WEGT (w/v) to A/J mice was shown to result in significant protection against DEN-induced lung and forestomach tumorigenicity.

In a recently completed study (S.K. Katiyar, R. Agarwal, and H. Mukhtar, unpublished data) we also compared the protective effects of WEGT against BP- and DEN-induced tumor initiation, tumor promotion, and complete carcinogenesis in lung and forestomach of A/J mice. Compared to the non-WEGT-fed control group of animals (BP or DEN was used to induce the lung and forestomach tumors), oral feeding of WEGT (2.5%,, w/v) as the sole source of drinking water resulted in significantly fewer tumors per mouse in both lung and forestomach. In case of BP-induced tumorigenesis, WEGT feeding resulted in significant protection in total number of tumors per mouse in both lung (25-51%) and forestomach (61-71%) at all the stages of carcinogenesis. Similarly, in the case of DEN-induced tumorigenesis, 43-62 and 80-85% fewer tumors per mouse in lung and forestomach, respectively, were observed at all the stages of carcinogenesis in WEGT-fed animals compared to the group of mice administered DEN alone.

B. Protection of Lung and Forestomach Tumorigenicity by GTP

Utilizing the protocol as described above for WEGT, we also assessed the protective effects of GTP against BP- and DEN-induced tumor initiation, tumor promotion, and complete carcinogenesis in lung and forestomach of A/J mice (S.K. Katiyar, R. Agarwal, and H. Mukhtar, unpublished data). Oral feeding of 0.2% GTP as a sole source of drinking water to A/J mice before tumor initiation and during tumor promotion and the entire tumorigenesis protocol resulted in significant protection against lung and forestomach neoplasia development. In case of BP-induced tumorigenesis, GTP resulted in 25-46 and 39-66% protection in total number of tumors per mouse in lung and forestomach, respectively, at all the stages of carcinogenesis. Similarly, in case of DEN-induced tumorigenesis, feeding of GTP resulted in 38-43 and 68-82% protection in total number of tumors per mouse in lung and forestomach, respectively, during initiation, promotion, and the complete carcinogenesis protocol.

V. Chemoprevention of Cancer in Other Tumor Models by Green Tea

In addition to cancer chemopreventive effects of green tea in experimental models utilizing skin, forestomach, and lung as the target organs for tumorigenesis (summarized in Table 3), several other investigators working in this area have also employed other experimental tumor models. In a recent study, Wang *et al.*[17] showed the cancer chemopreventive effects of WEGT against 4-(methylnitrosamino)-1-(3-pyridyl)-1-butanone (NNK)-induced lung tumorigenesis in A/J mice. Compared to non-WEGT-fed mice, oral feeding of 0.6% WEGT prior to or after the challenge with NNK up to the end of the experiment resulted in significant protection in terms of tumor incidence as well as tumor multiplicity.[17] Similar results were also reported in Xu *et al.*[18] where chemopreventive effects of WEGT and EGCG against NNK-induced lung tumorigenicity in A/J mice were demonstrated. Fujita *et al.*[19] reported the chemopreventive effects of EGCG against *N*-ethyl-*N'*-nitro-N-nitrosoguanidine (ENNG)-induced duodenum tumorigenicity in C57BL/6 mice. In this study, compared to non-EGCG-fed animals, oral feeding of 0.005% EGCG after the challenge with carcinogen up to the termination of the experiment resulted in significant protection against ENNG-caused tumor promotion in duodenum as observed by a decrease in tumor incidence, the total number of tumors, and the number of tumors per mouse.[19] Chen[20] reported the chemopreventive effects of different varieties of Chinese tea including green tea against N-nitrosomethylbenzylamine (NMBzA)-induced esophageal tumorigenicity in Wistar rats. Oral feeding of 2.0% WEGT during the whole tumorigenicity protocol was shown to result in significantly smaller tumors and fewer tumors per animal compared to a non-WEGT-fed group of rats.[20] Yamane *et al.*[21] used a rat colon tumorigenicity model and showed that oral feeding of 0.01 or 0.1% GTP in drinking water after the challenge with azoxymethane (AOM) up to the termination of the experiment resulted in significant protection against AOM-induced colon cancer in Fisher rats as observed by decrease in tumor incidence, tumor number, and tumor size in GTP-fed animals compared to non-GTP-fed control rats.

Table 3. Protective Effects of Oral Feeding in Drinking Water of WEGT, EGCG, and GTP against Chemical Carcinogen-Induced Tumorigenesis in Other Tumor Models

Animal and tumor model	Carcinogen used	Tumorigenesis protocol	Dose of WEGT/ EGCG/GTP	% Protection (tumors/animal)	Reference
A/J mice, lung	NNK	Anti-tumor-initiating and -promoting	0.6% WEGT	67-85	17
A/J mice, lung	NNK	Complete carcinogenesis	2.0% WEGT	45	18
A/J mice, lung	NNK	Complete carcinogenesis	560 ppm EGCG	30	18
C57BL/6 mice, duodenum	ENNG	Anti-tumor-promoting	0.005% EGCG	63-75	19
Wistar rats, esophageal	NMBzA	Complete carcinogenesis	2.0% WEGT	40-60	20
Fisher rats, colon	AOM	Anti-tumor-promoting	0.01 and 0.1% GTP	53-60	21

VI. Mechanistic Studies

A. Inhibition of Metabolic Activation and Carcinogen-DNA Binding by GTP

In general, the initial step in carcinogenesis is the metabolic activation of chemical carcinogens, which requires cytochrome P_{450}-dependent biotransformation reactions.[22,23] For example, the ubiquitous environmental pollutant BP is known to cause cancer of the skin, lung, and forestomach (and other organs) in experimental animals only after its metabolic activation to highly reactive molecules.[22,23] The ultimate carcinogenic metabolite of BP is BPDE-2, the formation of which is catalyzed by three successive enzymatic steps catalyzed by cytochrome P_{450}-dependent monooxygenases and epoxide hydrolase.[22-24] During the course of this metabolic activation, the reactive intermediates may be detoxified by conjugating mechanisms mediated by glutathione S-transferase (GST), quinone reductase (QR), and other pathways.[24] The binding of BPDE-2 to cellular DNA leading to the formation of BPDE-2-dGua adduct is essential for the tumor-initiating activity of BP.[22-24] Similar metabolic pathways leading to specific DNA adduct formation are also described for other polycyclic aromatic hydrocarbons (PAHs), including DMBA.[22-24] The extent of specific DNA adduct formation and its persistence in target tissue appears to correlate with the tumor development risk for that tissue.[22,23] Consistent with this knowledge, it is believed that agents that decrease the risk of cancer may include substances that inhibit carcinogenesis by modifying cellular metabolism of pro-carcinogens in a manner that ultimately prevents carcinogen formation, increases detoxification of carcinogens and/or their metabolites, or blocks the interaction of the ultimate carcinogens with cellular DNA.[2]

Our studies have shown that GTP and epicatechins derived from green tea interact with cytochrome P_{450} and inhibit cytochrome P_{450}-dependent aryl hydrocarbon hyrdroxylase, 7-ethoxycoumarin O-deethylase, and 7-ethoxyresorufin O-deethylase activities.[25] Based on the structure of different epicatechin derivatives, we suggested that their inhibitory effects may be due to the galloyl or hydroxyl groups attached to the parent skeleton of catechin.[25] GTP was also found to inhibit hepatic and epidermal microsomal enzyme-mediated binding of [^{3}H]BP to calf thymus DNA.[5] In additional studies, the effect of GTP on *in vivo* binding of carcinogens to epidermal DNA was assessed.[8] Oral feeding of GTP in drinking water (0.5 g/liter, w/v) for 50 days or its topical application (10 mg GTP/0.2 ml acetone per mouse per day) for 7 days to SENCAR mice prior to a single topical application of [^{3}H]BP or [^{3}H]-DMBA showed that GTP significantly decreased the binding of topically applied polycyclic aromatic hydrocarbons (PAHs) to epidermal DNA. These observations provided evidence that GTP is capable of inhibiting PAH-DNA binding in the epidermis.[8]

B. Inhibition of Biochemical Markers of Tumor Promotion by GTP

The morphological, biochemical, and physiological changes occurring in mouse skin as a result of topical application of phorbol esters like TPA include induction of ornithine decarboxylase (ODC) activity followed by an increase in the levels of polyamines, epidermal hyperplasia, and inflammation, and an increase in the number of dark basal keratinocytes.[22,26] However, it is difficult to establish which of these effects or many other effects of phorbol ester tumor promoters are obligatory or sufficient components of the tumor promotion process. The induction of ODC activity is considered to be closely associated with the tumor-promoting activity of a variety of tumor promoters.[22,26] ODC is involved in polyamine biosynthesis, which plays an essential role in cell proliferation and differentiation.[22,26] The induction of inflammation in skin mediated by TPA is believed to be governed by cyclooxygenase- and lipoxygenase-catalyzed metabolites of arachidonic acid, specifically prostaglandins (PGs) and hydroxy eicosatetraenoic acids (HETEs), respectively.[22,26] The importance of induction of epidermal ODC, cyclooxygenase, and lipoxygenase activities in skin tumor promotion is evident from the fact that several inhibitors of these enzymes inhibit the tumor promotion in murine skin.[26-29] These inhibitory studies support the involvement of arachidonic acid metabolism pathways in skin tumor promotion.[26-29]

We assessed the effect of skin application of GTP to SENCAR mice on the induction of ODC activity caused by TPA and other skin tumor promoters, and TPA-caused induction of cyclooxygenase and lipoxygenase activities. Topical application of GTP to mouse skin was found to inhibit TPA-induced epidermal ODC activity in a dose-dependent manner.[30] The inhibitory effect of GTP was also dependent on the time of its application relative to TPA treatment. Maximum inhibitory effect was observed when GTP was applied 30 min prior to that of TPA.[30] GTP application to SENCAR mice also inhibited the induction of epidermal ODC activity caused by several structurally different mouse skin tumor promoters.[30] As quantitated by the formation of prostaglandin and hydroxy eicosatetraenoic acid metabolites from cyclooxygenase- and lipoxygenase-catalyzed metabolism of arachidonic acid, respectively, skin application of GTP to SENCAR mice also resulted in significant inhibition of TPA-caused effects on these two enzymes.[10] Prior application of GTP to mouse skin also resulted in 30-46% inhibition of TPA-induced epidermal edema and hyperplasia.[10] The results of these studies suggested that the anti-skin tumor-promoting effects of GTP involve inhibition of tumor promoter-caused induction of epidermal ODC, cyclooxygenase and lipoxygenase activities, edema, and hyperplasia.

C. Induction of Antioxidant and Phase II Enzyme Activities by GTP

The basic functions of antioxidant enzymes are to protect tissue against toxic effects mediated by reactive oxygen species (ROS) generated during normal metabolism or under oxidative stress.[27,28,31] Several studies have shown that antioxidant enzymes play an important role in modulating the induction and development of tumorigenesis.[27,28,31] It has been shown that topical application of TPA results in a 20-40% decrease in glutathione peroxidase (GSH-Px) activity and reduces the GSH:GSSG ratio significantly in epidermal homogenate.[31-34] Catalase activity in epidermis is also known to be decreased significantly following single or multiple application of TPA to murine skin.[35,36] Xenobiotic compounds and/or their metabolites, formed by the phase I reaction catalyzed by cytochrome P_{450}, are removed by phase II reactions from the cell by their conjugation with reduced glutathione, a reaction catalyzed by a family of enzymes known as GST.[2] QR, a flavoenzyme, catalyzes the two-electron reduction of quinones and quinonoid compounds to hydroquinones, facilitating their glucuronidation and ultimate excretion.[37] QR also reduces mutagenicity, covalent binding, and oxygen radicals resulting from quinone metabolites,[38] which indicates it may protect cells against oxidative stress.

Several lines of evidence have shown that the administration of cancer chemopreventive agents to experimental animals induces both antioxidant and phase II enzyme activities in target organs,[2] and have suggested that this may be the mechanism of cancer chemoprevention by these agents.[2] We recently have shown that oral feeding of GTP (0.2%, w/v) in drinking water for 30 days to female SKH-1 hairless mice results in the enhancement of antioxidant and phase II enzyme activities in small bowel, liver, and lung.[39] The relationship of these effects to anticarcinogenicity of GTP remains to be elucidated.

VII. Epidemiological Studies Related to Cancer Chemopreventive Effects of Green Tea

The epidemiological studies related to the effects of tea consumption on human cancer, though inconclusive, have suggested a protective effect.[16] A case-control study by Kono *et al.*[40] has indicated that individuals consuming green tea tend to have a lower risk for gastric cancer. Studies by Oguni *et al.*[41] in Shizuoka Prefecture in Japan showed that the stomach cancer death rate in this tea-producing and -consuming area was lower than the national average.

It is important to emphasize here that the International Agency for Research on Cancer recently concluded that "there is no evidence to implicate tea or tea component(s) as human carcinogens."[42] However, based largely on recently published studies from our laboratory and from other investigators, a suggestion was made that GTP contains an interesting and potentially important group of chemicals with anticarcinogenic effect.

VIII. Future Outlook

Based on a wide range of cancer chemopreventive studies showing chemopreventive effects of those chemicals present in trace amounts in our daily diet, it has been suggested by several investigators working in this area of research that changes in lifestyle and dietary habits, with the intake of more cancer chemopreventive agents, appear to be the most practical approach for cancer prevention. On a visit to grocery and drug stores in Japan, one can find that gums, candies, tobacco products, beverages, toothpastes, mouthwashes, topically used creams, several commonly consumed food products, etc., are supplemented with "Sunphenon," a 70% enriched polyphenolic preparation from green tea. The Japanese population consume such items with the hope that they might be protected from the cancer. Although at this stage such consumption may be considered premature, it is quite possible that the polyphenols present in green tea may find a place in supplementing various items routinely consumed by humans once their long-term safety and beneficial effects have been established.

Acknowledgments

The green tea work from the author's laboratory is supported by American Institute for Cancer Research grant 90A47.

Correspondence should be addressed to Hasan Mukhtar at the Department of Dermatology Case Western Reserve University, 2074 Abington Road, Cleveland, OH 44106.

References

1. Anonymous, Effects to prevent cancer are on the increase, *Science* 253:613 (1991).
2. L.W. Wattenberg, Inhibition of carcinogenesis by naturally occurring and synthetic compounds, *in: Antimutagenesis and Anticarcinogenesis, Mechanisms II*, Y. Kuroda, D.M. Shankel, M.D. Waters, eds., Plenum Publishing Corp., New York, pp. 155-166 (1990).
3. I.B. Weinstein, Cancer prevention: Recent progress and future opportunities, *Cancer Res* 51(Suppl):5080s (1991).
4. H.J. Prochaska, A.B. Santamaria, P. Talalay, Rapid detection of inducers of enzymes that protect against carcinogens, *Proc Natl Acad Sci USA* 89:2394 (1992).
5. Z.Y. Wang, S.J. Cheng, Z.C. Zhou, M. Athar, W.A. Khan, D.R. Bickers, H. Mukhtar, Antimutagenic activity of green tea polyphenols, *Mutat Res* 223:273 (1989).
6. A.K. Jain, K. Shimoi, Y. Nakamura, T. Kada, Y. Hara, I. Tomita, Crude tea extracts decrease the mutagenic activity of N-methyl-N′-nitro-N-nitrosoguanidine *in vitro* and in intragastric tract of rats, *Mutat Res* 210:1 (1989).
7. R.J. Ruch, S.-J. Cheng, J.E. Klaunig, Prevention of cytotoxicity and inhibition of intercellular communication by antioxidant catechins isolated from chinese green tea, *Carcinogenesis* 10:1003 (1989).
8. Z.Y. Wang, W.A. Khan, D.R. Bickers, H. Mukhtar, Protection against polycyclic aromatic hydrocarbon-induced skin tumor initiation in mice by green tea polyphenols, *Carcinogenesis* 10:411 (1989).
9. W.A. Khan, Z.Y. Wang, M. Athar, D.R. Bickers, H. Mukhtar, Inhibition of the skin tumorigenicity of (+)-7β,8α-dihydroxy-9α,10α-epoxy-7,8,9,10-tetrahydrobenzo(a)pyrene by tannic acid, green tea polyphenols and quercetin in SENCAR mice, *Cancer Lett* 42:7 (1988).
10. S.K. Katiyar, R. Agarwal, Z.Y. Wang, A.K. Bhatia, H. Mukhtar, (−)-Epigallocatechin-3-gallate in *Camellia sinensis* leaves from Himalayan region of Sikkim: Inhibitory effects against biochemical events and tumor initiation in SENCAR mouse skin, *Nutr Cancer* 18:73 (1992).
11. S.K. Katiyar, R. Agarwal, G.S. Wood, H. Mukhtar, Inhibition of 12-*O*-tetradecanoylphorbol-13-acetate-caused tumor promotion in 7,12-dimethylbenz[a]anthracene-initiated SENCAR mouse skin by a polyphenolic fraction isolated from green tea, *Cancer Res* 52:6890 (1992).
12. M.-T. Huang, C.-T. Ho, Z.Y. Wang, T. Ferraro, T. Finnegan-Olive, Y.-R. Lou, J.M. Mitchell, J.D. Laskin, H. Newmark, C.S. Yang, A.H. Conney, Inhibitory effect of topical application of a green tea polyphenol fraction on tumor initiation and promotion in mouse skin, *Carcinogenesis* 13:947 (1992).
13. Z.Y. Wang, M.-T. Huang, T. Ferraro, C.-Q. Wong, Y.-R. Lou, M. Iatropoulos, C.S. Yang, A.H. Conney, Inhibitory effect of green tea in the drinking water on tumorigenesis by ultraviolet light and 12-O-tetradecanoylphorbol-13-acetate in the skin of SKH-1 mice, *Cancer Res* 52:1162 (1992).
14. S. Yoshizawa, T. Horiuchi, H. Fujiki, T. Yoshida, T. Okuda, T. Sugimura, Antitumor promoting activity of (−)-epigallocatechin gallate, the main constituent of "tannin" in green tea, *Phytotherapy Res* 1:44 (1987).
15. Z.Y. Wang, R. Agarwal, D.R. Bickers, H. Mukhtar, Protection against ultraviolet B radiation-induced photocarcinogenesis in hairless mice by green tea polyphenols, *Carcinogenesis* 12:1527 (1991).

16. Z.Y. Wang, R. Agarwal, W.A. Khan, H. Mukhtar, Protection against benzo(a)pyrene and N-nitrosodiethylamine-induced lung and forestomach tumorigenesis in A/J mice by water extracts of green tea and licorice, *Carcinogenesis* 13:1491 (1992).
17. Z.Y. Wang, J.Y. Hong, M.-T. Huang, K.R. Reuhl, A.H. Conney, C.S. Yang, Inhibition of N-nitrosodiethylamine- and 4-(methylnitrosamino)-1-(3-pyridyl)-1-butanone-induced tumorigenesis in A/J mice by green tea and black tea, *Cancer Res* 52:1943 (1992).
18. Y. Xu, C.-T. Ho. S.G. Amin, C. Han, F.L. Chung, Inhibition of tobacco-specific nitrosamine-induced lung tumorigenesis in A/J mice by green tea and its major polyphenol as antioxidants, *Cancer Res* 52:3875 (1992).
19. Y. Fujita, T. Yamane, M. Tanaka, K. Kuwata, J. Okuzumi, T. Takahashi, H. Fujiki, T. Okuda, Inhibitory effect of (–)-epigallocatechin gallate on carcinogenesis with N-ethyl-N′-nitro-N-nitrosoguanidine in mouse duodenum, *Jpn J Cancer Res (Gann)* 80:503 (1989).
20. J. Chen, The effects of chinese tea on the occurrence of esophageal tumors induced by N-nitrosomethylbenzylamine in rats, *Prev Med* 21:385 (1992).
21. T. Yamane, N. Hagiwara, M. Tateishi, S. Akachi, M. Kim, J. Okuzumi, Y. Kitao, M. Inagake, K. Kuwata, T. Takahashi, Inhibition of azoxymethane-induced colon carcinogenesis in rat by green tea polyphenol fraction, *Jpn J Cancer Res (Gann)* 82:1336 (1991).
22. R. Agarwal, H. Mukhtar, Cutaneous chemical carcinogenesis, *in*: *Pharmacology of the Skin*, H. Mukhtar, ed., CRC Press, Boca Raton, FL, pp. 371-387 (1991).
23. A.H. Conney, Induction of microsomal enzymes by foreign chemicals and carcinogenesis by polycyclic aromatic hydrocarbons: GHA Clowes Memorial Lecture, *Cancer Res* 42:4875 (1982).
24. H. Mukhtar, R. Agarwal, D.R. Bickers, Cutaneous metabolism of xenobiotics and steroid hormones, *in*: *Pharmacology of the Skin*, H. Mukhtar, ed., CRC press, Boca Raton, FL, pp. 89-110 (1991).
25. Z.Y. Wang, M. Das, D.R. Bickers, H. Mukhtar, Interaction of epicatechins derived from green tea with rat hepatic cytochrome P-450, *Drug Metab Dispos* 16:98 (1988).
26. J. DiGiovanni, Modification of multistage skin carcinogenesis in mice, *in*: *Modification of Tumor Development in Rodents*, N. Ito, H. Sugano, eds., *Prog Exp Tumor Res*, Basel, Karger, vol. 33 pp. 192-199 (1991).
27. R. Agarwal, H. Mukhtar, Oxidative stress in skin chemical carcinogenesis, *in*: *Oxidative Stress in Dermatology*, L. Packer, J. Fuchs, eds., Marcel Dekker, NY, pp. 207-242 (1993).
28. J.-P. Perchellet, E.M. Perchellet, Antioxidants and multistage carcinogenesis in mouse skin, *Free Radical Biol Med* 7:377 (1989).
29. H.U. Gali, E.M. Perchellet, J.P. Perchellet, Inhibition of tumor promoter-induced ornithine decarboxylase activity by tannic acid and other polyphenols in mouse epidermis in vivo, *Cancer Res* 51:2820 (1991).
30. R. Agarwal, S.K. Katiyar, S.I.A. Zaidi, H. Mukhtar, Inhibition of tumor promoter-caused induction of ornithine decarboxylase activity in SENCAR mice by polyphenolic fraction isolated from green tea and its individual epicatechin derivatives, *Cancer Res* 52:3582 (1992).
31. Y. Sun, Free radicals, antioxidant enzymes, and carcinogenesis, *Free Radical Biol Med* 8:583 (1990).
32. E.M. Perchellet, E.C. Maata, N.L. Abney, J.-P. Perchellet, Effects of diverse intracellular thiol delivery agents on gluthathione peroxidase activity, the ratio of reduced/oxidized glutathione, and ornithine decarboxylase induction in isolated mouse epidermal cells treated with 12-O-tetradecanoylphorbol-13-acetate, *J Cell Phys* 131:64 (1987).
33. J.-P. Perchellet, N.L. Abney, R.M. Thomas, Y.L. Guislain, E.M. Perchellet, Effects of combined treatments with selenium, glutathione and vitamin E on glutathione peroxidase activity, ornithine decarboxylase induction and complete and multistage carcinogenesis in mouse skin, *Cancer Res* 47:477 (1987).
34. J.-P. Perchellet, N.L. Abney, R.M. Thomas, E.M. Perchellet, E.A. Maatta, Inhibition of multistage tumor promotion in mouse skin by diethyldithiocarbamate, *Cancer Res* 47:6302 (1987).

35. V. Solanki, R.S. Rana, T.J. Slaga, Diminution of mouse epidermal superoxide dismutase and catalase activities by tumor promoters, *Carcinogenesis* 2:1141 (1981).
36. T.W. Kensler, D.M. Bush, W.J. Kozumbo, Inhibition of tumor promotion by a biomimetic superoxide dismutase, *Science* 221:75 (1983).
37. A.D. Jassim, V.L. Sparnins, L.W. Wattenberg, Long term effects of benzo(a)pyrene on the activity of NAD(P)H:quinone reductase in the forestomach and glandular stomach of ICR/Ha mice, *Cancer Res* 50:2064 (1990).
38. H. Wefers, H. Sies, Hepatic low level chemiluminescence during redox cycling of menadione and the menadione-glutathione conjugate: Relation to glutathione and NAD(P)H: quinone reductase (DT-diaphorase) activity, *Arch Biochem Biophys* 224:568 (1983).
39. S.G. Khan, S.K. Katiyar, R. Agarwal, H. Mukhtar, Enhancement of antioxidant and phase II enzymes by oral feeding of green tea polyphenols in drinking water to SKH-1 hairless mice: Possible role in cancer chemoprevention, *Cancer Res* 52:4050 (1992).
40. S. Kono, M. Ikeda, S. Tokudome, M. Kuratsune, A case-control study of gastric cancer and diet in Northern Kyushu, *Jpn J Cancer Res* 79:1067 (1988).
41. I. Oguni, K. Nasu, S. Yammamoto, T. Nomura, On the antitumor activity of fresh green tea leaf, *Agric Biol Chem* 52:1879 (1988).
42. International Agency for Research on Cancer Working Group, Monographs on the evaluation of the carcinogenic risk to humans: Coffee, tea, mate, methylxanthines and methylglyoxal, IARC, Vol. 51 (1991).

Chapter 10

Potential Role of Dietary Isoflavones in the Prevention of Cancer

STEPHEN BARNES, GREG PETERSON, CLINTON GRUBBS and KEN SETCHELL

I. Introduction

The death rate from cancer in the United States is unacceptably high. However, examination of time-related changes in death rates from individual cancers reveals that deaths from some cancers (e.g., stomach cancer) have fallen dramatically over the past 50 years, whereas others (principally lung cancer) have risen sharply.[1] Since lung cancer is strongly associated with cigarette smoking, new public health policies with regard to smoking should lead to a reduction in the rate of this cancer.

The death rates from two other cancers, breast cancer and prostatic cancer, have remained essentially unchanged over this same period.[1] It is well known that there are wide variations in the international death rates for each of these cancers;[2] these variations have been attributed to environmental differences (particularly in the diet) from country to country, rather than to ethnic susceptibilities.[3,4] Asians have consistently lower rates of death from breast (two- to three-fold) and prostatic cancer (six-fold) than Americans.[2] However, Asians who emigrate to the United States are at higher risk for these cancers; the risk of cancer in their offspring approaches that of other men and women born in the United States.[5] The most well known hypothesis to explain these data is the much higher levels of fat in the U.S. diet compared to the Asian one;[6] however, this hypothesis has been subject to considerable debate.[7,8] We may, like Alice in *Through the Looking Glass*, be viewing the problem back to front, i.e., Asians living in the United States, as they incorporate American eating habits, may be eating decreasing amounts of a key, naturally occurring, dietary anti-cancer component normally present in the Asian diet.

Stephen Barnes • Departments of Biochemistry and Pharmacology, University of Alabama at Birmingham, Birmingham, AL 35294; Greg Peterson • Department of Biochemistry, University of Alabama at Birmingham, Birmingham, AL 35294; Clinton Grubbs • Department of Nutrition Sciences, University of Alabama at Birmingham, Birmingham, AL 35294; Ken Setchell • Clinical Mass Spectrometry Center, Children's Hospital, Cincinnati, OH 45229

Diet and Cancer: Markers, Prevention, and Treatment
Edited by M.M Jacobs, Plenum Press, New York, 1994

A common feature of breast and prostatic cancer is that they are sex hormone-dependent cancers. Treatment of breast and prostatic cancer patients often involves administration of estrogen-like drugs. In the case of breast cancer, the anti-estrogen tamoxifen is used to prevent the metastatic growth of estrogen-dependent primary tumors. It may also have an application as an anti-promoter and is currently being tested in a chemoprevention trial in women at high risk for breast cancer. In androgen-dependent prostatic cancer, the nonsteroidal synthetic estrogen, diethylstilbestrol, is used in some patients to reduce the synthesis of testosterone. Therefore, we have asked the question: could the Asian diet contain a substance that interferes with the action of physiologic sex hormones on tumor cells?

Soybean-based foods are common components of Asian, but not American, diets. Soybeans contain large (mg per g) quantities of the phytoestrogens—the isoflavones, genistein and daidzein (as their β-glucosides).[10] Since these isoflavones are weak estrogen agonists,[11] they may interfere with the promotive effects on the tumors by physiologic estrogens and androgens. Thus, it has been proposed that the low rates of breast cancer in Asians result from eating soy and, in particular, the phytoestrogens contained in soy.[12,13]

Epidemiological data support the hypothesis that consumption of soy lowers the risk of breast and prostatic cancer. In the most systematic study of the effects of soy carried out so far, Lee *et al.*[14] recently showed that increased soy protein consumption was significantly correlated with a reduction in the risk of breast cancer. In the case of prostatic cancer, Severson *et al.*[15] reported that daily consumption of tofu, but not shoyu (soy sauce) or miso soup, decreased the relative risk of prostatic cancer three-fold in Japanese men living in Hawaii. These data suggest that the anti-cancer effect of soy may lie in the protein fraction.

Using animal models of breast cancer, we have confirmed the hypothesis that consumption of soy reduces the risk of breast cancer.[13] Our results with diets including autoclaved or nonautoclaved soy demonstrated that the protease inhibitors in soy were not the anti-cancer components, as had been previously suggested by Troll and his colleagues.[16,17]

In the present study, we tested the anti-cancer capability of several soy fractions that contain different amounts of the isoflavones. In addition, we examined the effects of the isoflavones, genistein and daidzein, on the growth of human breast and prostatic cancer cells in tissue culture. The data obtained support the hypothesis that it is the isoflavones in soy that are responsible for the reduction in cancer risk; however, their mechanism of action is unclear.

II. Materials and Methods

A. Animals, Diets, and Treatments

Female Sprague-Dawley rats were obtained from Harland Sprague-Dawley (Indianapolis, IN) at 22 days of age and housed in polycarbonate cages (five per cage) in an environmentally controlled room maintained at 72 ± 2°F with a 12-hr light-dark cycle. They were placed on the test diets at 25 days of age and remained on the diet for the rest of the study. Rats were allowed to feed and drink *ad libitum*. Casein in the standard AIN-76A diet (20% w/w) was totally or partially replaced in the diets containing soy protein isolate or soy protein concentrate. Diets were prepared in 8-kg lots by mixing a basal mix concentrate (Teklad, Madison, WI) that contained no casein, soy protein isolate, or soy protein concentrate; casein (as required to give a final protein concentration of 20%); and sucrose (as required to give a final carbohydrate concentration of 65%) in a tumbling blender. Diets containing soy molasses were prepared in a similar manner. The soy molasses was added to 10% by weight. Casein was the only protein added. Final compositions of each of the diets tested are given in Tables 1 and 2. Soy protein isolate (Ardex F), aqueous alcohol extracted soy protein concentrate (Arcon F), and soy molasses were donated by the Archer Daniels Midland Co. (Decatur, IL). They were used without further treatment.

Table 1. Composition of Diets (g/100g) in Experiment 1

Group	No. of Rats	Agent	AIN-76A concentrate	Soy protein isolate	Casein	Sucrose
1	30	DMBA	60	0	20	20
2	30	DMBA	60	11	10	19
3	30	DMBA	60	22	0	18
4	15	None	60	0	20	20
5	15	None	60	22	0	18

Table 2. Composition of Diets (g/100g) in Experiment 2

Group	No. of Rats	Agent	AIN-76A concentrate	Soy protein concentrate	Soy molasses	Casein	Sucrose
6	30	DMBA	60	0	0	20	20
7	30	DMBA	60	38	0	0	2
8	30	DMBA	56	0	9	19	16
9	15	None	60	0	0	20	20
10	15	None	60	38	0	0	2
11	15	None	56	0	9	19	16

7,12-Dimethylbenz[a]anthracene (DMBA) was administered by gavage as a single dose (15 mg) dissolved in sesame oil. Animals in control groups were given sesame oil alone. Animals were weighed and palpated twice weekly to detect tumors beginning 4 weeks after the administration of carcinogen. Vaginal smears were taken for a period of 10 days at 2 monthly intervals. Three to five animals from each group were placed in metabolic cages for 3 days to monitor food and water intake and fecal and urine outputs. This was repeated at six weekly intervals beginning 1 week prior to the administration of DMBA. At the end of the study, animals were killed by CO_2 asphyxiation. During the study, animals with necrotic tumors that had broken through the skin layer were sacrificed immediately. All tumors detected by palpation and others found at necropsy were resected. Biopsy specimens were taken for histological analysis.

B. Analysis of Soy Isoflavones in Diets

The isoflavones in each soy fraction were recovered by exhaustive extraction with hot, 80% aqueous methanol and delipidated by extraction with n-hexane. Analysis of both conjugated and unconjugated isoflavones was carried out by reversed-phase HPLC. Aliquots of each sample were injected onto a 30 x 0.45 cm Aquapore C_8 reversed-phase column (Brownlee Labs, Santa Clara, CA). The eluting medium was a gradient of 0–45% acetonitrile in water; the concentration of acetonitrile rose at 1.125% per min. The flow rate of the eluting solvent was 1.5 ml/min. Eluted compounds were detected by their absorbance at 262 nm.

Isoflavones in urines were analyzed by HPLC-thermospray mass spectrometry[18] and by capillary gas-liquid chromatography-mass spectrometry.[19]

C. Cell Culture Studies

MCF-7 and MCF-7-D40 human breast cancer cells (William Dalton, University of Arizona) were maintained on modified Eagle's medium supplemented with 7% (v/v) fetal bovine serum and antibiotics; MCF-7-40 cells also received 10^{-8} M doxorubicin to maintain their multiple drug-resistance phenotype. MDA-468 cells (Jeff Kudlow, UAB) were maintained on Dulbecco's Modified Eagles medium low glucose, with 10% (v/v) fetal bovine serum and antibiotics. LNCaP and DU-145 human prostate cancer cells (ATCC, Rockville, MD) were maintained in RPMI 1640 medium supplemented with 7% (v/v) fetal bovine serum and antibiotics. Cells were cultured as monolayers in a 95% air: 5% CO_2, water-saturated atmosphere at 37°C. In assays of the effect on cell growth of the isoflavones and tyrphostin A25 (LC Services Woburn, MA), a specific EGF receptor tyrosine kinase inhibitor, cells were plated into 96-well microtiter plates at densities of $2\text{-}10 \times 10^3$ cells/well in 198 µl of medium. The cells were allowed to attach for 2 days, the test compounds were added (in 2-µl solutions in DMSO; controls were given 2-µl DMSO alone), and incubation continued for a further 4 days. Growth stimulation of 2-day quiesced cells by EGF was studied in a similar manner, using serum-free media containing insulin, transferrin, selenium, and bovine serum albumin. The viable cell mass in each well was determined by the MTT assay of Carmichael *et al.*[20] Each experiment consisted of at least three plates (18 replicates).

To determine the effects of isoflavones and tyrphostin on EGF receptor tyrosine phosphorylation, cells were quiesced in serum-free media for 48 hr. After addition of the isoflavones or tyrphostin, cells were cooled to 4°C and exposed to EGF for 15 min. Following a brief warming at 37°C, cells were lysed with a NP-40 buffer containing phosphatase and protease inhibitors. Proteins in the lysate were immunoprecipitated with a specific anti-EGF receptor monoclonal antibody (B1D8; Jeff Kudlow), separated by SDS-PAGE, and transferred to nitrocellulose. Tyrosine-phosphorylated proteins were detected using an anti-phosphotyrosine monoclonal antibody (UBI, Lake Placid, NY) and either a peroxidase-linked or alkaline phosphatase-linked secondary antibody. Bands on the immunoblot were quantitated by reflectance spectrophotometry or by transmission.

III. Results

Vaginal smears showed that there was no difference in the estrous cycling between any of the groups tested (data not shown). Food intake, weight gain, and fecal output were unaffected by the replacement of casein by soy protein isolate in the diet (Table 3). Water intake and urine output were significantly higher as a function of the amount of soy protein isolate in the diet.

The total isoflavone content of soy protein isolate (0.4 mg/g) was lower than that in powdered soybean chips (1.5 mg/g). Unconjugated isoflavones constituted one half of the total. The aqueous alcohol-extracted soy protein concentrate had an even lower total isoflavone content (<0.1 mg/g). Soy molasses, a concentrate of the aqueous alcohol fraction, had high levels of isoflavones (total content 5 mg/g), mostly as the β-glucoside conjugates.

Based on the concentration of isoflavones in each diet and the amount of diet consumed by the rats, daily intake of isoflavones ranged from 0.35 mg/g in the 20% soy protein concentrate group (group 7) to 1.1 mg/day in the 20% soy protein isolate group (group 3) to 6 mg/day in the soy molasses group (group 8).

Isoflavones were not detectable in the urine of control animals except in the first urine collection (10 days after transferring them from a standard lab chow diet; data not shown). The combined urinary excretion of isoflavones in animals in the 10% soy protein group represented 12.3% of the ingested isoflavones, whereas in the 20% soy protein group urinary

Table 3. Effects of Diets on Physiological Parameters[a]

Group[b]	Soy (%)	Body wt(g)	Food intake (g/day)	Feces (g/day)	Water intake (ml/day)	Urine (ml/day)
Experiment 1						
1	0	288 ± 15	12.9 ± 0.2	2.0 ± 0.2	16.3 ± 0.9	7.1 ± 1.0
2	10	293 ± 10	12.8 ± 0.2	2.3 ± 0.1	20.5 ± 0.5	11.4 ± 1.0
3	20	290 ± 06	13.5 ± 0.5	2.4 ± 0.1	21.0 ± 1.6	12.1 ± 1.2
4	0	298 ± 07	14.0 ± 0.4	2.3 ± 0.1	21.0 ± 1.8	11.4 ± 1.1
5	20	295 ± 06	12.9 ± 0.4	2.4 ± 0.2	28.3 ± 3.5	17.5 ± 2.3
Experiment 2						
6	0	272	–	–	–	–
7	20	268	–	–	–	–
8	10	270	15.8 ± 0.9	2.5 ± 0.2	16.0 ± 0.3	8.0 ± 0.4
9	0	284	–	–	–	–
10	20	262	13.9 ± 0.4	2.3 ± 0.2	17.7 ± 0.7	8.9 ± 0.6
11	10	271	–	–	–	–

[a] Mean ± SEM of six collections taken throughout the experiment.
[b] See Tables 1 and 2 for composition of diets.

excretion was 14.1% (Table 4). In this study, equol was a minor metabolite, rarely exceeding 10% of the total urinary isoflavone excretion. Genistein and daidzein (as conjugates) were present in approximately equal amounts.

Diets containing soy protein isolate reduced the number of histologically confirmed mammary tumors compared to the control group in a dose-dependent manner (Figure 1). There was also an effect on the latency period before tumors were detected for animals on the 20% soy protein isolate diet (Figure 1). However, by the end of the study most animals had at least one mammary tumor. No mammary tumors were detected in control or test diet animals that had not received carcinogen.

The diet containing alcohol-extracted soy protein concentrate had no effect on the appearance of mammary tumors (Figure 2), whereas the diet containing soy molasses increased the latency period before tumors were detected and, in addition, inhibited the total number of mammary tumors compared to the control group (Figure 2).

Since the anti-cancer activity of soy in the animal model of breast cancer was correlated with the isoflavone content of the soy fraction, the isoflavones, genistein and daidzein, were isolated from soy molasses[21] and tested for their effects on the serum and EGF-stimulated proliferative growth of human breast cancer cells in culture. Genistein inhibited not only the growth of MCF-7 (estrogen receptor-positive) cells, but also the growth of MDA-468 (estrogen receptor-negative) cells (Figure 3). The IC_{50} values were similar (10.5 and 6.5 µg/ml, respectively) in each cell line. Daidzein was a much weaker inhibitor of the growth of both cell types (IC_{50} values of 28 and 34 µg/ml, respectively). Genistein also inhibited the serum-stimulated growth of MCF-7-D40 cells that overexpress the multidrug-resistance gene product, whereas these cells were insensitive to doxorubicin (Figure 4).

Table 4. Dietary and Urinary Isoflavones

		Dietary intake		Urinary output			
Group	Age (days)	Genistein (mg/day)	Daidzein (mg/day)	Genistein (mg/day)	Daidzein (mg/day)	Equol (mg/day)	Urinary output (%)
MNU-treated							
10% soy protein	36	0.61	0.21	0.039	0.044	0.006	10.9
	77	0.66	0.23	0.044	0.052	0.006	11.7
	117	0.64	0.22	0.046	0.056	0.007	12.6
	155	0.62	0.22	0.051	0.060	0.006	14.0
20% soy protein	36	0.80	0.33	0.083	0.081	0.010	15.5
	71	0.85	0.35	0.062	0.068	0.006	11.4
	113	0.66	0.27	0.076	0.073	0.007	16.7
	155	0.76	0.32	0.064	0.070	0.006	13.0
Controls							
20% soy protein	36	0.81	0.34	0.045	0.049	0.007	8.8
	71	0.74	0.30	0.057	0.062	0.007	12.2
	113	0.78	0.32	0.065	0.074	0.005	13.0
	155	0.85	0.35	0.072	0.080	0.008	13.4

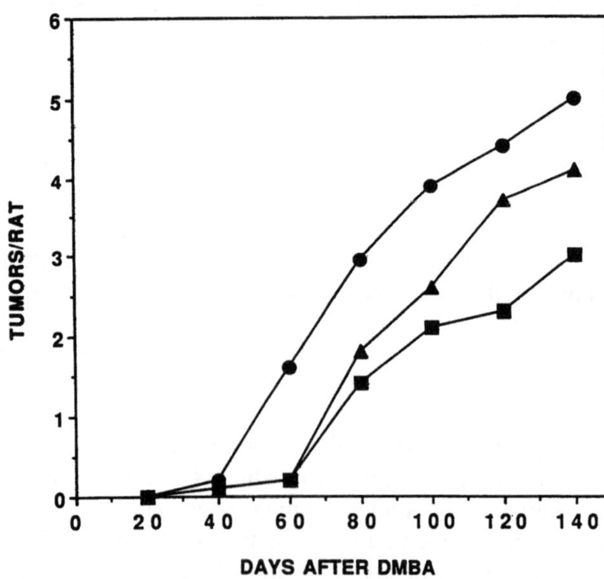

Figure 1. Rate of appearance of histologically confirmed mammary carcinomas in female rats treated with DMBA consuming a control diet (0% soy protein) (●), or test diets containing 10% soy protein isolate (▲) and 20% soy protein isolate (■).

Genistein also inhibited the serum-stimulated growth of the human prostate cancer cell lines, LNCaP and DU-145, although its IC_{50}values (22 and 27 µg/ml, respectively) were lower than for the breast cancer cell lines (Figure 5). It had a more potent inhibitory effect on EGF-stimulated growth of these prostate cancer cells (IC_{50} values, 15 and 4.3 µg/ml, respectively; Figure 6). Tyrphostin A25 had no effect on serum-stimulated growth of the

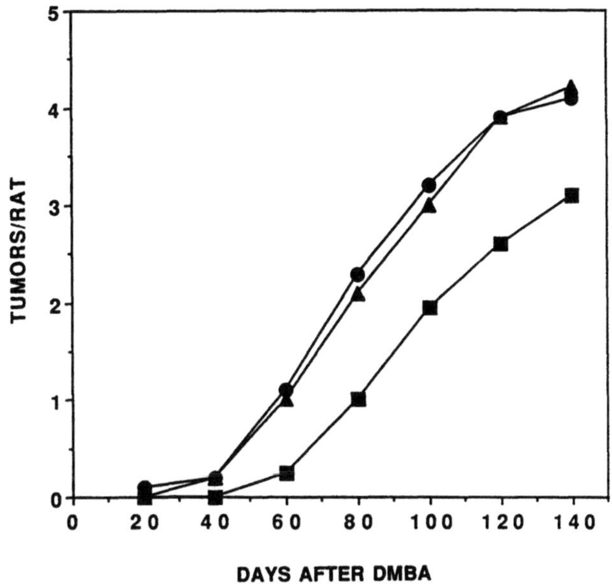

Figure 2. Rate of appearance of histologically confirmed mammary carcinomas in female rats treated with DMBA consuming a control diet (0% soy protein) (●), or test diets containing 20% soy protein concentrate (▲) and 10% soy molasses (■).

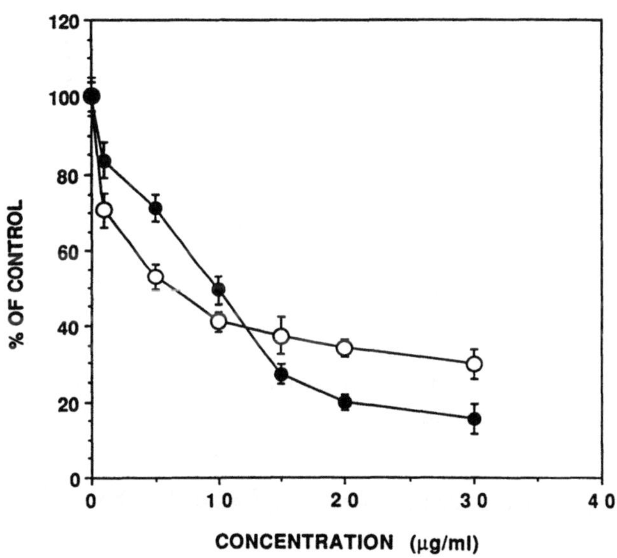

Figure 3. Inhibition of genistein of serum-stimulated growth of MCF-7 (●) and MDA-468 (O) human breast cancer cells.

prostate cancer cells, but strongly inhibited EGF-stimulated growth (IC_{50}values of 5.2 and 6.0 µg/ml, respectively).

When used at its IC_{50} value (6 µg/ml), tyrphostin strongly inhibited EGF receptor tyrosine phosphorylation in both prostate cancer cell types. In contrast, even at concentra-

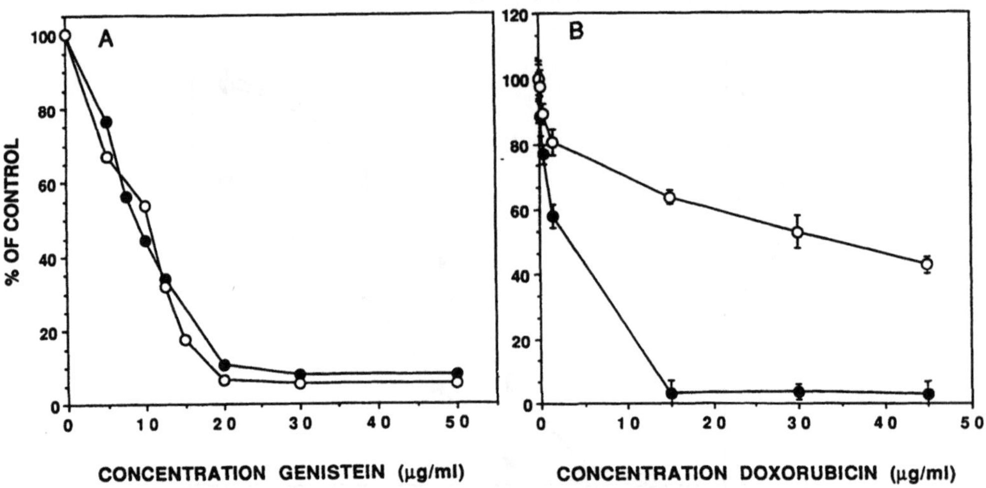

Figure 4. Inhibition of serum-stimulated growth by genistein (A) and doxorubicin (B) on wild-type MCF-7 cells (○) and MCF-7 cells overexpressing the multidrug-resistance gene product (●).

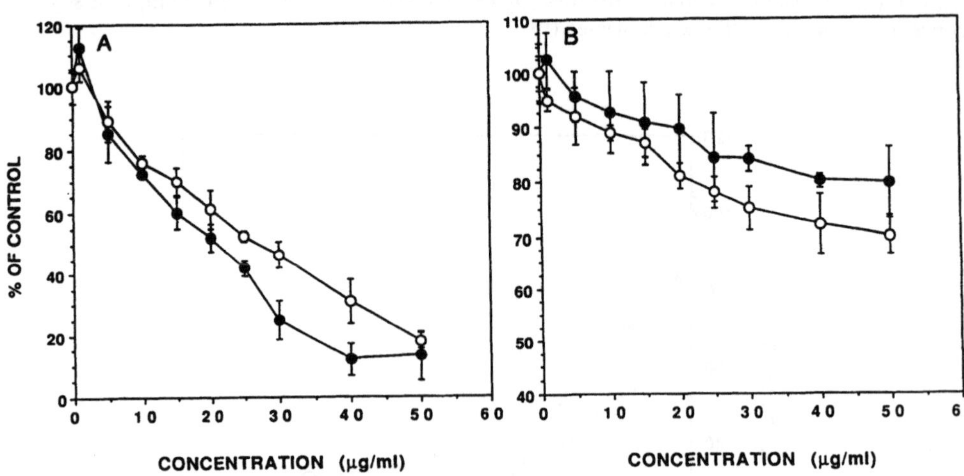

Figure 5. Inhibition by genistein (A) and tyrphostin A25 (B) of serum-stimulated growth of LNCaP (●) and DU-145 (○) human prostate cancer cells.

tions above its IC_{50}, genistein had no effect on tyrosine phosphorylation of the EGF receptor in these cells.

IV. Discussion

The potential for anti-cancer agents in foods is eminently reasonable. Not only are the health benefits of nonnutrient vitamins and minerals in the diet well understood, but many plants and animals are sources of pharmacologically active agents. And just as some authors have argued that we live in a sea of naturally occurring dietary toxins and carcinogens that far exceed the synthetic toxins produced by our industrial society,[22] it is equally likely that our diets contain anti-cancer substances.

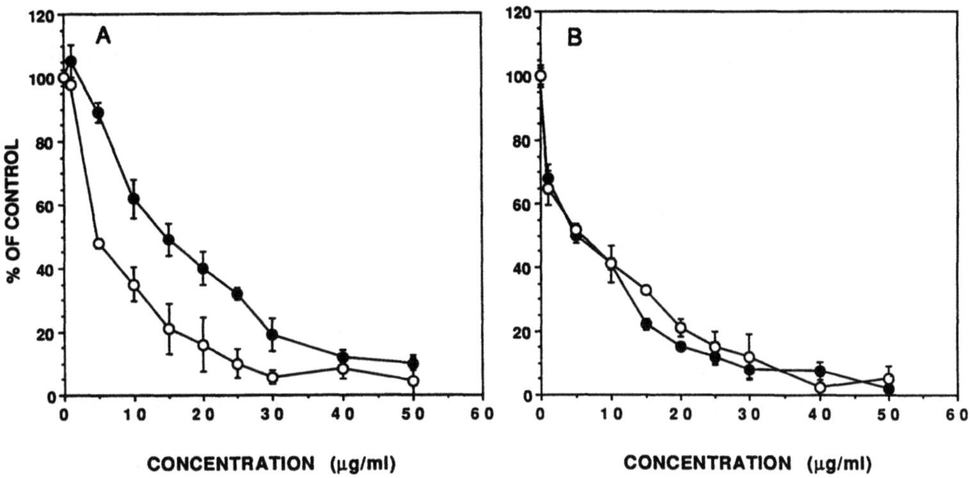

Figure 6. Inhibition by genistein (A) and tyrphostin A25 (B) of EGF-stimulated growth of LNCaP (●) and DU-145 (O) human prostate cancer cells.

In the present study, soy protein isolate, widely used in foods consumed in small quantities by most U.S. consumers, has a chemopreventive effect in the DMBA rat model of breast cancer although it was weaker when compared on a protein basis to unprocessed, powdered soybean chips.[13] The data confirm those obtained by Hawrylewicz *et al.* who showed that soy protein isolate inhibits the appearance of mammary tumors in a MNU-induced model of breast cancer in rats.[23] These investigators introduced soy protein isolate into the diet after injection of the MNU, indicating that the effect of the soy was on the promotional phase of tumor development.

Previous data obtained in tissue culture experiments led to the suggestion that protease inhibitors are alternative chemopreventive agents in soybeans.[16,17] However, using the MNU model of breast cancer, we have shown that both autoclaved and nonautoclaved soybean chips caused the same degree of inhibition of appearance of mammary tumors.[13] Since there was no measurable protease inhibitor in the autoclaved powdered soybeans (autoclaved for 60 min), it was concluded[13] that protease inhibitors or soybean enzymes were not responsible for the chemopreventive effect. The absence of measurable protease inhibitors in the soy protein isolate used in the present study also makes it unlikely that the chemopreventive effect could be explained by this mechanism.

Since the daily intake of isoflavones for the soy protein isolate diets (up to 1 mg/day per rat) was substantially lower than for the powdered soybean chip diets (up to 5 mg/day per rat), the data from the present study are consistent with our hypothesis that isoflavones are the active chemopreventive agent in soybeans. This concept was further supported by our observation that a soy protein concentrate, prepared by extraction of soy flour with 65% aqueous alcohol, has no chemopreventive effect in the rat model, whereas soy molasses, a concentrate of the aqueous alcohol extract, did.

The isoflavone composition of soy protein isolate differs from that of unprocessed, powdered soybean chips. In the latter, the isoflavones were present as their β-glucoside conjugates, daidzin and genistin, whereas in soy protein isolate there was an almost equal proportion of the unconjugated and conjugated isoflavones. This may be a reflection of the processing of the soybeans to make soy protein isolate. In the manufacturing process, defatted soy flour is first treated with alkali to solubilize the proteins. This also leads to the dissociation of both the conjugated and unconjugated isoflavones from the protein. Subsequent acidification to precipitate the proteins results in the co-precipitation of the unconjugated isoflavones (they become fully protonated below pH 7—their solubility falls to less than 10 µg/ml—H. Armstrong and G. Peterson, unpublished observations), but not the isoflavone β-glucoside conjugates.

Only 9–17% of the isoflavones in the diets consumed appeared in the urine. One explanation for these data is that only a small fraction of dietary isoflavones is absorbed from the intestine. However, using rats with exteriorized biliary cannulas, we have shown that 43% of duodenally administered genistein is recovered in bile within 4 hr (H. Armstrong and S. Barnes, unpublished data). Thus, there is substantial biliary as opposed to renal excretion of isoflavone metabolites. Further experiments are needed to define the enterohepatic circulation and metabolism of isoflavones in rats.

Our experiments with human breast cancer cell lines have demonstrated that genistein is an effective inhibitor of cell growth, independently of whether or not the cells express an estrogen receptor,[24] suggesting that the mechanism of action of the isoflavones in preventing appearance of mammary tumors may not be a direct antiestrogenic effect as we previously postulated.[13] The growth of human prostate cancer cell lines was also inhibited by genistein, again largely independently of whether these cells contain an androgen receptor. These data suggest that the cellular biochemical target(s) of genistein include those which are distal to the action of steroids on their receptors. Both in breast tissue and the prostate, gonadal steroids stimulate the production of growth factors and lead to increased numbers of receptors to these growth factors.[25-29] Therefore, the target of isoflavones may be signal transduction pathways induced by interaction of growth factors with their receptors on tumors.

It has been shown that in cell-free preparations genistein is a potent inhibitor of the EGF receptor tyrosine kinase activity.[30] However, we were unable in the present study to show significant inhibition by genistein at its IC_{50} value on the tyrosine phosphorylation of the EGF receptor in genistein-sensitive human prostate cancer cells; in contrast, tyrphostin A25, an *in vitro*-specific inhibitor of the EGF receptor tyrosine kinase,[31] fully blocked EGF receptor tyrosine phosphorylation in these cells. These data suggest that the critical cellular target of genistein in human prostate cancer cells is not the EGF receptor tyrosine kinase. Indeed, Linassier *et al.*[32] have presented evidence that ribosomal S6-kinase, rather the EGF receptor tyrosine kinase, is the cellular target of genistein in NIH 3T3 cells. Therefore, the precise cellular targets for genistein in cells remain to be discovered and could depend on phenotypic expression in an individual cell.

Genistein has been shown to inhibit signal transduction events triggered by many receptors with intrinsic or tightly associated tyrosine protein kinase activity. These include platelet-derived growth factor receptor (PDGF-R),[33,34] the insulin receptor,[35] c-*kit*,[36] interleukin-2 receptor,[37] interleukin-3 receptor,[38] nerve growth factor receptor,[39,40] surface Ig receptor in B-cells,[41] and CD-7.[42] Only in the cases of PDGF-R and c-*kit* has genistein been shown to directly inhibit receptor autophosphorylation. However, even in these cases, the concentration of genistein used (30 µg/ml) was much higher than the IC_{50} values for inhibition of cell growth. Thus, not all tyrosine kinases are inhibited by genistein; rather, it must have a selective action dependent on the chemical nature of the site of inhibition on the kinase. Indeed, a tyrosine kinase (pp^{40}) identified in bovine thymocytes is potently inhibited by apigenin (the flavone isomer of genistein) and not by genistein.[43]

In addition to their effects on growth factor-induced signal transduction, other mechanisms for the action of isoflavones have been suggested. Genistein inhibits mammalian DNA topoisomerase I[44] and DNA topoisomerase II,[45,46] and tumor promoter-induced hydrogen peroxide formation in neutrophils and in mouse skin.[47] Each of these inhibitory effects would be important in the transformation and proliferation of tumor cells.

Acknowledgment

This study was funded in part by a grant-in-aid from the American Cancer Society (BC-599) and by grants from the American Institute for Cancer Research (87B64 and 91B58). We would also like to acknowledge the generosity of the Archer Daniels Midland Co. and Protein Technologies International in providing the soybean materials for our experiments. William Dalton, University of Arizona, and Jeff Kudlow, University of Alabama at Birmingham, kindly provided human cancer cells; Dr. Kudlow also supplied the anti-EGF receptor monclonal antibody. Mark Messina is especially thanked for his enthusiasm and for providing data on soybean consumption around the world.

References

1. G.R. Newell, Epidemiology of cancer, *in: Cancer. Principles and Practice of Oncology*, 2nd ed, V.T. DeVita, Jr., S. Hellman, and S.A. Rosenberg, eds., Lippincott, Philadelphia, pp. 151-195 (1985).
2. American Cancer Society, Cancer facts & figures—1991, ACS, Atlanta (1991).
3. J.E. Dunn, Cancer epidemiology in populations of the United States—with emphasis on Hawaii and California—and Japan, *Cancer Res* 35:3240-3245 (1975).
4. G.E. Gray, M.C. Pike, B.E. Henderson, Breast cancer incidence and mortality rates in different countries in relation to known factors and dietary practice, *Br J Cancer* 39:1-7 (1979).
5. H. Shimizu, R.K. Ross, L. Bernstein, R. Yatani, B.E. Henderson, T.M. Mack, Cancer of the prostate and breast among Japanese and white immigrants in Los Angeles County, *Br J Cancer* 63:963-966 (1991).
6. P. Greenwald, Principles of cancer prevention: Diet and nutrition, *in: Cancer. Principles and Practice of Oncology*, 3rd ed, V.T. DeVita, Jr., S. Hellman, and S.A. Rosenberg, eds., Lippincott, Philadelphia, pp. 151-195 (1989).
7. W.C. Willett, M.J. Stampfer, G.A. Colditz, B.A. Rosner, C.H. Hennekens, F.E. Speizer, Dietary fat and the risk of breast cancer, *N Engl J Med* 316:22-28 (1987).
8. W.C. Willett, D.J. Hunter, M.J. Stampfer, G.A. Colditz, J.E. Manson, D. Spiegelman, B.A. Rosner, C.H. Hennekens, F.E. Speizer, Dietary fat and fiber in relation to the risk of breast cancer, *JAMA* 268:2037-2044 (1992).
9. Unpublished data compiled by Soyatech Inc., Bar Harbor, Maine.
10. L. Coward, N. Barnes, S. Barnes, K.D.R. Setchell, Genistein and daidzein and their β-glucoside conjugates: Antitumor isoflavones in soybean foods from American and Asian diets, *J Agric Food Sci*, (in press).
11. B.Y. Tang, N.R. Adams, Effect of equol on oestrogen receptors and on synthesis of DNA and protein in the immature rat uterus, *J Endocrin* 85:291-297 (1980).
12. K.D.R. Setchell, S.P. Borriello, P. Hulme, D.N. Kirk, M. Axelson, Non-steroidal estrogens of dietary origin: possible roles in hormone-dependent disease, *Am J Clin Nutr* 40: 569-578 (1984).
13. S. Barnes, C. Grubbs, K.D.R. Setchell, J. Carlson, Soybeans inhibit mammary tumors in models of breast cancer, *in: Mutagens and Carcinogens in the Diet*, M. Pariza, ed., Liss, New York, pp. 239-253 (1990).
14. H.P. Lee, L. Gourley, S.W. Duffy, J. Esteve, J. Lee, N.E. Day, Dietary effects on breast cancer risk in Singapore, *Lancet* 337:1197-1200 (1991).
15. R.K. Severson, A.M.Y. Nomura, J.S. Grove, G.N. Stemmerman, A prospective study of demographics, diet, and prostate cancer among men of Japanese ancestry in Hawaii, *Cancer Res* 49:1857-1860 (1989).
16. W. Troll, R. Wiesner, C.J. Shellabarger, S. Holtzman, J.P. Stone, Soybean diet lowers breast tumor incidence in irradiated rats, *Carcinogenesis* 1:469-472 (1980).
17. W. Troll, K. Frenkel, R. Wiesner, Protease inhibitors as anticarcinogens, *J Natl Canc Inst* 73:1245-1250 (1984).

18. K.D.R. Setchell, M.B. Welsh, C.K. Lim, HPLC analysis of phytoestrogens in soy protein preparations with ultraviolet, electrochemical, and thermospray mass spectrometric detection, *J Chromatogr* 368:315-323 (1987).
19. M. Axelson, J. Sjovall, B. Gustafsson, K.D.R. Setchell, Soya—a dietary source of the non-steroid oestrogen equol in humans and animals, *J Endocr* 10Z:49-56 (1984).
20. J. Carmichael, W.G. DeGraff, A. Gazdar, J. Minna, J.B. Mitchell, Evaluation of a tetrazolium-base semiautomated colorimetric assay: Assessment of chemosensitivity testing, *Cancer Res* 47:936-942 (1987).
21. E.D. Walter, Genistin (an isoflavone glucoside) and its aglucone, genistein from soybean, *J Am Oil Chem Soc* 63:3273-3276 (1941).
22. L.S. Gold, T.H. Slone, B.R. Stern, N.B. Manley, B.N. Ames, Rodent carcinogens: Setting priorities, *Science* 258:261-265 (1992).
23. E.J. Hawrylewicz, H.H. Huang, W.H. Blair, Dietary soybean isolate and methionine supplementation affect mammary tumor progression in rats, *J Nutr* 121:1693-1698 (1991).
24. T.G. Peterson, S. Barnes, Genistein inhibition of the growth of human breast cancer cells: Independence from estrogen receptors and the multi-drug resistance gene, *Biochem Biophys Res Commun* 179:661-667 (1991).
25. S.E. Bates, N.E. Davidson, E.M. Valverius, C.E. Freter, R.B. Dickson, J.P. Tam, J.E. Kudlow, M.E. Lippman, D.S. Soloman, Expression of transforming growth factor alpha and its messenger ribonucleic acid in human breast cancer: Its regulation by estrogen and its possible functional significance, *Molec Endocrinol* 2:543-555 (1988).
26. Y. Berthois, X.F. Dong, P.M. Martin, Regulation of epidermal growth factor receptor by oestrogen and anti-oestrogen in the human breast cancer cell line MCF-7, *Biochem Biophys Res Commun* 159:126-131 (1988).
27. J.M. Connally, D.P. Rose, Production of epidermal growth factor receptor and transforming growth factor by the androgen-responsive LNCaP human prostate cancer cell line, *Prostate* 16:1235-1241 (1990).
28. J.M. Connally, D.P. Rose, Secretion of epidermal growth factor and related polypeptides by the DU-145 human prostate cancer cell line, *Prostate* 15:177-186 (1989).
29. A.L.G. Schuurmans, J. Bolt, E. Mulder, Androgens stimulate both growth rate and epidermal growth factor receptor activity of the human prostate tumor cell LNCaP, *Prostate*12:55-68 (1988).
30. T. Akiyama, J. Ishida, S. Nakagawa, H. Ogawara, S. Watanabe, N. Itoh, M. Shibuya, Y. Fukami, Genistein, a specific inhibitor of tyrosine-specific protein kinases, *J Biol Chem* 262:5592-5595 (1987).
31. A. Gazit, P. Yaish, C. Gilon, A. Levitski, Tyrophostins I: Synthesis and biological activity of protein tyrosine kinase inhibitors, *J Med Chem* 32:2344-2352 (1989).
32. C. Linassier, M. Pierre, J.B. LePecq, J. Pierre, Mechanisms of action in NIH-3T3 cells of genistein, an inhibitor of EGF receptor tyrosine protein kinase activity, *Biochem Pharmacol* 39:187-193 (1990).
33. T.D. Hill, N.M. Dean, L.J. Mordan, A.F. Lau, M.Y. Kanemitsu, A.L. Boynton, PDGF-induced activation of phospholipase C is not required for induction of DNA synthesis, *Science* 248:1660-1663 (1990).
34. G. Davidai, A. Lee, I. Schvartz, E. Hazum, PDGF induced tyrosine phosphorylation in osteoblast-like cells: Relevance to mitogenesis, *Am J Physiol* 263:E205-209 (1992).
35. A. Abler, J.A. Smith, P.A. Randazzo, P.L. Rothenburg, L. Jarrett, Genistein differentially inhibits postreceptor effects of insulin in rat adipocytes without inhibiting the insulin receptor kinase, *J Biol Chem* 267:3946-3951 (1992).
36. A. Kuriu, H. Ikeda, Y. Kanakura, J.D. Griffin, B. Druker, H. Yagura, H. Kitayama, J. Ishikawa, T. Nishiura, Y. Kanayama, T. Yonezawa, S. Tarui, Proliferation of human myeloid leukemia cell line associated with tyrosine-phosphorylation and activation of the proto-oncogene c-*kit* product, *Blood* 78:2834-2840 (1991).
37. I. Merida, E. Diez, G.N. Gaulton, IL-2 binding activates a tyrosine-phosphorylated phosphatidylinositol-3-kinase, *J Immunol* 147:2202-2207 (1991).

38. R.A. Mufson, J. Szabo, D. Eckert, Human IL-3 induction of c-*jun* in normal monocytes is independent of tyrosine kinase and involves protein kinase C, *J Immunol* 148:1129-1135 (1991).
39. E.D. Lloyd, M.W. Wooton, pp$^{42/44}$MAP kinase is a component of the neurogenic pathway utilized by nerve growth factor in PC12 cells, *J Neurochem* 59:1099-1109 (1992).
40. K. Muroya, Y. Hashimoto, S. Hattori, S. Nakamura, Specific inhibition of NGF receptor tyrosine kinase activity by K-252a, *Biochem Biophys Acta* 1135:353-356 (1992).
41. S. Takagi, M. Daibata, T.J. Last, R.E. Humphreys, D.C. Parker, T. Sairenji, Intracellular localization of tyrosine kinase substrates beneath crosslinked surface immunoglobulins in B cells, *J Exp Med* 174:381-388 (1991).
42. L.K.L. Jung, A.K. Roy, H.R. Chakkalath, CD7 augments T cell proliferation via the interleukin-2 autocrine pathway, *Cell Immunol* 141:189-199 (1992).
43. R.L. Geahlen, N.M. Koonchanok, J.L. McLaughlin, D.E. Pratt, Inhibition of protein-tyrosine kinase activity by flavanoids and related compounds, *J Nat Prod* 52:982-986 (1989).
44. A. Okura, H. Arakawa, H. Oka, T. Yoshinari, Y. Monden, Effect of genistein on topoisomerase activity and on the growth of [val 12]Ha-*ras*-transformed NIH 3T3 cells, *Biochem Biophys Res Commun* 157:183-189 (1988).
45. J. Markowits, J. Couprie, P. Fosse, J.M. Saucier, J. Perre, C. Linassier, A.K. Larsen, J.B. Lepecq, A. Jacqueminsablon, Inhibitory effects of the tyrosine kinase inhibitor genistein on mammalian DNA topoisomerase II, *Cancer Res* 49:5111-5117 (1989).
46. Y. Yamashita, S.Z. Kawada, H. Nakano, Induction of mammalian topoisomerase II dependent cleavage by nonintercalative flavanoids, genistein and orobol, *Biochem Pharmacol* 39:737-744 (1990).
47. H. Wei, L. Wei, K. Frenkel, R. Bowen, S. Barnes, Inhibition of tumor promoter-induced hydrogen peroxide formation *in vitro* and *in vivo* by genistein, *Nutr Cancer* 20:1-12 (1993).

Chapter 11

Carcinogen-DNA and Protein Adducts as Intermediate Biomarkers for Human Chemoprotection Trials

JOHN D. GROOPMAN, BILL D. ROEBUCK
and THOMAS W. KENSLER

I. Introduction to Macromolecular Adducts as Biomarkers

Our rapidly expanding understanding of the progressive processes of carcinogenesis provides opportunities for the identification of molecular biological markers reflecting events from carcinogen exposure through clinical cancer. A joint committee of the National Academy of Sciences and the National Research Council[1] has developed a useful conceptual framework for these processes and is shown in an adapted form in Figure 1. This model groups molecular biomarkers into sets reflective of internal dose, biologically effective dose (dose to critical macromolecules), early biological effect, altered structure/function, and clinical disease as well as those reflecting underlying susceptibility factors. In more general terms, molecular biological markers can be considered to fall into broad categories of markers of exposure, biological effect, and susceptibility. Markers of exposure reflect exposure levels to toxic agents, markers of effect indicate a biological response to an exposure, and markers of susceptibility provide information about the inherent sensitivity of an individual to a toxic agent. By definition some of these markers are chemical-agent specific, such as a carcinogen-DNA or protein adduct, while others are biological-process specific, such as the altered expression of an oncogene protein.

This chapter highlights the development and validation of molecular biomarkers of aflatoxins, compounds that are classified as group I carcinogens by the International Agency for Research on Cancer. This classification of an agent as a definitive human carcinogen has led to a need for accurate means to relate exposure to these agents to an individual's risk

---

John D. Groopman and Thomas W. Kensler • Department of Environmental Health Sciences, The Johns Hopkins University School of Hygiene and Public Health, Baltimore, MD 21205; Bill D. Roebuck • Department of Pharmacology and Toxicology, Dartmouth Medical School, Hanover, NH 03755

Diet and Cancer: Markers, Prevention, and Treatment
Edited by M.M Jacobs, Plenum Press, New York, 1994

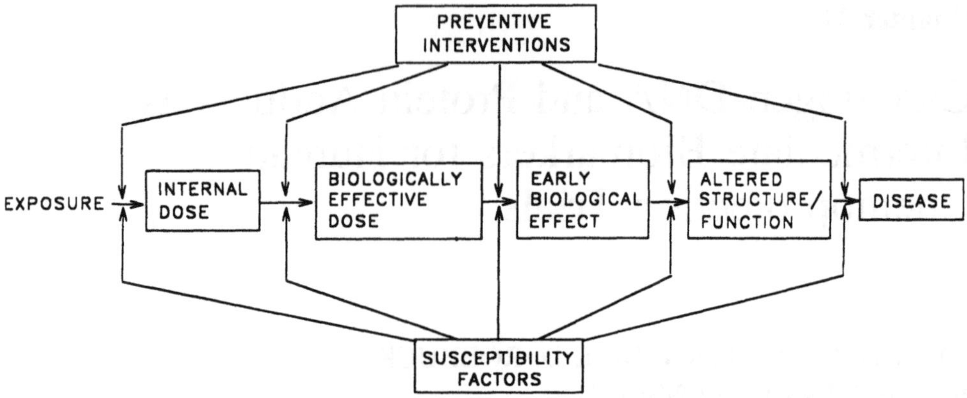

Figure 1. Paradigm for the development of molecular biomarkers.

of developing disease. The characterization of molecular biomarkers for aflatoxins is based upon the extensive research on their metabolism and macromolecular adduct formation.[2,3] In addition to using these biomarkers as measures of exposure to aflatoxins, we propose that these molecular biomarkers might serve as intermediate endpoints for assessing the efficacy of cancer prevention interventions. This goal will be facilitated by first establishing the aflatoxin biomarker-cancer link.

The validation of any biomarker-cancer link requires parallel experimental animal and human studies (Figure 2). Ideally, an appropriate animal model is used to determine the role (associative or causal) or the marker in the disease pathway and to establish dose-response relationships. The developed marker can then be used in pilot human studies where sensitivity, specificity, accuracy, and reliability parameters are established. The critical next step is the use of the marker in transitional epidemiological studies[4] to help validate the marker. These studies can assess intra- or interindividual variability, background levels, relationship of the marker to external dose or to disease status, as well as issues of feasibility for use in larger population-based studies. It is important to establish a connection, in humans, between the biological marker and the exposure or the outcome of interest. To fully interpret the information that the marker can provide, prospective epidemiological studies may be necessary to demonstrate the role that the marker plays in the overall pathogenesis of the disease. The aflatoxins are among the very few agents that have been extensively studied to date using the validation scheme shown in Figure 2. Thus, the studies described herein could serve as a template for the development, validation and application of aflatoxin biomarkers to preventive interventions.

II. Markers of Biologically Effective Dose

A. Experimental Animal Studies

Among the various possible biomarkers of aflatoxin exposure, the measurement of carcinogen-DNA and protein adducts are of major interest because they are direct products of or surrogate markers for damage to a critical macromolecular target. The chemical structures of the major aflatoxin-DNA, aflatoxin-N^7-guanine (AFB-N^7-gua), and serum albumin adducts have been identified[5,6] and both the DNA and albumin adducts are formed in a linear dose-dependent manner following single-dose exposures.[7,8] The AFB-N^7-gua adduct is rapidly removed from DNA with a half-life in rat liver of between 8 and 10 hr,[9] and is excreted exclusively in urine of exposed rats.[10] The serum albumin adduct exhibits

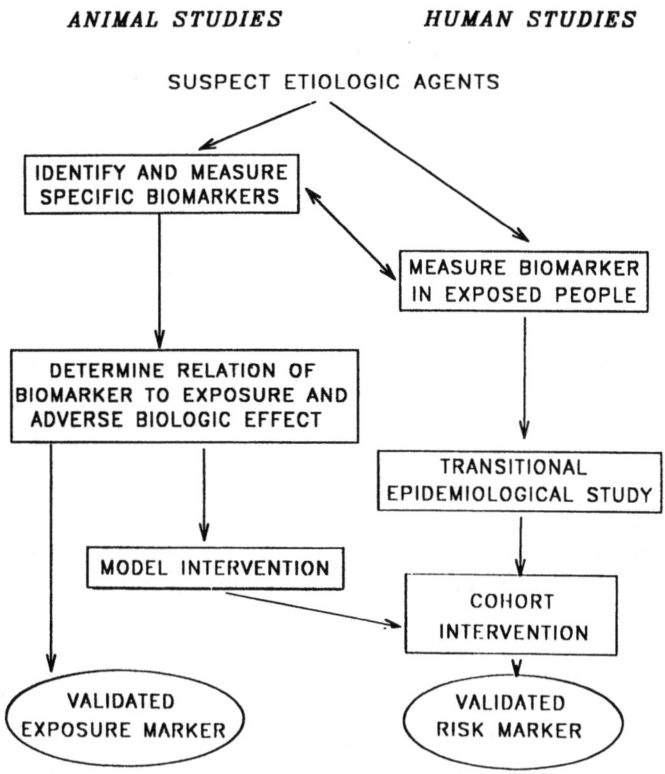

Figure 2. Paradigm for the validation process linking biomarkers and disease outcomes.

a longer half-life reflective of the biological lifetime of serum albumin (about 3 weeks in humans).[8] Thus, the serum albumin adduct can measure exposures integrated over longer time periods.

Validation studies for the dose-dependent excretion of urinary aflatoxin biomarkers were done in rats following a single exposure to AFB_1.[11] The relationship between AFB_1 dose and the excretion of the AFB-N^7-gua adduct over the initial 24-hr period following exposure demonstrated an excellent linear correspondence between oral dose and excretion of a biologically relevant metabolite in urine. In contrast, other oxidative metabolites such as aflatoxin P_1 (AFP_1), revealed no linear excretion characteristics. Residual levels of hepatic AFB_1-DNA adducts at 24 hr postdosing directly correspond to AFB-N^7-gua excretion in urine, supporting the concept that measurement of the AFB-N^7-gua adduct in urine reflects DNA damage in the primary target organ.

There have been many studies conducted over the years to examine the relation between a single AFB_1 dose and the formation of DNA and protein adducts and a comprehensive review of these investigations is beyond the scope of this chapter; however, there are relatively few studies examining the more critical and relevant relationship between chronic exposure to AFB_1 and macromolecular adduct formation. These studies form an important basis for future human analyses and the reader is urged to examine these efforts.[8,12-14] Later in this review some of these multiple dose experiments will be described in the context of chemoprevention studies.

B. Transitional Epidemiological Studies

In parallel with the development and validation of aflatoxin-adduct biomarkers in animal models, a systematic evaluation of these molecular biomarkers in several human populations has been undertaken. Early studies by Autrup *et al.*[15,16] used synchronous fluorescence spectroscopy for the analysis of AFB-DNA adducts in human urine samples in Kenya. Over 1000 urine samples were analyzed and nearly 13% of the urines were positive for excreted AFB-N^7-gua. These data provided important evidence that humans had the metabolic capacity to produce the aflatoxin-DNA adducts previously detected in experimental animals.

Subsequent studies conducted in the Guangxi Autonomous region, a high area of liver cancer incidence in the People's Republic of China, determined both dietary intake of aflatoxin as well as the levels of urinary aflatoxin biomarkers. The average dietary intake of aflatoxins, primarily from contaminated corn consumption, was approximately 275 μg and 540 μg in males and females, respectively, over a 7-day monitoring period.[17] Total 24-hr urine samples were collected starting on day 4 of the study as consecutive 12-hr fractions. Total aflatoxin metabolites in the urine samples were measured and the relationship between AFB_1 intake per day and total aflatoxin metabolite excretion per day revealed a correlation coefficient of 0.26 that was not statistically significant at the 0.05 level. Thus, levels of total aflatoxin equivalents in urine did not correspond with aflatoxin intake, indicating that measurement of total aflatoxin metabolites is not an appropriate dosimeter for determining individual exposure status.

These results prompted a sequential immunoaffinity-HPLC analysis of the 273 urine samples to quantitate levels of individual aflatoxins. AFB-N^7-gua, aflatoxin M_1 (AFM_1), AFP_1, and AFB_1 were the aflatoxins most commonly detected. The linear regression analyses for the urinary levels of each of these individual aflatoxins was compared to aflatoxin intake and the correlation coefficients for AFB-N^7-gua, AFM_1, AFP_1, and AFB_1 were 0.65, 0.55, 0.02, and –0.10, respectively. Thus, both AFB-N^7-gua and AFM_1 appeared to be useful biomarkers of exposure.

One objective of this study was to determine the number of samples required from an individual and the time frame for sample collection necessary for a biomarker to reflect exposure status of AFB_1. There was particular interest in characterizing the molecular dosimetry of AFB-N^7-gua because of the putative relationship of this metabolite with the cancer initiation process and the potential to use this marker of the biologically effective dose to link exposure to risk status. Figure 3 shows total AFB-N^7-gua excretion in the urine of the male and female subjects over the complete urine collection period plotted against the total AFB_1 exposure in the diet for each of the subjects. This analysis smoothed day-to-day variations in both intake and excretion of AFB-N^7-gua and revealed a highly statistically significant correlation coefficient of 0.80. This analysis demonstrated that a summation of excretion and exposure status provides a stronger association between exposure and a molecular dosimetry marker than was seen in prior statistical analyses and supported the concept that quantitation of the AFB-N^7-gua adduct in urine can be a reliable biomarker for AFB_1 exposures.

Using the same study population, Gan *et al.*[18] monitored levels of aflatoxin-serum albumin adducts. A highly significant correlation of adduct level with intake ($r = 0.69$, $P < 0.000001$) was observed. From the slope of the regression line for adduct level as a function of intake, it was calculated that 1.4-2.3% of ingested AFB_1 became covalently bound to serum albumin, a value very similar to that observed when rats are administered AFB_1. When the data for DNA adduct excretion in urine[17] and serum albumin were compared, a statistically significant relationship was seen with a correlation coefficient of 0.73. Thus, both of these markers appear to be useful in human monitoring studies.

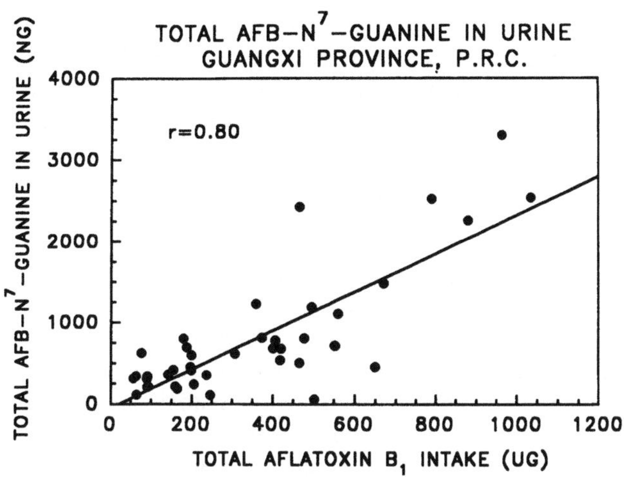

Figure 3. Linear regression analysis of the association between total AFB-N^7-gua adduct excretion in urine from males and females with total dietary aflatoxin in exposure during the study period in Guangxi Province, P.R.C.[17].

Similar studies examining the relationship between aflatoxin exposure and DNA adduct excretion and serum albumin adduct formation have been conducted in The Gambia, West Africa. These studies revealed relationships between dietary exposure and adduct levels[19,20] similar to those seen in China. Thus, the utility of these dosimetry markers as effective monitors of exposure has been confirmed in two populations at high risk for liver cancer that have different dietary sources of aflatoxins. From a practical perspective pertinent to epidemiological studies, the measurement and quantification of the aflatoxin-serum albumin adduct offers a rapid, facile approach that can be used to screen very large numbers of people. This method has been extensively validated in experimental and human sample analyses and the technique is described in detail in Wild *et al.*[21] Extensive geographic surveys have been carried out on hundreds of samples.[21,22]

III. Traditional Epidemiological Studies

Since the discovery of the aflatoxins in the early 1960s, many traditional epidemiological studies have been done to explore the aflatoxin-human liver cancer link. These early studies have been recently reviewed[3] and all of these investigations suffered from a lack of good aflatoxin exposure data and/or poor cancer incidence information. In addition, the studies done before 1980 suffered from the lack of an available accurate test for hepatitis B virus (HBV) status. For these reasons, the aflatoxin-human liver cancer link remained unclear until the application of both molecular dosimetry methods for aflatoxins and accurate HBV assessment methods to epidemiological studies.

Molecular dosimetry markers for aflatoxin exposure can be applied in a variety of epidemiological investigations, including cross-sectional, case-control and cohort or prospective investigations, and may facilitate efforts to define exposure-disease relationships. A recent report of a cross-sectional study by Campbell *et al.*[23] revealed no association between total aflatoxin metabolites in human urine and liver cancer incidence. In the context of human data described earlier in this chapter indicating that the composite measurement of aflatoxin equivalents in urine is not a good marker for exposure to aflatoxins, it is difficult to assess the significance of this null relationship between exposure and risk.

Recently, a case-control study of potential risk factors for hepatocellular carcinoma was reported for northeast Thailand.[24] In this study, 65 cases from three hospitals with matched controls were examined. Infection with HBV was the major risk factor identified and chronic carriers of hepatitis surface antigen had an estimated relative risk of 15.2. No increase in risk was found with recent aflatoxin intake, as estimated by consumption of possibly contaminated foods, or by measuring aflatoxin-albumin adducts in serum. Regular use of alcohol (two or more glasses of spirits per week) was associated with a nonsignificant elevation in risk (odds ratio = 3.4; 95% confidence interval = 0.8-14.6), but the number of regular drinkers in the population was small. Thus, in this case-control study, the contribution of recent exposure to aflatoxin and liver cancer was not significant.

In general, the most rigorous test of an association between an agent and disease outcome is found in prospective epidemiological studies where healthy people are followed until the diagnosis of disease. A nested case-control study initiated in 1986 in Shanghai has started to examine the relation between markers for aflatoxin and HBV and the development of liver cancer.[25] In this study, 18,244 urine samples were collected from healthy males between the ages of 45 and 64. In the subsequent 4 years, 22 of these individuals developed liver cancer. The urine samples for each case were age-matched and residence-matched with 5 to 10 controls and analyzed for both aflatoxin biomarkers and HBV surface antigen status. The data revealed a highly significant increase in the relative risk ($RR = 4.9$) for those liver cancer cases where AFB-N^7-gua was detected. There were also elevated risks for other aflatoxin urinary markers. The relative risk for people who tested positive for the HBV surface antigen was also about 5, but individuals with both urinary aflatoxins and positive HBV surface antigen status had a relative risk for developing liver cancer of about 60. These results show for the first time a relationship between presence of carcinogen-specific biomarkers and cancer risk. Moreover, these findings provide the first demonstration of a multiplicative interaction between two major risk factors for liver cancer: HBV and AFB_1 exposure.

IV. Experimental Chemoprotection Studies Using Dithiolethiones, Including Oltipraz

It is possible to modify risk for aflatoxin hepatocarcinogenesis in animals using chemoprotective interventions, and a number of classic chemprotective agents (notably BHA, BHT, and ethoxyquin) inhibit AFB_1 hepatocarcinogenesis in rats when fed simultaneously with the carcinogen.[26,27] These agents also powerfully protect against aflatoxin-DNA adduct formation in the livers of AFB_1-exposed animals.[28] More recently we have observed that a substituted dithiolethione, oltipraz, was a considerably more potent inhibitor of AFB_1-induced tumorigenesis in rats than the aforementioned food antioxidants.[29]

These chemoprotection studies with oltipraz have been extended to a cancer endpoint.[30] Male F344 rats received 10 intragastric doses of AFB_1 (25 μg/rat per day) and this subchronic dose exposure to AFB_1 produced an 11% incidence of hepatocellular carcinoma at 23 months with an additional 9% of the rats exhibiting hyperplastic nodules in their livers. By contrast, feeding rats a diet supplemented with 0.075% oltipraz for a 4-week period surrounding the time of AFB_1 exposure afforded complete protection against AFB_1-induced hepatocellular carcinomas and hyperplastic nodules. None of these lesions were observed in the oltipraz-fed, AFB_1-treated animals. Further, no tumors were found at secondary, extrahepatic sites for AFB_1 carcinogenesis such as the colon and kidney.

These findings also led to the study of the possible role of the dithiolethiones in the protection against the acute hepatotoxicity of AFB_1 in the rat,[31] since hepatotoxicity (with resultant cell proliferation) may be a requisite toxicological effect for complete carcinogens like AFB_1. In these hepatotoxicity experiments, male F344 rats were fed purified diets supplemented with either 0.075% oltipraz or 0.01% 1,2-dithiole-3-thione starting 1 week before treatment with AFB_1 and throughout the experimental period.[31] Pretreatment with oltipraz reduced the mortality produced by a single p.o. dose of 10 mg AFB_1/kg from 83% to 36%. In a subchronic study, treatment of rats with 500 µg AFB_1/kg per day for 10 days produced 100% mortality within 1 week following cessation of dosing. However, animals fed a diet supplemented with 1,2-dithiole-3-thione were completely protected from the lethal effects of AFB_1. In fact, growth rates in this group were comparable to the non-AFB_1-treated controls.

Since tissue damage with cell killing and consequent cell proliferation is thought to be important in the promotion of liver cancers, we investigated the effects of oltipraz on preventing cell death. As shown in Figure 4, the subchronic AFB_1 treatment regimen resulted in the loss of nearly 60% of hepatocytes prelabeled with [^{3}H]-thymidine, while oltipraz supplementation largely prevented this loss. This loss of DNA from the liver is

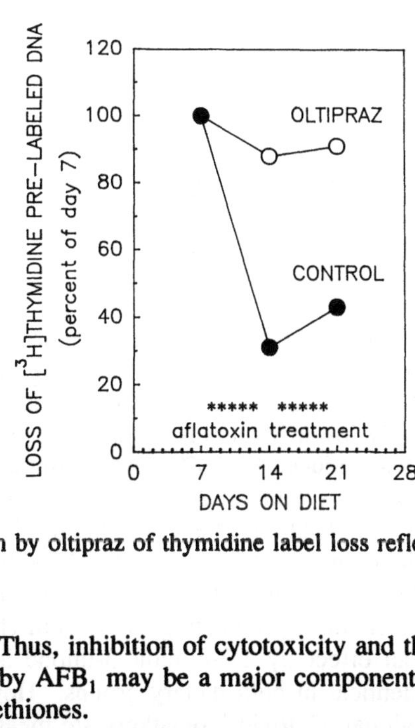

Figure 4. Protection by oltipraz of thymidine label loss reflecting liver cell death.

equated with cell death. Thus, inhibition of cytotoxicity and the subsequent compensatory cell proliferation elicited by AFB_1 may be a major component of the protective actions of oltipraz and other dithiolethiones.

The AFB_1-dithiolethione chemoprotection model is a useful experimental system to examine the relationships between specific aflatoxin biomarkers in biological fluids and cancer risk in animals. Recently completed experiments indicated that the unsubstituted congener of oltipraz, 1,2-dithiole-3-thione, is an extremely effective inhibitor of AFB_1-induced tumorigenesis, as determined by analyses for preneoplastic foci.[32] Thus, the impact of chemoprotection by 1,2-dithiole-3-thione on the molecular dosimetry of AFB-*N*^{7}-gua in urine and the modulation of levels of hepatic AFB_1-DNA adducts have been examined. The effects of 1,2-dithiole-3-thione on the kinetics of hepatic AFB_1-DNA adduct formation and removal in rats receiving 250 µg AFB_1/kg by gavage on each of days 0-4 and 7-11 are shown in Figure 5A. Maximal levels of carcinogen binding were achieved following the third dose in the control group and declined thereafter despite continued exposure to AFB_1. This diminution of binding, particularly during the second dosing cycle, has been observed

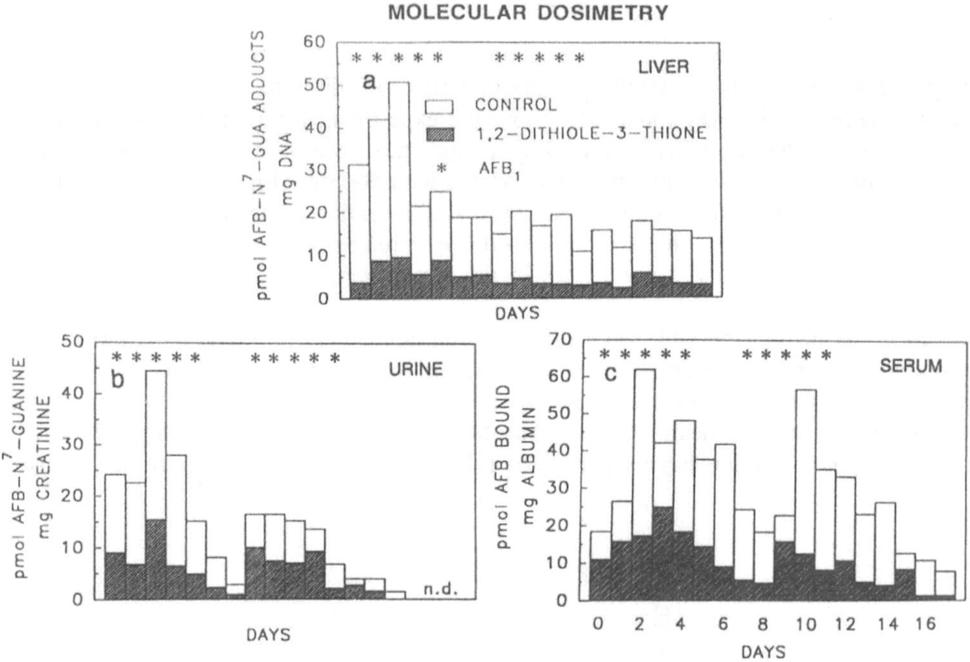

Figure 5. Effect of 1,2-dithiole-3-thione on the time-course of aflatoxin binding in rats. (a) Liver DNA adduct formation and removal. (b) Excretion of AFB-N^7-guanine in urine. (c) Formation of aflatoxin-serum albumin adducts.

previously[8,14,33] and may be a consequence of the induction of glutathione S-transferases and/or other enzymes involved in AFB_1 detoxication following chronic exposure to AFB_1.[32] Inclusion of 0.03% 1,2-dithiole-3-thione in the diet, beginning 1 week prior to dosing with AFB_1, resulted in substantially lower levels of hepatic AFB_1-DNA adducts throughout the exposure period. Binding was reduced by 76% over the initial 18-day period.

The levels of total AFB_1 equivalents in 24-hr urine samples collected over the 2-week exposure period showed no remarkable differences in the levels of AFB_1 metabolites in rats fed the control AIN-76A diet compared to those fed the 1,2-dithiole-3-thione-supplemented diet. Urinary aflatoxin levels rise rapidly following dosing with AFB_1 and drop equally quickly following cessation of dosing, reflecting the overall short *in vivo* half-life of this carcinogen. The lack of an effect by 1,2-dithiole-3-thione is not surprising, given that exposures to AFB_1 were identical in both dietary groups. However, a distinctly different pattern emerges when individual aflatoxin biomarkers are measured. Shown in Figure 5B are the levels of AFB-N^7-gua in serial 24-hr urine samples collected from rats undergoing the chemoprotective intervention. The highest level of AFB-N^7-gua excretion occurred on day 2 in both groups following the third dosage of AFB_1. This outcome is identical to that observed with hepatic levels of AFB_1-DNA adducts and with serum albumin adduct formation (Figure 5C). Over the 15-day collection period in which AFB-N^7-gua adducts were detectable in the urine, feeding of 1,2-dithiole-3-thione produced an overall reduction of 62% in the elimination of this AFB_1-DNA adduct excision product, mirroring the data on the overall levels of hepatic AFB_1-DNA adducts. The amount of AFB-N^7-gua in urine represents only 1% of the total aflatoxin metabolites in urine and accounts as to why the dramatic differences seen between groups in AFB-N^7-gua levels are not reflected in the levels of total urinary aflatoxin metabolites. Thus, these data indicate that the excreted DNA adduct in urine and the formation of the serum albumin adduct accurately reflects the amount of genotoxic damage at the target organ site in the liver. In addition, these data indicate that the measurement of these adducts may reflect risk for disease development in the animal.

V. Molecular Biomarkers as Intermediate Endpoints in Future Human Chemoprotection Trials

Primary hepatocellular carcinoma is one of the most lethal and common cancers in the world. A number of epidemiological studies have associated the exposure status of people to AFB_1 as being important in the etiology of liver cancer. Until the recent report by Ross *et al.*,[25] these studies have relied upon the criteria of presumptive intake data, rather than relying upon quantitative analyses of aflatoxin biomarkers obtained by monitoring biological fluids from exposed people. While further work needs to be done, both experimental animal studies and human analyses support the concept that measurement of the major, rapidly excised AFB-N^7-gua adduct in urine and the more persistent aflatoxin albumin adduct are appropriate dosimeters for estimating exposure status in individuals consuming this mycotoxin. A further finding of this recent study was the strong multiplicative interaction between HBV status and aflatoxin biomarkers. The interaction between a potent cell proliferative agent (HBV) and a strong initiating agent (aflatoxin) could lead to the type of multiplicative effect seen in this study. These data also provide support for two different approaches for prevention of liver cancer. A first approach is vaccination against HBV. Unfortunately, people living in high-risk areas for liver cancer acquire the HBV infection before 3 years of age. Thus, an immunization program for total population protection would minimally have to occur over a 20- to 30-year period. Despite this problem, vaccination programs for HBV should be implemented with the required long-term follow-up. A second approach for cancer prevention would be the elimination of aflatoxin exposures. Primary prevention of aflatoxin exposures could be accomplished through large expenditures of resources for proper crop storage and handling; however, this approach is not economically feasible in many areas of the world. Thus, chemoprotective interventions may provide a near-term alternative for some very high-risk populations. Indeed, clinical trials of chemoprotective agents in such populations are presently underway.

The development of biomarkers as intermediate endpoints in chemoprotection trials is an important and necessary goal. Cancer as the endpoint of a chemoprotection intervention ensures very slow progress, be it in experimental models or man. Intermediate biomarkers can reflect several different intermediary stages of the neoplastic process.[34] These markers might include organ site-specific genomic, proliferation, and differentiation markers. Although very few such markers have been rigorously validated to date, techniques of modern biology allow for the detection of specific genetic and phenotypic changes in small populations of cells. As an adjunct to these process-dependent markers it may also be possible to devise markers specific to interventions in selected high-risk groups. These agent-dependent approaches would be primarily based upon knowledge of the etiologic agent in the study population. In this context the aflatoxin specific biomarkers discussed in this chapter should be useful and objective intermediate endpoints for assessing the efficacy of chemoprotective interventions. For example, a large reduction in the urinary levels of AFB-N^7-gua in a group dosed with a chemoprotective agent compared to levels in a control group could be a presage for success of the intervention. Clearly these initial studies would be of modest size to demonstrate feasibility of the intermediate biomarker strategy and would not include follow-up to a disease endpoint. Thus, these adduct markers might be particularly useful in optimizing dose and schedule parameters of chemoprotective agents in advance of longer-term clinical trials utilizing actual disease endpoints.

The long-term goal of the research described in this chapter is the application of biomarkers to the development of chemoprevention agents for use in human populations at high risk for cancer. Several of the aflatoxin specific biomarkers have been validated in epidemiological studies and are now available to use as intermediate biomarkers in chemoprotection trials. As described herein the development of these aflatoxin biomarkers has been based upon the knowledge of the biochemistry and toxicology of aflatoxin gleaned from both experimental and human studies. These biomarkers have been subsequently utilized in experimental chemoprotection models to provide data on the modulation of these markers under different disease risk situations. This systematic approach should serve as a template for the development, validation, and application of other chemical-specific biomarkers to cancer or other chronic diseases.

Acknowledgments

This work was supported by grants from the USPHS CA39416, CA48409, CA54114, ES06052, NIEHS Center Grant ES03819 and KO4 CA 01517.

References

1. National Research Council Committee on Biological Markers, Biological markers in environmental health research, *Env Hlth Persp* 74:3-9 (1987).
2. W.F. Busby, G.N. Wogan, Aflatoxins, *in*: *Chemical Carcinogens*, 2nd Ed., C.E. Searle, ed., pp. 945-1136, American Chemical Society, Washington, D.C. (1985).
3. J.D. Groopman, G. Sabbioni, C.P. Wild, Molecular dosimetry of aflatoxin exposures, *in*: *Molecular Dosimetry of Human Cancer: Epidemiological, Analytical and Social Considerations*, J.D. Groopman, P. Skipper, eds., pp. 302-324, CRC Press, Boca Raton, FL (1991).
4. B. Hulka, Epidemiological studies using biological markers: Issues for epidemiologists, *Cancer Epidemiology, Biomarkers and Prevention* 1:13-19 (1991).
5. J.M. Essigmann, R.G. Croy, A.M. Nadzan, W.F. Busby, Jr., V.N. Reinhold, G. Buchi, G.N. Wogan, Structural identification of the major DNA adduct formed by aflatoxin B_1 *in vitro*, *Proc Natl Acad Sci USA* 74:1870-1874 (1977).
6. G. Sabbioni, P. Skipper, G. Buchi, S.R. Tannenbaum, Isolation and characterization of the major serum albumin adduct formed by aflatoxin B_1 *in vivo* in rats, *Carcinogenesis* 8:819-824 (1987).
7. S. Appleton, M.P. Goetchius, T.C. Campbell, Linear dose-response curve for hepatic macromolecular binding of aflatoxin B_1 in rats at very low exposures, *Cancer Res* 42:3659-3662 (1982).
8. C.P. Wild, R.G. Garner, R. Montesano, F. Tursi, Aflatoxin B_1 binding to plasma albumin and liver DNA upon chronic administration to rats, *Carcinogenesis* 7:853-858 (1986).
9. J.D. Groopman, W.F. Busby, G.N. Wogan, The nuclear distribution of aflatoxin B_1 and its interaction with histones *in vivo*, *Cancer Res* 40:4343-4351 (1980).
10. R.A. Bennett, J.M. Essigmann, G.N. Wogan, Excretion of an aflatoxin-guanine adduct in the urine of aflatoxin B_1-treated rats, *Cancer Res* 41:650-654 (1981).
11. J.D. Groopman, J. Hasler, L.J. Trudel, A. Pikul, P.R. Donahue, G.N. Wogan, Molecular dosimetry in rat urine of aflatoxin-N^7-guanine and other aflatoxin metabolites by multiple monoclonal antibody affinity chromatography and HPLC, *Cancer Res* 52:267-274 (1992).
12. R.G. Croy, G.N. Wogan, Temporal patterns of covalent DNA adducts in rat liver after single and multiple doses of aflatoxin B_1, *Cancer Res* 41:197-203 (1981).
13. D.E. Goeger, D. Shelton, J.D. Hendricks, C. Pereira, G.S. Bailey, Comparative effect of dietary butylated hydroxyanisole and β-naphthoflavone on aflatoxin B_1 metabolism, DNA adduct formation and carcinogenesis in rainbow trout, *Carcinogenesis* 9:1793-1800 (1988).

14. T.W. Kensler, P.A. Egner, N.E. Davidson, B.D. Roebuck, A. Pikul, J.D. Groopman, Modulation of aflatoxin metabolism, aflatoxin-N^7-guanine formation, and hepatic tumorigenesis in rats fed ethoxyquin: Role of induction of glutathione S-transferase, *Cancer Res* 46:3924-3931 (1986).
15. H. Autrup, K.A. Bradley, A.K.M. Shamsuddin, J. Wakhisi, A. Wasunna, Detection of putative adduct with fluroescence characteristics identical to 2,3-dihydro-2-(7'-guanyl)-3-hydroxyaflatoxin B1 in human urine collected in Murang'a district, Kenya, *Carcinogenesis* 4:1193-1195 (1983).
16. H. Autrup, T. Seremet, J. Wakhisi, A. Wasunna, Aflatoxin exposure measured by urinary excretion of aflatoxin B_1-guanine adduct and hepatitis B virus infection in areas with different liver cancer incidence in Kenya, *Cancer Res* 47:3430-3433 (1987).
17. J.D. Groopman, J. Zhu, P.R. Donahue, A. Pikul, L.-S. Zhang, J.-S. Chen, G.N. Wogan, Molecular dosimetry of urinary aflatoxin DNA adducts in people living in Guangxi Autonomous Region, People's Republic of China, *Cancer Res* 52:45-51 (1992).
18. L.-S. Gan, P.L. Skipper, X.-C. Peng, J.D. Groopman, J.-S. Chen, G.N. Wogan, S.R. Tannenbaum, Serum albumin adducts in the molecular epidemiology of aflatoxin carcinogenesis: Correlation with aflatoxin B_1 intake and urinary excretion of aflatoxin M_1, *Carcinogenesis* 9:1323-1325 (1988).
19. J.D. Groopman, A. Hall, H. Whittle, G. Hudson, G.N. Wogan, R. Montesano, C.P. Wild, Molecular dosimetry of aflatoxin-N^7-guanine in human urine obtained in The Gambia, West Africa, *Cancer Epidemiology, Biomarkers and Prevention* 1:221-228 (1992).
20. C.P. Wild, G. Hudson, G. Sabbioni, G.N. Wogan, H. Whittle, R. Montesano, J.D. Groopman, Correlation of dietary intake of aflatoxins with the level of albumin bound aflatoxin in peripheral blood in The Gambia, West Africa, *Cancer Epidemiology, Biomarkers and Prevention* 1:229-234 (1992).
21. C.P. Wild, Y.Z. Jiang, G. Sabbioni, B. Chapot, R. Montesano, Evaluation of methods for quantitation of aflatoxin-albumin adducts and their application to human exposure assessment, *Cancer Res* 50:245-251 (1990).
22. C.P. Wild, Y.Z. Jiang, S.J. Allen, L.A. Jansen, A.J. Hall, R. Montesano, Aflatoxin-albumin adducts in human sera from different regions of the world, *Carcinogenesis* 11:2271-2274 (1990).
23. T.C. Campbell, J. Chen, C. Liu, J. Li, B. Parpia, Nonassociation of aflatoxin with primary liver cancer in a cross-sectional ecological survey in the People's Republic of China, *Cancer Res* 50:6881-6893 (1990).
24. P. Srivatanakul, D.M. Parkin, M. Khlat, D. Chenvidhya, P. Chotiwan, S. Insiripong, K.A. L'Abbe, C.P. Wild, Liver cancer in Thailand. II. A case-control study of hepatocellular carcinoma, *Int J Cancer* 48:329-332 (1991).
25. R. Ross, J.-M. Yuan, M. Yu, G.N. Wogan, G.-S. Qian, J.-T. Tu, J.D. Groopman, Y.-T. Gao, B.E. Henderson, Urinary aflatoxin in biomarkers and risk of hepatocellular carcinoma, *The Lancet* 339:943-946 (1992).
26. J.R. Cabral, G.E. Neal, The inhibitory effects of ethoxyquin on the carcinogenic action of aflatoxin B_1 in rats, *Cancer Lett* 19:125-132 (1983).
27. G.M. Williams, T. Tanaka, Y. Maeura, Dose-related inhibition of aflatoxin B_1-induced hepatocarcinogenesis by the phenolic antioxidants butylated hydroxytoluene and butylated hydroxyanisole, *Carcinogenesis* 7:1043-1046 (1986).
28. T.W. Kensler, P.A. Egner, M.A. Trush, E. Bueding, J.D. Groopman, Modification of aflatoxin B_1 binding to DNA *in vivo* in rats fed phenolic antioxidants, ethoxyquin, and a dithiothione, *Carcinogenesis* 6:759-764 (1985).
29. T.W. Kensler, P.A. Egner, P.H. Dolan, J.D. Groopman, B.D. Roebuck, Mechanism of protection against aflatoxin tumorigenesis in rats fed oltipraz and related 1,2-dithiol-3-thiones and 1,2-dithiol-3-ones, *Cancer Res* 47:4271-4277 (1987).
30. B.D. Roebuck, Y.-L. Liu, A.E. Rogers, J.D. Groopman, T.W. Kensler, Protection against aflatoxin B_1-induced hepatocarcinogenesis in F344 rats by 5-(2-pyrazinyl)-4-methyl-1,2-

dithiol-3-thione (oltipraz): Predictive role for short-term molecular dosimetry, *Cancer Res* 51:5501-5506 (1991).

31. Y.-L. Liu, B.D. Roebuck, J.D. Yager, J.D. Groopman, T.W. Kensler, Protection by 5-(2-pyrazinyl)-4-methyl-1,2-dithiol-3-thione (oltipraz) against the hepatotoxicity of aflatoxin B_1 in the rat, *Toxicol Appl Pharmacol* 93:442-451 (1988).
32. T.W. Kensler, J.D. Groopman, D.L. Eaton, T.J. Curphey, B.D. Roebuck, Potent inhibition of aflatoxin-induced hepatic tumorigenesis by the monofunctional enzyme inducer 1,2-dithiole-3-thione, *Carcinogenesis* 12:95-100 (1992).
33. R.G. Croy, J.M. Essigmann, V.N. Reinhold, G.N. Wogan, Identification of the principal aflatoxin B_1-DNA adduct formed *in vivo* in rat liver, *Proc Natl Acad Sci USA* 75:1745-1749 (1978).
34. S.M. Lippman, J.S. Lee, R. Lotan, W. Hittleman, M.J. Wargovich, W.K. Hong, Biomarkers as intermediate end points in chemoprevention trials, *J Natl Cancer Inst* 82:555-562 (1990).

Chapter 12

Cytokines, Iron Homeostasis, and Cancer

FRANK M. TORTI and SUZY V. TORTI

I. Introduction

Iron is of central importance to all, from bacteria to higher eukaryotes.[1-3] It is critical for such important cellular processes as DNA synthesis, cellular respiration, and oxygen transport. However, iron also is a catalytic participant in the generation of damaging oxygen free radicals. Given this balance of iron availability and cellular survival and iron availability and cell injury, the level and redox state of iron is very tightly regulated within cells by a number of important proteins. These proteins—transferrin, transferrin receptor, and ferritin—are highly conserved. In bacteria and lower eukaryotes, well-defined systems for uptake and regulation of iron show a high degree of similarity to those in mammals. [4]

Perhaps for similar reasons, uptake, distribution, and compartmentalization of iron are tightly regulated in man. This is primarily achieved through gastrointestinal uptake; the enterocyte possesses a highly specialized ability for uptake and cellular transport of ferrous iron and its conversion to ferric iron. This enterocyte iron either binds plasma transferrin and enters the circulation, or is complexed with ferritin and returns to the lumen as the intestinal mucosa is sloughed. This ability to partition iron provides the capacity to modulate intestinal iron uptake over a wide range, depending on the iron status of the individual.

Iron is important for the growth of both normal and neoplastic cells; in cell culture, the withdrawal of iron from culture media inhibits cell proliferation. This is due in large part to the dependence on iron of ribonucleotide reductase, an essential enzyme in DNA synthesis.

II. Compartments of Iron

In man, iron is partitioned into well-defined compartments. Two thirds (approximately 2500 mg in a 70-kg man) of all available iron is present as hemoglobin, an iron-containing protein that functions in oxygen transport. Twenty-five percent (approximately 1000 mg) is present in "storage" form, either as ferritin or hemosiderin. These iron stores accumulate

Frank M. Torti and Suzy V. Torti • Department of Medicine, Stanford University and Palo Alto Veterans Affairs Medical Center, Palo Alto, CA 94304

Diet and Cancer: Markers, Prevention, and Treatment
Edited by M.M Jacobs, Plenum Press, New York, 1994

with age,[5] and may participate in a number of pathophysiologic processes.[6] Thus ferritin and hemosiderin comprise the vast majority of iron outside the erythron. All other iron sources, in sum, account for less than 6% of total body iron: myoglobin accounts for an additional 130 mg, the "regulatory" or "labile" iron pools within cells, for 80 mg. Other tissue iron (in cytochromes, etc.) contributes 8 mg. Transferrin iron and other transport forms contribute less than one tenth of one percent (3 mg) of iron.[7]

III. Iron Sources in the Diet

Virtually all iron is absorbed, regardless of dietary source, as ferrous (Fe^{2+}) iron. The entire small intestine is capable of absorbing iron, although in health the duodenum appears to participate most efficiently in iron absorption. At low levels of luminal iron concentration, the extent of iron absorption is regulated by the level of the body iron stores. When these stores are low, iron absorption is increased; when high, most of the iron taken up by the intestinal mucosal cells is trapped and sloughed with the normal maturation of this proliferating epithelium at the villus tip. Thus a normal man will absorb approximately 1 mg of iron daily, an amount equivalent to that lost in the mucosal epithelium. A menstruating woman will absorb approximately two to ten times this amount. The signals that communicate this body iron stores information to the mucosal epithelial cell remain unknown. At high luminal concentrations of iron achieved during therapy with ferrous iron compounds, however, this regulation can be overcome. As intestinal luminal iron concentration increases with increasing doses of exogenous iron, the degree of iron absorption increases proportionally.[7]

The availability of iron for intestinal absorption is dependent on the dietary form of iron. Heme iron is thought to be more readily absorbed than non-heme sources. The acidic environment in the proximal small bowel is thought to enhance absorption. Further, the ingestion of food containing phytates or oxalates can retard iron absorption, presumably by complexing with iron in the intestinal lumen. Phosphates also reduce iron absorption. Alcohol and ascorbic acid enhance absorption.

IV. The Epidemiology of Cancer and Iron Body Stores

A number of studies have correlated iron stores in the body with cancer risk. In a study published in the *New England Journal of Medicine* in 1988, Stevens and co-workers showed that the risk of cancer increased with increasing serum iron and transferrin saturation.[3] Lung, colon, bladder, and esophageal cancers showed a particularly high correlation with increased transferrin saturation and serum iron. In patients with colon cancer, the relative risk of developing colon cancer increased from 1.0 in the lowest quartile of transferrin saturation to 4.69 in the highest quartile, although the biologic explanation for this observation is still uncertain. These recent results confirm and extend earlier studies in Solomon Islanders. In these studies, liver cancer risk was greater in those who had low transferrin and high ferritin.[2,8]

V. Iron Modulation and Cancer Growth

A. Animal Studies

The growth of L1210 leukemic cells has been investigated under conditions in which DBA-2 mice were treated with supplemental iron. Intraperitoneal injections of iron-dextran (12.5 mg or 125 mg elemental iron/kg body weight every 8 hr for two doses) were administered to mice that were injected with 1×10^6 L1210 cells 6 hr later. Iron-dextran treatment enhanced the number of L1210 cells as early as 2 days after tumor inoculation. This enhanced proliferation persisted for the 8 days of the study. At the higher levels of supplemental iron, the L1210 animals died 25% faster than controls.[9] The lower dose of supplemental iron was chosen to approximate the dose used therapeutically in humans. The investigators conclude that supplemental iron enhanced the growth of L1210 cells, resulting in a shorter lifespan for these mice.

In a study of subcutaneously implanted tumors in three strains of mice maintained on a low (5 mg iron/kg body weight) or normal (312 mg/kg) iron-supplemented diet, the low iron diet was shown to result in less rapid tumor growth.[10] Three tumor cell lines were studied—a colon cancer, a breast cancer, and a hepatoma cell line. In one of the mouse lines the low iron group demonstrated less weight than the controls, but in the other two strains the low iron group mean weight was greater, leading the authors to conclude that weight changes did not contribute to the slower tumor growth in the low dietary iron group. Similar observations have been made in spontaneous tumors in rats.[11]

In addition to investigations of the growth of tumors under different conditions of dietary iron supplementation or deprivation, a number of studies have been performed to examine the rate of carcinogenesis in animals whose iron content has been modulated. In a study by Thompson and associates,[12] low, excess, and normal iron-supplemented diets were administered to Sprague-Dawley rats 21 days old. All groups were treated with 1-methyl-1-nitrosourea (MNU), an experimental model for breast cancer. Although the initial rates of tumor formation were equivalent in the excess iron supplemention group and the normal iron supplemention group, the rats with the excess iron-supplemented diets continued to develop tumors for a longer time period, and eventually had the greatest number and extent of mammary cancers induced. In the iron-poor dietary groups, the results are more difficult to interpret. Those fed low iron diets had a substantial reduction in the rate of tumor formation, but these animals also had a low hematocrit and reduced weight gain, factors that can also influence tumor formation. The investigators conclude that excess dietary iron appears to be a more potent stimulus to modulation of tumor formation than does iron depletion in this model.[12]

Dimethylhydrazine (DMH) has been used as a tumor promoter in studies designed to assess the role of oral iron supplementation in colorectal tumor formation.[13] In male NMRI mice fed a diet supplemented with 3.5% Fe-fumarate, the rate of tumor formation after DMH treatment increased 6.5-fold (from 3.5 to 13.9 tumors/mouse). Desferrioxamine did not protect against tumor formation, however. DMH methylase activity was similar in iron-supplemented and control mice.

Using a model of liver carcinogenesis induced by hexachlorobenzene, a single iron-dextran injection in C57BL/10ScSn mice followed by feedings containing 0.01% HCB were shown to increase the likelihood of developing liver cancer.[14]

B. Human Studies

It has been suggested that in mammals the decrease in serum iron that occurs during inflammation is a host defense: low iron concentration in serum of septic mammals inhibits the growth of pathogenic bacteria, high iron concentration stimulates growth.[15] Many tumor cells require iron for growth. In cancer patients, serum iron and serum transferrin are frequently depressed, leading to the speculation that decreased serum iron is also a host defense against malignancy.

If this is true, modulation of host iron by dietary or other means might alter tumor growth characteristics. Therapy of cancer utilizing iron depletion to selectively target tumor cells has been used successfully in *in vivo* and *in vitro* animal models as discussed above, and has been tested in a limited number of clinical situations. These include the use of deferoxamine in the treatment of neuroblastoma,[16] deferoxamine in the treatment of neonatal acute leukemia,[17] I^{131} and Y^{90} antiferritin antibodies in Hodgkins disease,[18,19] gallium nitrate,[20] and transferrin receptor antibodies (currently in clinical trials).

VI. Serum Iron and Serum Ferritin in Cancer Patients

Three major questions are addressed in the multiple studies in lymphoid malignancies and solid tumors that evaluate serum ferritin: (1) Does the magnitude of elevation of serum ferritin correlate with disease extent or prognosis? (2) Does cytoreduction of tumor—either surgically or chemotherapeutically—result in normalization of serum ferritin levels? and (3) Is cancer recurrence preceded by elevation of serum ferritin?

A. Lymphoproliferative Diseases and Leukemia

Serum ferritin correlates closely with disease extent in the lymphoproliferative disorders. In Hodgkins disease, Bezwada and colleagues[21] demonstrated a decrease in the serum iron, serum transferrin saturation, and hemoglobin in patients with Hodgkins disease. These alterations were correlated with disease stage: In patients with minimal disease (stage IA), the mean serum iron for the group was 18.4 μmol/liter, while in patients with advanced and symptomatic disease (stage IVB), it was 7.2 μmol/liter. A reverse relationship was observed with serum ferritin, which increased from a mean of 80 μg/liter in stage IA patients to 2072 μg/liter in stage IVB patients. The rise in serum ferritin correlated closely with the content of non-heme iron in liver and marrow. Total body iron content remained relatively constant at 2g. However, there was a redistribution favoring storage iron in the liver and a reduction in heme iron content.[21]

A similar increase in serum ferritin concentration with increasing stage of Hodgkins disease was also demonstrated by Jacobs and co-workers.[22] These investigators also demonstrated a statistically significant increase in those patients with B symptoms. In addition, a decrease in serum iron and transferrin saturation was noted. However, in contrast to the report of Bezwada and colleagues,[21] there was no correlation with the degree of decrease in serum iron and transferrin saturation and more advanced clinical stage of disease. Hann confirmed the association of increased serum ferritin with a poor prognosis in childhood Hodgkins disease, and demonstrated that low serum transferrin conferred similar prognostic information.[23]

Acute leukemias, Hodgkins disease, and histocytic lymphoma show the most striking correlations with serum ferritin among the lymphoproliferative diseases. In one study of histocytic lymphoma, for example, values of serum ferritin ranged from 1000-10,000 μg/liter and normalized on treatment.[24] Serum ferritin elevations can be striking in these rapidly proliferating lymphoid disorders. Parry and co-workers[25] observed marked elevation in serum ferritin (10 times the normal range) in patients with acute leukemias. Jones and co-workers[26] also found marked elevations in serum ferritin in Hodgkins disease and leukemias; in addition, they found elevated transferrin-bound iron in the leukemia patients and reduced transferrin-bound iron in those patients with Hodgkins disease.

Elevated serum ferritin has been observed in patients with AIDS and the AIDS-related complex (ARC).[27] These elevated ferritin levels have been proposed to contribute to the profound immunosuppression seen in these conditions.[28]

B. Malignancies of Childhood

In a number of childhood malignancies, serum ferritin is elevated and frequently decreases with effective anti-cancer treatment. In addition, markedly higher serum ferritin has been seen in groups of patients with active disease and in those in remission. Nonetheless, some patients in remission have higher than normal serum ferritin. This has put into question the utility of the test in identifying complete remission in individual patients.[29] Whether this represents a failure of serum ferritin as a predictive test (patients actually free of disease when their serum ferritin was elevated) or whether those patients in remission with high serum ferritin eventually relapsed (the test was accurate but clinical staging was inaccurate) remains uncertain. This uncertainty highlights the major problem with many of these clinical studies—the lack of long-term patient follow up as an endpoint to correlate with serum ferritin values.

One exception is the study of Potaznik and colleagues,[30] which does provide population-based long-term follow up. These investigators demonstrated a difference in actuarial survival for pediatric acute lymphoblastic leukemia in relation to serum ferritin concentration. Patients with serum ferritin greater than 200 ng/ml had a substantially worse survival (as did patients with high transferrin saturation or high serum iron) than those with lower ferritin levels.

C. Solid Tumors in Adults

The correlations demonstrated in Hodgkins disease between degree of elevation of serum ferritin and advancing clinical stage of disease have not been found universally in solid tumors, although striking correlations exist in certain tumors.

In hepatocellular carcinoma, ferritin has been widely studied. Among Korean patients with chronic liver disease (most of whom had hepatitis B), those with persistently high serum ferritin had a high-rate probability of developing hepatocellular carcinoma (50% vs. 20% for men with hepatitis and lower serum ferritin).[31] Elevated serum ferritin has been found in 56.8% of hepatocellular carcinomas. In concert with alpha fetoprotein measurements, measurement of ferritin raised the diagnostic accuracy of liver lesions less than 3 cm over what either serum marker could provide singly.[32] However, although serum ferritin is elevated in many patients with hepatocellular carcinoma, Nagasue and co-workers[33] found inconsistent decreases after resection of hepatocellular cancer; these researchers suggested that serum ferritin would not be a useful tumor marker in resected patients.

Although increased serum ferritin has been reported in pancreatic carcinoma,[34-37] recent evidence[34] suggests that altered hepatic function, particularly cholestasis, makes a substantial contribution to the overall elevation of serum ferritin in this disease, as well as in liver cancer. The correlation between increased serum ferritin, serum bilirubin, and alkaline phosphatase led the authors to speculate that decreased reticuloendothelial function might decrease ferritin clearance by the liver and, as a consequence, increase serum ferritin.

Serum ferritin was found to be elevated in patients with a diagnosis of renal cell carcinoma when compared to serum ferritin in patients without cancer.[38] A correlation of ferritin with increasing stage in renal cell carcinoma was also found. Further, in patients with surgically resectable renal cell cancer, nephrectomy was associated with reduction of serum ferritin. Immunocytochemistry on fixed tissue specimens showed intense tissue staining of the renal cell cancer specimens, but limited staining of the surrounding normal tissues.

Breast cancer causes a relatively modest increase in serum ferritin. Nonetheless, Gaglia and colleagues[39] showed a correlation between an increase in serum ferritin and increasing tumor size, but not nodal status of patients. In a study of breast cancer patients with locally advanced or metastatic disease, serum ferritin was found to be substantially higher in these patients prior to receiving endocrine therapy than in disease-free breast cancer patients. In stage IV patients, 33% had serum ferritin of greater than 200 μg/liter. Of note, there was a strong association between response to endocrine therapy and ferritin decline.[40]

Changes in serum ferritin, serum iron, and other iron-binding proteins have been studied somewhat less intensively in other solid tumors. In small cell lung cancer patients, Cox and co-workers[41] were unable to correlate the elevated levels of serum ferritin with either the presence or absence of metastases or clinical course. However, it is difficult to evaluate this study because patient numbers were small and patients with small cell lung cancer tend to have occult metastases even with apparently localized disease. In gastric cancer, Saji and co-workers[42] demonstrated an increase in the mean value from 76.5 ng/ml in normal subjects to 169.8 µg/liter in all gastric cancer patients, but differences according to stage were unconvincing.

A complex relationship exists between serum ferritin and stomach cancer. Using stored serum samples from the Hiroshima and Nagasaki populations, Akiba and colleagues[43] found that low serum ferritin correlated with the risk of stomach cancer. The investigators speculated that the concordant association with acholorhydria and stomach cancer might affect iron levels, since achorhydria reduces the amount of ferrous iron absorbed in the duodenum and proximal jejunum. Nor could the investigators exclude the possibility that occult blood loss prior to the diagnosis of stomach cancer contributed to the low serum ferritin.

VII. Relationships Between Cytokines and Iron Homeostasis

Studies in our laboratory on the effect of cytokines on proteins of iron metabolism grew out of experiments on cachexia. These experiments utilized TA1 adipocytes, a cell line isolated in the laboratory of Dr. Gordon Ringold by treatment of mouse 10T 1/2 cells with 5-azacytidine followed by screening for a propensity for adipocyte differentiation. We observed that treatment of TA1 cells with tumor necrosis factor alpha (TNF-α) resulted in a loss of lipid content, and a decrease in the expression of adipose-specific genes.[44] However the expression of one gene was consistently elevated, rather than repressed, following treatment with TNF-α. The cDNA for this gene (which was initially termed XGS) was sequenced, and shown by sequence similarity to be the heavy chain of murine ferritin.[45] This was the first indication that cytokines might impact intracellular iron balance in normal cells.

In order to assess the generality of this observation, we measured the effects of TNF-α on ferritin H gene expression in cultured skeletal myoblasts and fibroblasts. We observed that in both predifferentiated myoblasts and fully differentiated myotubes, TNF-α induced ferritin H expression.[46] Fibroblasts also responded to TNF-α treatment with an increase in ferritin H synthesis. In all cases, increased levels of ferritin H mRNA were accompanied by increased levels of ferritin H protein. These results are depicted in Figure 1.

Interestingly, induction of ferritin H also ensued following treatment of cultured muscle cells with interleukin 1 alpha (1L-1α), another inflammatory cytokine,[47] but not following treatment with TNF-β or interferon beta or gamma. Thus induction of ferritin H is a specific, cytokine-dependent, response.

One of our early observations was that the induction of ferritin H occurred without any coordinate effect on ferritin L mRNA or protein. In this regard, cytokines appear to differ from iron, which has long been known to elevate synthesis of both ferritin subunits through translational mobilization of stored mRNA.[48] This led us to hypothesize that induction of ferritin H synthesis in the mesencyhymal cell lines we were studying might occur by a different mechanism than "classical" iron-mediated ferritin induction. In support of this conclusion, we observed that cytokine-dependent induction of ferritin H appeared independent of cellular iron levels, and occurred in the presence of actinomycin D.[49] More recently, transcriptional "run-off" experiments have demonstrated directly that TNF-α leads to a stimulation of ferritin H transcription (E. Kwak and F. Torti, unpublished observations).

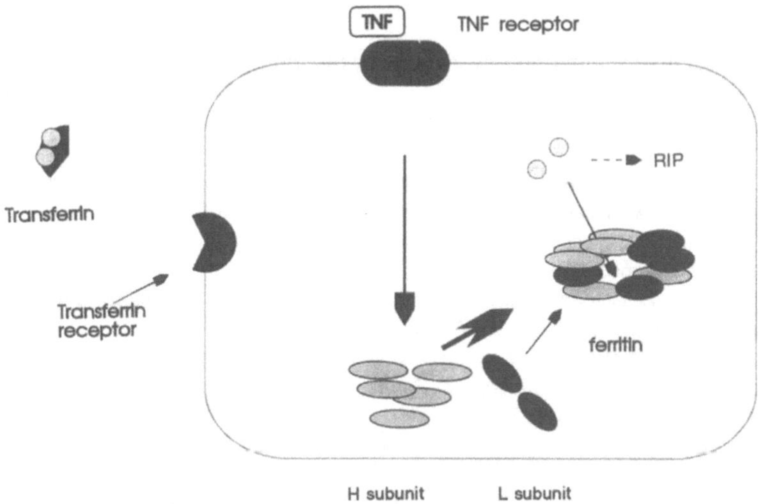

Figure 1. Cytokines modulate iron homeostasis: TNF-α (or IL-1α) binding to its receptor(s) induces a selective augmentation of ferritin H mRNA, altering the ratio of H to L mRNAs in target cells. For a given iron-mediated translational stimulus, the resultant newly synthesized ferritin proteins will have a relative preponderance of H-rich subunits compared to control cells not exposed to TNF-α. RIP = regulatory iron pool.

However, in other tissues, particularly liver, cytokines may affect ferritin by alternative, non-transcriptional mechanisms. Rogers and co-workers[50] have shown that in cultured hepatocytes, treatment with IL-1β induces ferritin H and L synthesis, and that this likely occurs at the translational level.

The changes in ferritin metabolism following cytokine treatment described above have a measurable impact on cellular iron balance. Following the early increase in ferritin H observed in normal human diploid fibroblasts treated with TNF-α or IL-1 α, we demonstrated a subsequent increase in synthesis of the transferrin receptor.[51] Since it is known that ferritin mlecules rich in H subunits take up and release iron more readily than L-rich feritins, this observation suggests that the alteration in ferritin composition that ensues following TNF-α or IL-1 α treatment is sufficient to deplete the "regulatory pool" of intracellular iron, leading to increased synthesis of transferrin receptor in much the same way that iron chelators induce transferrin receptor synthesis. However, confirmation of this hypothesis awaits more direct measurements of intracellular iron levels following TNF-α treatment.

What is the cellular rationale underlying the increase in ferritin H in response to TNF-α? One clue may lie in the response of L929 cells to TNF-α. Unlike the normal cell lines discussed above, these transformed, tumorigenic fibroblasts exhibit a cytotoxic response to TNF-α. In this cell line, treatment with TNF-α does not induce ferritin H synthesis.[52] This raises the possibility that increased levels of ferritin H serve as one of the protective mechanisms employed by normal cells to protect themselves from injury by this endogenous cytokine. This hypothesis is consistent with the role of iron in hydroxyl radical formation, and with the role of oxidative injury in cytotoxicity mediated by TNF-α. Experiments measuring the response to TNF-α in cells in which ferritin H synthesis has been elevated by transfection should allow this idea to be tested directly.

Acknowledgments

This research was supported by American Institute for Cancer Research Grant 87A104.

References

1. J.V. Selby, G.D. Friedman, Epidemiologic evidence of an association between body iron stores and risk of cancer, *Int J Cancer* 41:677-682 (1988).
2. R.G. Stevens, R.P. Beasley, B.S. Blumberg, Iron-binding proteins and risk of cancer in Taiwan, *JNCI* 76:605-610 (1989).
3. R.G. Stevens, D.Y. Jones, M.S. Micozzi, P.R. Taylor, Body iron stores and the risk of cancer, *N Engl J Med* 319:1047-1052 (1988).
4. J.B. Neilands, Parallels in the mode of regulation of iron assimilation in all living species, *in*: *Iron Transport and Storage*, P. Ponka, H.M. Schulman, and R.C. Woodworth, eds., CRC Press, Boca Raton (1990).
5. J.D. Cook, Evaluation of the iron status of a population, *Blood* 48:449-455 (1976).
6. R.B. Lauffer, Iron, aging and human disease: Historical background and new hypotheses, *in*: *Iron and Human Disease*, R.B. Lauffer, ed., CRC Press, Boca Raton (1992).
7. V.R. Fairbanks, E. Beutler, Iron metabolism, *in*: *Hematology*, 2nd ed, W.J. Williams, E. Beutler, A.J. Erslev, and R.W. Rundles, eds., McGraw-Hill, New York (1977).
8. R.G. Stevens, S. Kuvibidila, M. Kapps, J. Friedlaender, B.S. Blumberg, Iron-binding proteins, hepatitis B virus, and mortality in the Solomon Islands, *Am J Epidemiol* 118:550-561 (1983).
9. R.J. Bergeron, R.R. Streiff, G.T. Elliot, Influence of iron on *in vivo* proliferation and lethality of L1210 cells, *J Nutr* 115:369-374 (1985).
10. H.W.L. Hann, M.W. Stahlhut, B.S. Blumberg, Iron, nutrition and tumor growth: Decreased tumor growth in iron-deficient mice, *Cancer Res* 48:4168-4170 (1988).
11. H.W.L. Hann, M.W. Stahlhut, H. Menduke, Iron enhances tumor growth, *Cancer* 68:2407-2410 (1991).
12. H.J. Thompson, K. Kennedy, M. Witt, J. Juzefyk, Effect of dietary iron deficiency or excess on the induction of carcinogenesis by 1-methyl-1-nitrosourea, *Carcinogenesis* 12:111-114 (1991).
13. C.P. Siegers, D. Bumann, G. Baretton, M. Younes, Dietary iron enhances the tumor rate in dimethylhydrazine-induced carcinogenesis in mice, *Cancer Lett* 41:251-256 (1988).
14. A.G. Smith, J.R.P. Cabral, P. Carthew, J.E. Francis, M.M. Manson, Carcinogenicity of iron in conjunction with a chlorinated environmental chemical, hexachlorobenzene, in C57BL/10ScSn mice, *Int J Cancer* 43:492 (1989).
15. E.D. Weinberg, Iron in neoplastic disease, *Nutr Canc* 4:223-233 (1983).
16. A. Donfrancesco, G. Deb, C. Dominici, D. Pileggi, M.A. Castello, L. Helson, Effects of a single course of deferoxamine in neuroblastoma patients, *Cancer Res* 50:4929-4930 (1990).
17. Z. Estrov, A. Tawa, X. Wang, I. Dube, H. Sulh, A. Cohen, E.W. Gelfand, M.H. Freedman, *In vitro* and *in vivo* effects of deferoxamine in neonatal leukemia, *Blood* 69:757 (1987).
18. H.M. Vriesendorp, J.M. Herpst, M.A. Germack, J.L. Klein, P.K. Leichner, D.M. Loudenslager, S.E. Order, Phase I-II studies of Yttrium-labeled antiferritin treatment for end-stage Hodgkin's disease, including radiation therapy oncology group 87-01, *J Clin Oncol* 9:918 (1991).
19. S.E. Order, J.L. Klein, P.K. Leichner, J. Frincke, C. Lollo, D.J. Carlo, ^{90}Yttrium antiferritin—a new therapeutic radiolabeled antibody, *Int J Radiat Oncol Biol Phys* 12:277 (1986).
20. B.J. Foster, K. Clagett-Carr, D. Hoth, B. Leyland-Jones, Gallium nitrate: the second metal with clinical activity, *Cancer Treat Rep* 70:1311 (1986).
21. W.R. Bezwada, D.P. Derman, T.H. Bothwell, R.Baynes, C. Hesdorffer, A.P. MacPhail, Serum ferritin and Hodgkin's disease, *Scand J Haematol* 35:505-10 (1985).

22. A. Jacobs, A. Slater, J.A. Whittaker, G. Canellos, P.H. Wiernak, Serum ferritin concentration in untreated Hodgkin's disease, *Br J Cancer* 34:162-166 (1976).
23. H.W.L. Hann, B. Lange, M.W. Stahlhut, K.A. McGlynn, Prognostic importance of serum transferrin and ferritin in childhood Hodgkin's disease, *Cancer* 66:313-316 (1990).
24. Y. Matzner, A.M. Konijn, C. Hershko, Serum ferritin in hematologic malignancies, *Am J Hematol* 9:13-22 (1980).
25. D.H. Parry, M. Worwood, A. Jacobs, Serum ferritin in acute leukaemia at presentation and during remission, *Br Med J* 1:245-247 (1975).
26. P.A.E. Jones, F.M. Miller, M. Worwood, A. Jacobs, Ferritinaemia in leukaemia and Hodgkin's disease, *Br J Cancer* 27:212-217 (1973).
27. S. Gupta, A. Imam, K. Licorish, Serum ferritin in acquired immune deficiency syndrome, *Clin Lab Immunol* 20:11 (1986).
28. C. Moroz, S.L. Misrock. F.P. Siegal, Isoferritins in HIV infection: Relation to clinical stage, CD8 lymphocyte binding and the pathogenesis of AIDS, *AIDS* 3:11-16 (1989).
29. M.A. Siimes, W.C. Wang, P.R. Dallman, Elevated serum ferritin in children with malignancies, *Scand J Haematol* 19:153-158 (1977).
30. D. Potaznik, S. Groshen, D. Miller, R. Bagin, R. Bhalla, M. Schwartz, M. de Sousa, Association of serum iron, serum transferrin saturation, and serum ferritin with survival in acute lymphocytic leukemia, *Amer J Pediatr Hematol/Oncol* 9:350-355 (1987).
31. H.W. Hann, C.Y. Kim, T. London, B.S. Blumberg, Increased serum ferritin in chronic liver disease: A risk factor for primary hepatocellular carcinoma, *Int J Can* 43:376-379 (1989).
32. M. Tatsuta, H. Yamamura, H. Iishi, H. Kasugai, S. Okuda, Value of serum alpha-fetoprotein and ferritin in the diagnosis of hepatocellular carcinoma, *Oncology* 43:306-310 (1986).
33. N. Nagasue, H. Yukaya, Y.-C. Chang, Y. Ogawa, Serum ferritin level after resection of hepatocellular carcinoma, *Cancer* 57:1820-1823 (1986).
34. D. Basso, C. Fabris, G. Del Favero, T. Meggiato, M. Panozzo, D. Vianello, M. Plebani, R. Naccarato, Hepatic changes and serum ferritin in pancreatic cancer and other gastrointestinal diseases: The role of cholestasis, *Ann Clin Biochem* 28:34-38 (1991).
35. D. Nitti, C. Fabris, G. Del Favero, Serum ferritin in pancreatic disease. An accurate test of malignancy? *Digestion* 25:258-262 (1982).
36. C. Fabris, R. Farini, G. Del Favero, Combined evaluation of serum ribonuclease and ferritin: Any advantages in pancreatic cancer diagnosis? *Oncology* 41:393-395 (1984).
37. C. Fabris, R. Farini, A. Piccoli, CEA and ferritin in chronic pancreatic disease: A comparative evaluation, *Hepatogastroenterology* 32:146-148 (1985).
38. A. Esen, H. Ozen, A. Ayhan, A. Ergen, C. Tasar, F. Remzi, Serum ferritin: A tumor marker for renal cell carcinoma, *J Urol* 145:1134-1137 (1991).
39. P. Gaglia, B. Caldarola, R. Bussone, F. Potente, D. Lauro, A. Jayme, L. Caldarola, Prognostic value of CEA and ferritin assay in breast cancer; a multivariate analysis, *Eur J Cancer Clin Oncol* 24:1151-1155 (1988).
40. M.R. Williams, A. Turkes, D. Pearson, K. Griffiths, R.W. Blamey, Serum ferritin as a marker of therapeutic response in stage III and IV breast cancer, *Eur J Surg Oncol* 16:22-27 (1990).
41. R. Cox, O.H. Gyde, M.J. Leyland, Serum ferritin levels in small cell lung cancer, *Eur J Cancer Clin Oncol* 22:831 (1986).
42. S. Saji, Y. Yokoyama, H. Niwa, H. Takao, H. Kida, R. Kawata, H. Tanemura, K. Sakata, Clinical studies on serum immunosuppressive acidic protein (IAP) and ferritin in gastric cancer patients: With special reference to preoperative value and influence of surgical stress, *J Surg Oncol* 33:215-222 (1986).
43. M. Akiba, K. Neriishi, W.J. Blot, H. Kabuto, R.G. Stevens, H. Kato, C.E. Land, Serum ferritin and stomach cancer risk among a Japanese population, *Cancer* 67:1707 (1991).
44. F.M. Torti, B. Dieckmann, B. Beutler, A. Cerami, G.M. Ringold, A macrophage factor inhibits adipocyte gene expression: An *in vitro* model of cachexia, *Science* 229:867-869 (1985).

45. S.V. Torti, E. Kwak, S.C. Miller, G. Ringold, K. Myambo, F.M. Torti, The molecular cloning and characterization of murine ferritin heavy chain, a tumor necrosis factor-inducible gene, *J Biol Chem* 263:12638-12644 (1988).
46. S.C. Miller, H. Ito, H.M. Blau, F.M. Torti, Tumor necrosis factor inhibits human myogenesis *in vitro*, *Molec Cell Biol* 8:2295-2301 (1988).
47. Y. Wei, S.C. Miller, Y. Tsuji, S.V. Torti, F.M. Torti, Interleukin 1 induces ferritin heavy chain in human muscle cells, *Biochem Biophys Res Commun* 169:289-296 (1990).
48. J. Zahringer, B.S. Baliga, H.N. Munro, Novel mechanism for translational control in regulation of ferritin synthesis by iron, *Proc Natl Acad Sci USA* 73:857-861 (1976).
49. L.L. Miller, S.C. Miller, S.V. Torti, Y. Tsuji, F.M. Torti, Iron-independent induction of ferritin-H chain by tumor necrosis factor, *Proc Natl Acad Sci USA* 88:4946-4950 (1991).
50. J.T. Rogers, K.R. Bridges, G.P. Durmowicz, J. Glass, P.E. Auron, H.N. Munro, Translational control during the acute phase response, *J Biol Chem* 265:14572-14578 (1990).
51. Y. Tsuji, L.L. Miller, S.C. Miller, S.V. Torti, F.M. Torti, Tumor necrosis factor-α and interleukin-1-α regulate transferrin receptor in human diploid fibroblasts. Relationship to the induction of ferritin heavy chain, *J Biol Chem* 266:7257-7261 (1991).
52. S. Warren, S.C. Miller, S.V. Torti, Y. Tsuji, F.M. Torti, The role of iron in the cytotoxicity of tumor necrosis factor, *Lymphokine Cytokine Res* 12:75 (1993).

Chapter 13

Evidence for Nutrient Modulation of Tumor Phenotype: Impact of Tyrosine and Phenylalanine Restriction

CATHERINE A. ELSTAD, GARY G. MEADOWS, CHERYL J. ASLAKSON and JEAN R. STARKEY

I. Introduction

The link between nutrition and cancer is well documented, but also complex and oftentimes confusing. Nutritional factors are implicated in cancer progression, cancer prevention, and chemotherapy. Carbohydrates,[1,2] fats,[3-7] vitamins,[8-10] and other trace elements[11-13] alter not only tumor phenotype—including growth,[14] morphology,[15] membrane characteristics,[15,16] and metastatic potential[17,18]—but also host responses to tumor.[19-21] Dietary manipulation of amino acids, which exploits differences in nutritional requirements of tumor cells compared with normal cells,[14,15,22] can yield antitumor activity. For example, methionine dependence of several tumor systems has been utilized to preferentially inhibit tumor growth.[22-27] Additionally, L-asparaginase has achieved clinical success as an antileukemic drug.[28-30]

Our research has focused on the impact of two amino acids, tyrosine (Tyr) and phenylalanine (Phe), on pigmented melanomas, which have a dual requirement for these amino acids as precursors for protein synthesis and melanin production. The growth inhibitory activity of *in vivo* restriction of Tyr and Phe on pigmented S91 melanomas was first shown by Demopoulos in the 1960s.[31] Since then, the effectiveness of Tyr and Phe restriction against other murine and human tumor systems has been documented.[32-36]

Catherine A. Elstad and Gary G. Meadows • Department of Pharmaceutical Sciences, College of Pharmacy, and the Pharmacology/Toxicology Graduate Program, Washington State University Pullman, WA 99164; Cheryl J. Aslakson and Jean R. Starkey • Department of Microbiology, Montana State University, Bozeman, MT 59717

We are thoroughly investigating the antitumor effects of Tyr and Phe restriction, and previously have shown that dietary [37,38] or enzymatic[39,43] restriction of Tyr and Phe reduces plasma levels of these amino acids. Under these conditions, melanoma growth[37,38,40-42] is suppressed, survival of tumor-bearing mice is increased[37,38,44,45] and the effectiveness of levodopa against melanoma is augmented.[38,45] Metastasis, the more important clinical problem, is dramatically suppressed.[44,46-48] We have repeatedly shown that spontaneous metastasis of the highly invasive and metastatic B16-BL6 (BL6) murine melanoma is inhibited.[44,46-48] These antimetastatic effects of Tyr and Phe restriction are not limited to melanoma and are evident in other nonmelanoma tumor systems, such as Lewis lung carcinoma and RT74bs hepatocarcinoma[46] and L1210 leukemia.[35] Additionally, exposure of BL6 to low levels of Tyr and Phe *in vivo* and *in vitro* prior to intravenous inoculation into mice fed normal diet suppresses lung tumor colony formation and metastatic heterogeneity.[44,47,48] These antimetastatic effects are immediate, occurring after only one subcutaneous passage in mice fed diet low in Tyr and Phe[47] and after one *in vitro* passage in Tyr- and Phe-restricted medium.[48] Although stable *in vitro* for at least five passages,[48] the metastatic potential of the Tyr- and Phe-modulated tumor is unstable after prolonged *in vivo* passage in mice fed low Tyr and Phe diet and normal lung colonizing ability is regenerated after ten *in vivo* passages.[47] These antimetastatic effects are specific for Tyr and Phe and do not result from weight loss and sustained low body weight[44] or from restriction of other amino acids.[48]

Tyr and Phe restriction may induce changes in tumor phenotype directly, by impacting the tumor; indirectly, by modulating host factors that in turn affect tumor characteristics; or by affecting both tumor and host factors. In humans, Tyr and Phe restriction decreases platelet aggregation and platelet activating factor, increases natural killer, T-helper, and T-cytotoxic/suppressor lymphocyte numbers relative to neutrophils, and increases natural killer cell activity.[49] Conversely, in normal and *nude* mice, Tyr and Phe restriction impairs host cytotoxic T-cell-mediated immunity,[21] complement-independent cell-mediated cytotoxicity,[35] and natural killer cell activity.[50,51] Although these host immune responses are modulated by Tyr and Phe restriction, metastasis is still inhibited in mice. This evidence indicates that host immune responses may not play a significant role in the Tyr and Phe modulation of tumor phenotype and that altered tumor cell characteristics may be more important. We are presently investigating which tumor characteristics are modulated by Tyr and Phe restriction and which of these altered characteristics impacts metastatic phenotype. We have previously shown that Tyr and Phe modulation of metastasis is not due to differential tumor cell attachment to basement membrane components,[52] differential clearance or retention in the lung,[44,48] or decreased growth rate.[47] We are continuing to assess other mechanisms that may explain our observations. Data indicate that altered invasive capabilities,[52] protease activities, and growth factor production may by involved.

The antitumor effects of Tyr and Phe restriction most likely result from complex interactions between the tumor cell and host factors altered during amino acid restriction. In this chapter, we present additional evidence for the Tyr- and Phe-modulated suppression of metastasis, and demonstrate that these antimetastatic effects can result from a direct modulation of tumor cell phenotype, which is independent of modulated host effects.

II. Materials and Methods

A. Tumor Cells and Culture Conditions

The well characterized B16-BL6 (BL6) murine melanoma,[53,54] which is highly metastatic and highly invasive, was routinely cultured *in vitro* as a monolayer on tissue culture plastic in Dulbecco's modified Eagle's complete minimal essential medium supplemented with 10%

heat-inactivated fetal bovine serum (FBS), sodium pyruvate, nonessential amino acids, two-fold vitamin solution, L-glutamine, 100 U/ml penicillin, and 100 µg/ml streptomycin (DME). Cultures were incubated in a humidified atmosphere of 5% CO_2/95% air at 37°C. Cells were harvested with 4 mM EGTA to avoid the membrane-altering effects of trypsin. Cell viability was evaluated by trypan blue exclusion.

As described in greater detail below, *in vivo* and *in vitro* metastatic variants of BL6 were isolated during Tyr and Phe restriction. *In vivo* Tyr and Phe variants of BL6 were obtained from tumor-bearing mice fed one of two Tyr- and Phe-defined diets. *In vitro* Tyr and Phe variants of BL6 were obtained by passaging BL6 in Tyr- and Phe-defined media formulated from MEM Selectamine Kits.

B. Animals

Specific-pathogen-free, female B6D2F1 mice were purchased from Jackson Laboratories (Bar Harbor, ME) or Simonsen Laboratories (Gilroy, CA) at 6-8 weeks of age. Mice were housed singly or in groups of 5 mice/polycarbonate cage in the Wegner Hall Vivarium, which is accredited by the American Association for Accreditation of Laboratory Animal Care, at Washington State University. BALB/c *nude* and w/w^v mast cell-deficient mice were bred from colonies maintained at the Montana State University Animal Facility.

C. *In Vivo* Tyr and Phe Restriction and Isolation of *In Vivo* BL6 Variants

Mice were fed one of two fully characterized, defined crystalline amino acid diets (BioServ, Inc., Frenchtown, NJ), which differ in their content of Tyr and Phe,[43] for approximately two weeks and subsequently inoculated with BL6. The nutritionally complete diet (normal diet) contains 0.3% Tyr and 0.6% Phe. Approximately ten times lower in Tyr and Phe, the low Tyr and Phe diet contains 0.04% Tyr and 0.08% Phe. Both diets are isocaloric (~4 cal/g) and are consumed equally (~3 g/day). Mice consuming either diet are healthy.[44]

Mice were inoculated either subcutaneously into the pinna of the ear with 5×10^4 viable cells in 20 µl Ca^{2+}- and Mg^{2+}-free buffered saline solution (CMF) or subcutaneously into the dorsal hip with 1×10^6 viable BL6 cells in 200 µl CMF. Following injection into the pinna, BL6 variants were isolated from spontaneous cervical lymph node metastases and individual spontaneous lung metastases, as described previously.[44] Subcutaneous, dorsal hip, primary tumors were excised, enzymatically digested, and cultured *in vitro* as monlayers on plastic.[47] Some experiments utilized BL6 serially transplanted as a subcutaneous tumor for 10 *in vivo* passages in mice fed normal diet or a diet low in Tyr and Phe.[47] BL6 variants isolated from mice fed normal diet are designated "ND"; variants isolated from mice fed low Tyr and Phe diet are designated "LTP".

D. *In Vitro* Tyr and Phe Restriction and Isolation of *In Vitro* BL6 Variants

BL6 was cultured *in vitro* in media restricted in Tyr and Phe. Media were formulated using MEM Selectamine Kits (GIBCO Laboratories, Grand Island, NY) supplemented with FBS, which was dialysed to eliminate free amino acids. Both Tyr and Phe were restricted to either 8 µg/ml or to 4 µg/ml of medium. A level of 8 µg/ml for these two amino acids corresponds to plasma levels of Tyr and Phe in mice fed the low Tyr and Phe diet[37] and approximately equals a 75-78% decrease compared to normal levels found in serum-free DME. Levels of 4 µg/ml correspond to decreases of 88-89% for both amino acids, compared to normal levels found in DME. BL6 variants isolated in medium containing 8/8 µg/ml Tyr/Phe are designated "LTP 8/8." "LTP 4/4" denotes those variants of BL6 grown in medium containing 4/4 µg/ml Tyr/Phe.

E. Lung Colonizing Ability (Figure 1)

Experimental metastatic potential of tumor cells was determined as described previously.[41,44,46-48] Briefly, unanesthetized mice were inoculated intravenously into the dilated lateral tail vein with 2×10^4 viable tumor cells in 200 µl CMF. After approximately 3 weeks, mice were killed and examined for pulmonary and extrapulmonary metastatses. Lungs were removed and fixed in Bouin's solution to facilitate visualization of superficial tumor foci, which were counted twice with the aid of a dissecting microscope.

F. Survival of Mice Inoculated with Metastatic BL6 Melanoma (Table 1)

To determine the effect of Tyr and Phe restriction on survival, mice were fed normal diet or diet low in Tyr and Phe and inoculated subcutaneously into the foot pad with 2.5×10^4 viable tumor cells in a volume of 20 µl CMF. Legs with primary tumors of approximately 1 cm in diameter were amputated at midfemur in mice under equithesian anesthesia. Most tumor-bearing legs were removed approximately 40 days after tumor inoculation. This surgery removes both the primary tumor and metastases to the popliteal lymph node and facilitates long-term survival study, since mice do not die from lymph node involvement. Mice were sacrificed when moribund. Survival was determined from the date of tumor injection to the date of sacrifice.

G. Effect of Tyr and Phe Restriction on Tumor Growth, Spontaneous Metastasis, Angiogenesis, and Mast Cell Numbers in *nude* Mice (Tables 2-6)

To eliminate the possibility that T-cell-dependent systems may be involved in Tyr and Phe modulation of tumor growth and metastasis, *nude* mice were fed normal diet or low Tyr and Phe diet and inoculated subcutaneously into the ventral surface of the pinna of the ear with 1×10^5 viable BL6 tumor cells. Primary tumors were measured twice weekly and removed by electrocautery surgery 26 days postinoculation. Mice were sacrificed 34 days after tumor inoculation. Cervical lymph node metastases were weighed upon necropsy of mice. Mice were also evaluated twice weekly for tumor-associated angiogenesis, as described previously.[55] Extent of angiogenesis was scored on the scale of one "+" (minor blood vessel formation) to four "++++" (extensive blood vessel involvement). The number of mast cells per 1-mm length of skin overlying the primary tumor was determined by astra blue/eosin staining.[55]

H. Effect of Tyr and Phe Restriction on Tumor Growth and Angiogenesis in Mast Cell-Deficient w/w^v Mice (Tables 7-8)

To determine whether Tyr- and Phe-suppressed angiogenesis is mast-cell mediated, littermate w/w^v and control +/+ mice were fed either normal diet or low Tyr and Phe diet for 3 weeks prior to inoculation of 1×10^5 BL6 tumor cells into the pinna of the ear. Tumor growth and angiogenesis were determined as described above.

I. Involvement of Natural Killer (NK) Cells in Tyr and Phe Suppression of Lung-Colonizing Ability (Table 9)

To determine whether NK cells play a role in the Tyr/Phe suppression of experimental metastatic potential, lung-colonizing abilities of ND and LTP *in vivo* variants in normal mice fed low Tyr and Phe diet were determined. Mice were fed low Tyr and Phe diet 14 days prior to tumor inoculation and experimental metastatic potential of tumor cells was determined as described above. We have previously shown that mice fed low Tyr and Phe diet have decreased NK activities compared to mice fed normal diet.[50,51]

Table 1. Survival of Mice Fed Normal Diet or Low Tyr and Phe Diet and Inoculated Subcutaneously into the Footpad with BL6

Dietary treatment	N	Survival (days) Median (range)	Percent increase in survival[a]
Normal diet	13	64 (43-126)	
Low Tyr and Phe diet	15	126 (59-126)[b]	97%

[a] Percent increase = [(median survival on low Tyr and Phe diet — median survival on normal diet)/median survival on normal diet] × 100.
[b] Median significantly different from normal diet treatment ($p < 0.05$, Wilcoxon 2-sample test).

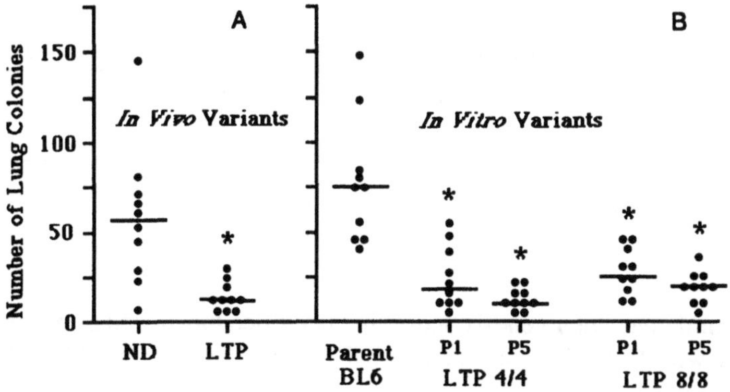

Figure 1. Lung-colonizing potentials of B16-BL6 and Tyr- and Phe-modulated *in vivo* and *in vitro* variants of B16-BL6. Each dot represents the number of pulmonary metastases in one mouse fed normal diet and inoculated intravenously with 2×10^4 viable tumor cells. Heavy lines represent the median number of lung colonies from 10 mice for each variant. Differences in lung colonization between the ND and LTP *in vivo* variants or between parent BL6 and the *in vitro* variants are statistically significant at $p < 0.05$ (Wilcoxon 2-sample test) and are designated by an asterisk.

Table 2. Volumes (mm^3) of Primary Tumors in *nude* Mice Fed Normal Diet or Low Tyr and Phe Diet

	Days after tumor inoculation			
Dietary treatment	16	20	23	26
Normal diet	3.21 ± 2.2	7.40 ± 5.9	17.40 ± 17.4	47.59 ± 61.1
Low Tyr and Phe diet	2.23 ± 2.0	3.85 ± 4.1	3.87 ± 4.3	3.47 ± 3.1

Table 3. Weight of Cervical Lymph Node Metastases following Subcutaneous Inoculation of B16-BL6 into the Ear of *nude* Mice Fed Normal Diet or Low Tyr and Phe Diet

Dietary treatment	Percentage of mice with lymph node metastases	Weight of cervical lymph node (g ± SD)
Normal diet	70%	2.55 ± 1.54
Low Tyr and Phe diet	70%	0.8 4 ± 0.44[a]

[a] Significantly different ($p < 0.05$) than normal diet treatment.

Table 4. Spontaneous Metastasis to the Lung of B16-BL6 following Subcutaneous Inoculation into the Ear of *nude* Mice Fed Normal Diet or Low Tyr and Phe Diet

Dietary treatment	Total number of lung metastases	Percentage of mice with tumor	Average number of metastases/mouse
Normal diet	280	90%	31
Low Tyr and Phe diet	17	60%	3[a]

[a] Significantly different ($p < 0.05$) from normal diet treatment.

Table 5. Cumulative Angiogenesis Scores in *nude* Mice Inoculated Subcutaneously into the Ear with B16-BL6 and Fed Normal Diet or Low Tyr and Phe Diet

	Days after tumor inoculation		
Dietary treatment	16	20	26
Normal diet	6	7	13
Low Tyr and Phe diet	1	1	1

Table 6. Number of Mast Cells per 1-mm Length of Skin Overlying the Primary Ear Tumor in *nude* Mice Fed Normal Diet or Low Tyr and Phe Diet

Dietary treatment	N	Number of mast cells (mean ± SD)
Normal diet	10	22 ± 3.7
Low Tyr and Phe diet	7	1 1 ± 3.4[a]

[a] Significantly different ($p < 0.05$) from normal diet.

Table 7. Volume (mm^3) of Primary Tumors in Littermate w/w^v and +/+ Mice Inoculated with B16-BL6 into the Pinna of the Ear and Fed Normal Diet or low Tyr and Phe Diet

			Days after tumor inoculation				
Group	Dietary treatment	N	8	12	16	20	23
w/w^v	Normal diet	9	3.8	10.0	63.4	207.7	347.0
w/w^v	Low Tyr and Phe diet	9	0.6	1.9	8.2	29.3	89.0
+/+	Normal diet	9	3.0	16.1	92.6	326.0	834.8

Table 8. Cumulative Angiogenesis Scores in Littermate w/w^v and +/+ Mice Inoculated Subcutaneously into the Ear with B16-BL6 and Fed Normal Diet or Low Tyr and Phe Diet

		Days after tumor inoculation			
Group	Dietary treatment	12	16	20	23
w/w^v	Normal diet	5.5	8.5	14.5	18.5
w/w^v	Low Tyr and Phe diet	4.0	5.0	5.0	10.0
+/+	Normal diet	8.0	15.5	20.5	23.5

Table 9. Experimental Metastatic Potential in Mice Fed Low Tyr and Phe Diet and Inoculated with ND and LTP *In Vivo* Variants

		Number of lung colonies	
In vivo variant	N	Median	Range
ND	10	166	42-205
LTP	10	63[a]	14-106

[a] Significantly different ($p < 0.05$) from ND variant by the Wilcoxon 2-sample test.

J. Attachment of Tumor Cells to Endothelium (Tables 10-11)

Attachment of Tyr- and Phe-modulated lymph node and lung colony metastases to bovine endothelial cell monolayers was determined by two methods. ND and LTP variants isolated from lymph nodes were seeded at 2.5×10^5 cells/ml into 75-cm^2 tissue culture flasks containing 3-day postconfluent bovine enthothelial cell monolayers and incubated at 37°C for 2 hr. Following incubation, cultures were gently agitated and washed with buffered saline. The number of unattached cells was enumerated. Lung colony variants were seeded at 1×10^4 cells/ml onto endothelium in 25-cm^2 tissue culture flasks and incubated at 37°C for 1 hr. Attachment was expressed as the number of cells attached per 10 microscopic fields.

Table 10. Attachment of B16-BL6 and Tyr- and Phe-Modulated Lymph Node Variants to Bovine Endothelium

Lymph node variant	Attachment (%)
B16-BL6	87.6 ± 2.19
ND	82.3 ± 1.69
LTP	80.8 ± 3.53

Table 11. Attachment to Bovine Endothelium of Tyr- and Phe-Modulated Lung Colony Variants with Different Experimental Metastatic Potentials[a]

Lung colony variant		Number of attached cells/10 microscopic fields[b]	Metastatic potential (median)
ND	1	301 ± 35	11
	2	226 ± 3	17
	3	289 ± 34	29
	4	450 ± 65	46
	5	394 ± 48	122
LTP	1	284 ± 38	6
	2	242 ± 42	10
	3	369 ± 33	10
	4	202 ± 47	13
	5	419 ± 6	43

[a] Experimental metastatic potentials of lung colony variants published elsewhere (see ref. 44).
[b] Numbers are averages of four counts.

K. Statistical Analyses

Differences in survival and lung-colonizing potentials were determined by the Wilcoxon 2-sample test. Differences in spontaneous metastasis and lymph node weights were determined by the Mann-Whitney U 2-tailed test.

III. Results

In vivo and *in vitro* restriction of Tyr and Phe suppresses the lung-colonizing ability of BL6 (Figure 1; see also refs. 44, 47, and 48). After one subcutaneous passage of BL6 in mice fed low Tyr and Phe diet, experimental metastatic potential is significantly reduced (Figure 1, Panel A). Similarly, passage of BL6 *in vitro* in media restricted in Tyr and Phe significantly suppresses lung colonization (Figure 1, Panel B). This suppression is evident after one *in vitro* passage and is stable for at least five passages *in vitro* in Tyr- and Phe-restricted media. This Tyr- and Phe-modulated suppression of lung colonization is also reflected in significantly increased median survival times of tumor-bearing mice fed low Tyr and Phe diet compared to mice fed normal diet (Table 1). Survival of mice fed low Tyr and Phe diet is increased 97%, from 64 days to 126 days.

Our results indicate that thymus-dependent immune responses are not involved in this observed Tyr- and Phe-mediated suppression of metastatic phenotype. Growth of primary tumors is slower in *nude* mice fed low Tyr and Phe diet compared to mice fed normal diet (Table 2), even though NK activity is reduced.[50,51] Twenty-six days after tumor inoculation, average tumor size in *nude* mice fed low Tyr and Phe diet is less than 10% of the tumor size measured in mice fed normal diet. While cervical lymph node involvement is not different between mice fed the two diets, lymph node tumor burden is less in *nude* mice fed low Tyr and Phe diet (Table 3). Tyr and Phe restriction further suppresses spontaneous metastasis to the lung (Table 4). *Nude* mice fed low Tyr and Phe diet have a 60% incidence of spontaneous lung metastases compared to a 90% incidence in the control group fed normal diet. Additionally, a 90% decrease in average number of lung metastases is evident in mice fed low Tyr and Phe diet. Tumor-associated angiogenesis is reduced in mice fed low Tyr and Phe diet (Table 5) and is accompanied by a 50% decease in mast cell numbers (Table 6).

Our research with mast cell-deficient mice indicates that the suppression of angiogenesis in *nude* mice fed low Tyr and Phe diet is not attributable to the lower numbers of mast cells in these mice. Tumor growth rate (Table 7) and tumor-associated angiogenesis (Table 8) are reduced in mast cell-deficient mice fed low Tyr and Phe diet compared to w/w^v mice and littermate +/+ mice fed normal diet, indicating that the effects of Tyr and Phe restriction are largely independent of mast cells.

The antimetastatic effects of Tyr and Phe restriction are likely not attributable to NK cells, as shown in Tables 2-4 and Table 9. Both *nude* mice and normal mice have decreased NK cell activity when fed the low Tyr and Phe diet.[50,51] Although lung colonization in mice fed low Tyr and Phe diet is greater compared to lung colonization in mice fed normal diet[47] due to decreased NK cell activity, the lung-colonizing potential of the LTP *in vivo* variant is suppressed compared to that of the ND *in vivo* variant, as shown in Table 9. The LTP variant forms 62% fewer lung tumor colonies after intravenous inoculation into mice fed low Tyr and Phe diet than does the ND variant.

Tyr and Phe modulation of the metastatic phenotype is not due to differences in attachment to endothelium (Tables 10 and 11). Approximately 80% of parent BL6 and ND and LTP lymph node variants seeded onto bovine endothelial cell monolayers attach (Table 10). Attachment of lung colony variants to endothelium is also not modulated by Tyr and Phe restriction (Table 11). Both ND and LTP lung colony variants demonstrate a heterogeneous response. Average numbers of ND lung colony variants attached range from 226-450 cells (range = 224); average numbers of attached LTP lung colony variants range from 202-419 cells (range = 217). Additionally, attachment of ND and LTP lung colony variants to endothelium is not associated with experimental metastatic potential.

IV. Discussion

Our research indicates that Tyr and Phe restriction suppresses the malignant phenotype of a variety of tumor systems, notably B16-BL6 murine melanoma.[44,47,48] We show in this report that *in vivo* and *in vitro* exposure of BL6 to low levels of Tyr and Phe suppresses metastatic heterogeneity and lung-colonizing ability of BL6 following intravenous inoculation into mice fed normal diet (Figure 1). Survival of mice bearing the metastatic BL6 variant of B16 melanoma and fed low Tyr and Phe diet is also increased (Table 1), confirming our results previously reported for mice bearing intraperitoneal parent B16 melanoma[38] and dorsal hip subcutaneous BL6 tumors.[45]

Tyr and Phe restriction may suppress metastasis and increase host survival by directly impacting the tumor, by indirectly modulating host response, or by affecting both tumor and host factors and their interaction. Our data strongly suggest that major antimetastatic effects of Tyr and Phe restriction are caused by direct effects on the tumor itself. We previously showed that Tyr and Phe restriction suppresses metastasis in the absence of host factors.[48] Prior exposure of BL6 to low levels of Tyr and Phe *in vitro* and *in vivo* suppresses lung-colonizing ability following intravenous inoculation into mice fed normal diet.[44,47,48] Evidence indicates that this suppressed metastatic potential is not due to changes in cell surface characteristics, since differences in attachment to endothelium (Tables 10 and 11) and basement membrane components such as laminin, fibronectin, collagen types I and IV, and Matrigel (data not shown; see also ref. 52) are not apparent.

These observations do not exclude involvement of host factors in the Tyr and Phe modulation of metastatic phenotype. Tyr and Phe restriction does modulate host responses by impairing cytotoxic T-cell-mediated immunity[21] and complement-independent cell-mediated cytotoxicity.[35] We have shown more recently that low levels of Tyr and Phe decrease NK activity in normal and *nude* mice.[50,51] Along with these data, the results presented in this report in *nude* and w/w^v mice support our hypothesis that the effect of Tyr and Phe restriction is independent of host immune responses.

T-cell-dependent, NK-dependent, and mast cell-dependent host immune responses do not play significant roles in the antimetastatic effect of Tyr and Phe restriction. Tumor growth (Tables 2 and 3) and spontaneous metastasis (Table 4) are inhibited in *nude* mice, which lack T lymphocytes. Tumor-associated angiogenesis is also inhibited in these mice (Table 5), which normally have large numbers of skin connective tissue mast cells to promote angiogenic responses. Although the number of mast cells is reduced in *nude* mice fed low Tyr and Phe diet (Table 6), the antiangiogenic effects of Tyr and Phe restriction are not mast cell mediated (Table 8). Tumor growth (Table 7) and angiogenesis are also inhibited in mast cell-deficient mice fed low Tyr and Phe diet compared to w/w^v mice fed normal diet. Mast cell-deficient mice are almost totally devoid of both mucosal and connective tissue mast cells. These results indicate that Tyr and Phe restriction may modulate the production of angiogenic factors by tumor cells and/or the response by the host to these factors. Recent studies indicate that Tyr- and Phe-modulated BL6 variants do not secrete basic fibroblast growth factor-like molecules or transforming growth factor β-like molecules.[56] These molecules have been implicated in angiogenesis.[57-59]

As mentioned previously, the antimetastatic effect of Tyr and Phe restriction most likely does not involve NK cells. Even though NK cell activity is decreased in mice fed low levels of Tyr and Phe,[50,51] metastasis is still suppressed in mice fed low Tyr and Phe diet (Table 9). In addition, we previously showed that Tyr- and Phe-modulated variants are not more sensitive to removal or clearance from the blood by NK cells.[44,48]

While Tyr and Phe restriction can suppress the metastatic phenotype by modulating tumor cell characteristics, this suppression of metastasis most likely involves a complex set of interactions among the amino acid restriction, the tumor, and the host environment. As the mechanisms for the antimetastatic effects of Tyr and Phe restriction are detailed, its potential as an adjuvant for effective cancer therapy will increase.

V. Summary

We have shown that Tyr and Phe restriction suppresses the malignant phenotype of the highly invasive and metastatic BL6 variant of B16 murine melanoma. Lung-colonizing abilities of Tyr- and Phe-modulated *in vivo* and *in vitro* variants of BL6 are inhibited following intravenous inoculation into mice fed normal diet. Although this antimetastatic effect of Tyr and Phe restriction is most likely not due to differences in attachment to endothelium, our data indicate that major impacts of Tyr and Phe restriction are at the level of the tumor, itself. Modulation of host immune responses, which in turn suppresses metastasis, does not appear to contribute significantly to the altered phenotype. Although numbers and function of T cells, mast cells, and NK cells are affected by Tyr and Phe restriction, they are not involved in the Tyr- and Phe-mediated suppression of tumor growth, metastasis, or angiogenesis. Our data do not rule out the importance of other host factors involved in the Tyr and Phe modulation of tumor phenotype. The outcome of this modulation results most likely from complex Tyr/Phe-tumor-host interactions.

Acknowledgments

These studies were supported in part by American Institute for Cancer Research grant 85B79 and National Cancer Institute grant CA42465.

References

1. G. E. Demetrakopoulos, M.F. Brennan, Tumoricidal potential of nutritional manipulations, *Cancer Res* 42:756s (1982).
2. A.E. Rogers, H.M. Nields, P.M. Newberne, Nutritional and dietary influences on liver tumorigenesis in mice and rats, *Arch Toxicol* 231:Suppl. 10 (1987).
3. S.H. Abou-El-Ela, K.W. Prasse, R.L. Farrell, R.W. Carroll, A.E. Wade, O.R. Bunce, Effects of D,L-2-difluoromethylornithine and indomethacin on mammary tumor promotion in rats fed high n-3 and/or n-6 fat diets, *Cancer Res* 49:1434 (1989).
4. D.F. Birt, L.T. White, B. Choi, J.C. Pelling, Dietary fat effects on the initiation and promotion of two-stage skin tumorigenesis in the SENCAR mouse, *Cancer Res* 49:4170 (1989).
5. E.B. Katz, E.S. Boylan, Effects of reciprocal changes of diets differing in fat content on pulmonary metastasis from the 13762 rat mammary tumor, *Cancer Res* 49:2477 (1989).
6. J. Silverman, J. Powers, P. Stromberg, J.A. Pultz, S. Kent, Effects on C3H mouse mammary cancer of changing from a high fat to a low fat diet before, at, or after puberty, *Cancer Res* 49:3857 (1989).
7. K. Sundram, H.T. Knor, A.S.H. Ong, R. Pathmanathan, Effect of dietary palm oils on mammary carcinogenesis in female rats induced by 7,12-dimethylbenz[*a*]anthracene, *Cancer Res* 49:1447 (1989).
8. N.S. Gardiner, J.R. Duncan, Inhibition of murine melanoma growth by sodium ascorbate, *J Nutr* 119:586 (1989).
9. R.C. Jones, S. Sugie, J. Braley, J.H. Weisburger, Dietary β-carotene in rat models of gastrointestinal cancer, *J Nutr* 119:508 (1989).
10. V. Noto, H.S. Taper, J. Yi-hua, J. Janssens, J. Bonte, W. DeLoecker, Effects of sodium ascorbate (vitamin C) and 2-methyl-1,4-naphthoquinone (vitamin K_3) treatment on human tumor cell growth *in vitro*, *Cancer* 63:901 (1989).
11. D. Medina, C.J. Oborn, Differential effects of selenium on the growth of mouse mammary cells *in vitro*, *Cancer Lett* 13:333 (1981).
12. U. Reinhold, H. Biltz, W. Bayer, K.H. Schmidt, Serum selenium levels in patients with malignant melanoma, *Acta Dermatovenereol (Stockholm)* 69:132 (1989).
13. J. van Eys, Nutrition and cancer: physiological interrelationships, *Ann Rev Nutr* 5:435 (1985).
14. T.J. Miller, R.S. Franco, W.T. Chance, J. Martelo, M.B. Popp, Amino acid requirements of a rat sarcoma as determined by a stem cell assay, *J Parenteral Enteral Nutr* 11:223 (1987).
15. J. Xiang, A.K. Kimura, *In vitro* modulation of the metastatic phenotype I. Analysis of differentiation forms of the B16 melanoma expressing Met-72 determinants and metastatic activity, *Clin Exp Met* 4:293 (1986).
16. M.K. Patterson, Uptake and utilization of amino acids by cell cultures, *in*: *Growth, Nutrition, and Metabolism of Cells in Culture*, G.H. Rothblast, V.J. Cristofalo, eds., Academic Press, New York (1972).

17. E.S. Boylan, L.A. Cohen, The influence of dietary fat on mammary tumor metastasis in the rat, *Nutr Cancer* 8:193 (1986).
18. R.F. Branda, J.J. McCormack, C.A. Perlmutter, L.A. Mathews, S.H. Robison, Effects of folate deficiency on the metastatic potential of murine melanoma cells, *Cancer Res* 48:4529 (1988).
19. G. Bounous, P.A.L. Kongshavn, The effect of dietary amino acids on immune reactivity, *Immunology* 35:257 (1978).
20. G. Bounous, C. Sandarangani, K.C. Pang, P.A.L. Kongshavn, Effect of dietary amino acids on tumor growth and cell mediated immune responses, *Clin Invest Med* 4:109 (1981).
21. D.G. Jose, R.A. Good, Quantitative effects of nutritional essential amino acid deficiency upon immune responses to tumor in mice, *J Exp Med* 137:1 (1973).
22. P.H. Stern, R.M. Hoffman, Enhanced *in vitro* selective toxicity of chemotherapeutic agents for human cancer cells based on a metabolic defect, *J Natl Cancer Inst* 76:629 (1986).
23. F. Breillout, M.F. Poupon, P. Blanchard, V. Lascaux, P. Echinard-Garin, M. Robert-Gero, Association of SIBA treatment and a Met-depleted diet inhibits *in vitro* growth and *in vivo* metastatic spread of experimental tumor cell lines, *Clin Exp Med* 6:3 (1988).
24. R.M. Hoffman, Methionine dependence in cancer cells—A review, *In Vitro* 18:421 (1982).
25. R.M. Hoffman, Altered methionine metabolism and transmethylation in cancer, *Anticancer Res* 5:1 (1985).
26. J.G. Judde, M. Ellis, P. Frost, Biochemical analysis of the role of transmethylation in the methionine dependence of tumor cells, *Cancer Res* 49:4859 (1989).
27. R.G. Liteplo, Altered methionine metabolism in metastatic variants of a human melanoma cell line, *Cancer Lett* 44:23 (1989).
28. B.L. Asselin, D. Ryan, C.N. Frantz, S.D. Bernal, P. Leavitt, S.E. Sallan, H.J. Cohen, *In vitro* and *in vivo* killing of acute lymphoblastic leukemia cells by L-asparaginase, *Cancer Res* 49:4363 (1989).
29. D.W. Ross, R.L. Capizzi, Differentiation versus cytoreduction during remission induction in acute nonlymphoblastic leukemia treated with sequential high-dose ara-C and asparaginase, *Cancer* 53:1651 (1984).
30. S.E. Sallan, S. Hitchcock-Bryan, R. Gelber, J.R. Cassady, E. Frei, III, D.G. Nathan, Influence of intensive asparaginase in the treatment of childhood non-T-cell acute lymphoblastic leukemia, *Cancer Res* 43:5601 (1983).
31. H.B. Demopoulos, Effects of low phenylalanine-tyrosine diets on S91 mouse melanomas, *J Natl Cancer Inst* 37:185 (1966).
32. C.W. Abell, D.S. Hodgins, W.J. Stith, An *in vivo* evaluation of the chemotherapeutic potency of phenylalanine ammonia-lyase, *Cancer Res* 33:2529 (1973).
33. A.B. Lorincz, R.B. Kuttner, M.B. Brandt, Tumor response to phenylalanine-tyrosine limited diets, *J Amer Dietetic Assoc* 54:198 (1969).
34. M.J. Pine, Effect of low phenylalanine diet on murine leukemia L1210, *J Natl Cancer Inst* 60:633 (1978).
35. M.J. Pine, Improved host defense against L1210 leukemia by deprivation of dietary phenylalanine, *Nutr Cancer* 3:94 (1981).
36. R.C. Theuer, Effect of essential amino acid restriction on the growth of female C57BL mice and their implanted BW10232 adenocarcinomas, *J Nutr* 101:223 (1971).
37. G.G. Meadows, D.E. Oeser, Response of B16 melanoma-bearing mice to varying levels of phenylalanine and tyrosine, *Nutr Rep Int* 28:1073 (1983).
38. G.G. Meadows, H.F. Pierson, R.M. Abdallah, P.R. Desai, Dietary influence of tyrosine and phenylalanine on the response of B16 melanoma to carbidopa-levodopa methyl ester chemotherapy, *Cancer Res* 42:3056 (1982).
39. G.W. Elmer, G.G. Meadows, C. Linden, J. DiGiovanni, J.S. Holcenberg, Influence of tyrosine phenol-lyase and phenylalanine ammonia-lyase on growth of B16 melanoma, *Pigment Cell* 2:339 (1976).
40. G.W. Elmer, C. Linden, G.G. Meadows, Influence of L-tyrosine phenol-lyase on the growth and metabolism of B16 melanoma, *Cancer Treat Rep* 63:1055 (1979).

41. G.G. Meadows, R.M. Abdallah, J.R. Starkey, Interaction between specific dietary factors and experimental chemotherapy of metastic melanoma, *Cancer Chemother Pharmacol* 16:229 (1986).
42. G.G. Meadows, J. DiGiovanni, L. Minor, G.W. Elmer, Some biological properties and an *in vivo* evaluation of tyrosine phenol-lyase on the growth of B16 melanoma, *Cancer Res* 36:167 (1976).
43. H.F. Pierson, G.G. Meadows, Properties of stearylamine liposomes containing tyrosine phenol-lyase, *Biochim Biophys Acta* 656:177 (1981).
44. C.A. Elstad, G.G. Meadows, R.M. Abdallah, Specificity of the suppression of metastatic phenotype by tyrosine and phenylalanine restriction, *Clin Exp Med* 8:393 (1990).
45. H.F. Pierson, G.G. Meadows, Sodium ascorbate enhancement of carbidopa-methyl ester antitumor activity against pigmented B16 melanoma, *Cancer Res* 43:2047 (1983).
46. R.M. Abdallah, J.R. Starkey, G.G. Meadows, Dietary restriction of tyrosine and phenylalanine: Inhibition of metastasis of three rodent tumors, *J Natl Cancer Inst* 78:759 (1987).
47. C.A. Elstad, G.G. Meadows, Phenotypic stability of B16-BL6 melanoma exposed to low levels of tyrosine and phenylalanine, *Anticancer Res* 10:1313 (1990).
48. C.A. Elstad, G.G. Meadows, Modulation of B16-BL6 murine melanoma metastatic phenotype by tyrosine and phenylalanine restriction in the absence of host selection pressures, *Anticancer Res* 13:523 (1993).
49. J.R. Norris, G.G. Meadows, L.K. Massey, J.R. Starkey, D.M. Sylvester, S.-Y. Liu, Tyrosine- and phenylalanine-restricted formula diet augments immunocompetence in healthy humans, *Am J Clin Nutr* 51:188 (1990).
50. R.M. Abdallah, J.R. Starkey, G.G. Meadows, Alcohol and related dietary effects on mouse natural killer-cell activity, *Immunology* 50:131 (1983).
51. G.G. Meadows, R.M. Abdallah, J.R. Starkey, C.J. Aslakson, Response of natural killer cells from dietary tyrosine and phenylalanine restricted mice to biological response modifiers, *Cancer Immunol Immunother* 26:67 (1988).
52. N.P. Robertson, J.R. Starkey, S. Hamner, G.G. Meadows, Tumor cell invasion of three-dimensional matrices of defined composition: Evidence for a specific role for heparan sulfate in rodent cell lines, *Cancer Res* 49:1816 (1989).
53. I.R. Hart, The selection and characterization of an invasive variant of the B16 melanoma, *Am J Pathol* 97:587 (1979).
54. G. Poste, J. Doll, I.R. Hart, I.J. Fidler, *In vitro* selection of murine B16 melanoma variants with enhanced tissue-invasive properties, *Cancer Res* 40:1636 (1980).
55. J.R. Starkey, P.K. Crowle, S. Taubenberger, Mast-cell-deficient w/w^v mice exhibit a decreased rate of tumor angiogenesis, *Int J Cancer* 42:48 (1988).
56. V.D. Blanckaert, M.E. Schelling, C.A. Elstad, G.G. Meadows, Differential growth factor production, secretion and response by high and low metastatic variants of B16BL6 melanoma, *Cancer Res* 53:4075 (1993).
57. D. Benharrock, D. Birnbaum, Biology of the fibroblast growth factor gene family, *Israel J Med Sci* 26:212 (1990).
58. J. Folkman, M. Klagsbrun, Angiogenic factors, *Science* 235:442 (1987).
59. V. Mahadevan, I.R. Hart, Metastasis and angiogenesis, *Acta Oncologica* 29:97 (1990).

Chapter 14

Possible Role of Neuropeptide Y in Experimental Cancer Anorexia

WILLIAM T. CHANCE, AMBIKAIPAKAN BALASUBRAMANIAM, SULAIMAN SHERIFF and JOSEF E. FISCHER

I. Introduction

Anorexia and the concomitant depletion of host tissue (cachexia) are commonly observed in cancer patients.[1] As many as 75% of cancer patients will exhibit anorexia,[2] and this was recognized as early as 1930 as a major contributor to death of cancer patients.[3] These aspects of cancer present very difficult problems for both the patient and physician in that the aggressive use of various therapies is severely limited in the cachectic patient.[4] In addition, the development of anorexia in cancer patients is associated with increased morbidity and mortality.[5]

Although anorexia may be caused by direct invasion of the tumor tissue into the digestive system, the incidence of anorexia in patients with primary tumors located in nondigestive sites suggests that cancer need not invade the gastrointestinal (GI) system to produce anorexia. Various experimental studies in animals suggest humoral or chemical mediation of anorexia, since food intake returns rapidly toward normal following tumor resection.[6] In addition, when tumor-bearing (TB) and normal rats are connected to allow sharing of circulating blood, a degree of anorexia is transferred to the normal rats.[7] This observation suggests that a circulating anorexigenic factor is either secreted by the tumor or is formed due to tumor-induced biochemical alterations in host metabolism. Although the search for an anorexigenic factor in cancer anorexia has produced many candidates, no single factor that can explain the anorexia adequately has been identified. Thus, circulating levels of cachectin are not elevated until well after the onset of anorexia.[8] Tumor-bearing rats also respond normally to the anorectic agent cholecystokinin (CCK)[9] and exhibit reduced levels of CCK in brain and plasma.[10] Anorectic TB rats also feed normally following the injection

---

William T. Chance • Department of Surgery, University of Cincinnati Medical Center, Veterans Affairs Medical Center, Cincinnati, OH 45267; Ambikaipakan Balasubramaniam, Sulaiman Sheriff, and Josef E. Fisher • Department of Surgery, University of Cincinnati Medical Center, Cincinnati, OH 45267

Diet and Cancer: Markers, Prevention, and Treatment
Edited by M.M Jacobs, Plenum Press, New York, 1994

of norepinephrine into the hypothalamus,[11] the systemic injection of insulin,[12] exposure to cold,[2] and lesioning of the ventromedial hypothalamic area.[2] Recent reports suggest increased turnover of serotonin and dopamine in several brain regions of anorectic TB rats.[6,13] The importance of these changes in neurochemistry for the development of anorexia are questionable, however, since their depletion or receptor blockade did not alleviate the anorexia.[14,15] Therefore the central nervous system molecular mechanism of cancer anorexia remains elusive.

Recent feeding research has been dominated by reports of the potency of neuropeptide Y (NPY) as a feeding stimulus. Neuropeptide Y is a 36 amino acid-residue peptide of the pancreatic polypeptide family[16] that is found co-localized with norepinephrine throughout the central and peripheral nervous system.[17,18] Following the intrahypothalamic (int) injection of as little as 500 ng of NPY, rats will consume over 50% of their daily intake within 60 min.[19] In addition, the chronic injection (3 × day) of NPY led to hyperphagia and weight gain in normal rats.[20] Thus, NPY is the most potent orexigenic agent examined experimentally to date. Therefore, in the present experiments we investigated the feeding response of TB rats to the iht administration of NPY. In addition, to permit assessment of alteration of endogenous NPY concentration, we also measured NPY-like immunoradioactivity in plasma and brains of control and anorectic TB rats. Preliminary results are also presented examining NPY receptor changes in hypothalamus taken from TB and control rats. In addition, since several studies from our laboratory indicate that hyperammonemia may be involved in the etiology of experimental cancer anorexia,[13,21,22] we investigated whether chronic infusion of ammonium salts would reduce NPY-induced feeding.

II. Materials and Methods

A. Feeding Response of TB rats to NPY

The possible role of NPY in the normal regulation of feeding and satiety led us to investigate its effectiveness as an endogenous feeding stimulus in TB rats. In this experiment cannulae (24 gauge) were surgically implanted into the perifornical hypothalamus (PFH) of 27 male 350- to 400-g Fischer 344 rats (Charles River Laboratories, Wilmington, MA). The rats were housed individually within a temperature- and humidity-controlled vivarium. One week later 15 of these rats were inoculated subcutaneously (sc) with 50 mg of viable fresh methylcholanthrene (MCA) sarcoma tissue, taken from a donor rat maintained in our animal tumor colony, while the remaining 12 rats were sham inoculated using the empty trocar. This tumor exhibits 100% take with no regression or metastases and causes anorexia in about 21 days postinoculation at approximately 10% of the host's total body weight (BW).[23] Therefore, the feeding response to iht NPY was investigated prior to (days 14, 18, and 21), during mild (days 25 and 28), and during severe (day 35) anorexia. On these test days TB rats were treated with 2 µg NPY ($n = 8$) or 1 µl artificial cerebrospinal fluid (CSF; $n = 7$). Non-TB rats also received identical treatments ($n = 7$ and 5, respectively). Intake of rat chow pellets and water was monitored for 4 hr following each injection. The rats were maintained on rat chow ad libitum throughout the experiments, and 24-hr intake of rat chow was also monitored daily. At the conclusion of the experiment, the rats were euthanized with an overdose of sodium pentobarbital and their brains were removed and stored in 10% buffered formalin prior to gross localization of cannulae placements.

B. Assessment of Blood and Brain NPY Levels in TB Rats

In order to determine whether blood and brain concentrations of NPY were altered in anorectic TB rats, a second experiment was conducted to measure NPY levels in TB and control rats. Methylcholanthrene sarcomas were transplanted into 8 Fisher 344 rats, while an additional 16 rats were subjected to sham inoculations. These sham-inoculated rats were divided into a freely feeding (FF) group of 8 rats having *ad libitum* access to rat chow, and a matched carcass weight group (MCW), whose daily food was rationed to reduce their BWs by 33% at the conclusion of the experiment, as compared to the FF group. The MCW control was employed to control for and permit analysis of the nutritional effects of the large tumor burden and anorexia exhibited by TB rats. All rats were euthanized by decapitation in the morning 32 days after tumor inoculation. Blood was collected from the cervical wound, centrifuged, and frozen (–70°C). The rats' brains were rapidly removed and the hypothalami were dissected free over ice as described previously[13] and frozen in liquid nitrogen.

The concentration of NPY was determined in plasma and hypothalamus using radioimmunoassay (RIA) methodology, according to our published procedures.[24] Plasma was extracted with 1.6 volumes of 95% ethanol and centrifuged. The resulting supernatants were dried and the residues reconstituted in 1 ml of assay buffer. Brain samples were extracted in 10 volumes of boiling 0.2 M acetic acid for 10 min. After homogenization over ice and centrifugation, the acetic acid extracts were lyophylized and resuspended in 1 ml of assay buffer. The assay mixture, consisting of 100 μl sample or standard, 100 μl assay buffer or NPY-free plasma, and 100 μl NPY antiserum, was incubated overnight at 4°C. Next, 100 μl of ^{125}I-NPY tracer was added and the mixture was incubated overnight again at 4°C. Then, 100 μl of anti-rabbit gamma globulin and 100 μl 10% polyethylene glycol were added, with the mixture being incubated for 2 hr prior to the addition of 500 μl of 1% borine serum albumin (BSA) assay buffer. Bound and free NPY were separated by centrifugation (20 min), with the supernatant being discarded and residue counted for 5 min.

C. Developmental Analysis of NPY Concentrations in TB Rats

This experiment was designed to determine whether blood and hypothalamic NPY concentrations are altered prior to the onset of anorexia and if altered NPY levels parallel the anorexia. Thirty male F344 rats were inoculated sc with fresh MCA sarcoma, while an additional 36 rats received sham inoculations. Twenty-seven of the control rats were matched in weight with TB rats to form three MCW and three TB groups. The remaining 9 rats were maintained on *ad libitum* rat chow and formed a FF control group. On days 14, 21, and 28, groups of MCW and TB rats were euthanized for the determination of NPY levels in blood and hypothalamus. The FF control group was sacrificed on day 28, with blood and hypothalamus being taken for the RIA of NPY.

D. NPY Receptor-Binding Studies

Methylcholanthrene sarcomas were inoculated sc into 16 male F344 rats. An additional 16 rats received sham inoculations and formed FF and MCW control groups. Tumor-bearing rats were euthanized 21 and 28 days after inoculation, while control rats were sacrificed along with the day 28 TB rats. Blood was collected for the analysis of plasma NPY by RIA and the hypothalamus was dissected free for the preparation of membranes for the binding studies.

Hypothalamic membranes were prepared according to the published procedures of Unden *et al.*[25] Tissue from FF, MCW, and TB groups was pooled and homogenized in 10 volumes of cold sucrose (0.32 M) buffered with 5-mM Hepes. The homogenates were centrifuged (1000 × *g*, 4°C) for 5 min, with the resulting supernatants being centrifuged again (10,000 × *g*, 4°C) for 45 min. The resulting pellet, representing the crude membrane preparation, was suspended in the assay buffer (20-mM Hepes containing 2% BSA, 100 μM

PMSF, 4 μg/ml leupeptin, 4 μg/ml chymostatin, 5 KIU/ml aprotinin, and 0.1% bacitracin) to prepare for binding. Membrane protein concentration was estimated by a modification of the Lowry procedure.[26] Binding of ^{125}I-NPY was conducted according to our published procedures[27] in a total volume of 0.25 ml of assay buffer, with 200 μg of membrane proteins being added to each assay tube. These samples were incubated for 2 hr at 18°C in a shaking water bath with various concentrations of ^{125}I-NPY (total binding). Nonspecific binding was determined by incubating the samples with a 500-fold excess of unlabeled NPY. After the initial incubation, 150-μl aliquots were transferred to tubes containing 250 μl of cold assay buffer. Free ligand was separated from bound ^{125}I-NPY by centrifugation (10,000 × *g*, 4°C) for 10 min. The resulting pellets were counted for bound radioactivity, with specific binding being obtained by subtracting the nonspecific binding from total binding. The values for Kd and B*max* were obtained by analyzing the binding data with the LIGAND® computer program for one- and two-site models.

E. Ammonia Infusion Studies

Following anesthetization [pentobarbital 45 mg/kg, intraperitoneal (ip)], cannulae (24 gauge) were implanted into the PFH of 21 male F 344 rats. Two weeks later, the rats were anesthetized again and silastic catheters (#602-155, Dow Corning, Midland, MI) were surgically implanted into the external jugular vein, according to our published procedures.[28] In 12 of these rats the catheters exited the dorsum of the rats through a stainless steel button (Harvard Biosciences, South Natick, MA) sutured to the midscapular skin. The catheter was protected by a spring and connected to a Harvard peristaltic infusion pump (Model 420) by a feed-through swivel (Harvard Biosciences), allowing the rat freedom to move about the cage. In the remaining rats, the catheter was occluded just beyond its exit from the skin. The rats with functional catheters were infused with sterile normal saline (2 ml/hr) for 3 days as normal feeding was returning following the surgical trauma. On day 4, infusion (2.0 ml/hr) of 0.2 M ammonium salts (0.1 M ammonium acetate + 0.1 M ammonium bicarbonate, pH = 7.6) was initiated in 7 of the rats, with the concentration being increased to 0.3 M on day 6. Infusion of saline was continued in the remaining 5 rats having functional catheters. Three days later, the 4-hr feeding effect of 2 μg of NPY or 1 μl of artificial CSF was assessed in all rats under *ad libitum* feeding conditions.

F. Statistical Evaluation

The results were evaluated for statistical significance using analysis of variance (ANOVA) techniques, with individual means being compared post hoc using Tukey's corrected *t* tests. Food intake results were subjected to a repeated-measures ANOVA. All procedures were approved by the University of Cincinnati Animal Care Committee.

III. Results

A. Feeding Response of TB Rats to NPY

As indicated in Figure 1, the TB rats exhibited consistent significant ($p < 0.01$) anorexia beginning 25 days after tumor inoculation. Table 1 shows that the feeding response to NPY was reduced significantly, as compared to the response of non-TB rats, every test day except day 14 and the 1 hr measurement period of day 21. Significant ($p < 0.01$) main effects for NPY vs. CSF were obtained for every test day, while main effects for TB vs. control were first observed on day 18 and consistently thereafter. Thus, even prior to the onset of overt anorexia, food intake in response to this potent stimulus of feeding was decreased. In addition, the feeding response was reduced even further as the TB rats became severely anorectic, decreasing by over 70% by day 35. Gross histological examination of brain slices using a dissecting microscope revealed cannulae terminations within 1 mm of the PFH.

Table 1. Mean (± SEM) Intake of Rat Chow by TB and Control (C) Rats 1, 2, and 4 hr after the iht Injection of 1 μl Artifical CSF or 2 μg NPY[a]

	Group	1 hr	2 hr	4 hr
Day 14	C-CSF	0.1 ± 0.1	0.5 ± 0.4	0.9 ± 0.3
	C-NPY[b]	5.2 ± 0.6	7.1 ± 0.8	7.8 ± 1.1
	T-CSF	0.1 ± 0.1	0.4 ± 0.2	0.5 ± 0.2
	T-NPY[b]	3.7 ± 0.9	5.2 ± 0.9	6.9 ± 1.2
Day 18	C-CSF	1.2 ± 0.3	1.4 ± 0.4	1.7 ± 0.5
	C-NPY[b]	6.7 ± 1.0	8.6 ± 0.3	9.0 ± 0.4
	T[c]-CSF	0.2 ± 0.1	0.8 ± 0.3	0.9 ± 0.3
	T[c]-NPY[b]	4.1 ± 1.3[d]	4.5 ± 1.3[d]	5.5 ± 1.2[d]
Day 21	C-CSF	0.4 ± 0.2	0.7 ± 0.3	1.2 ± 0.4
	C-NPY[b]	8.1 ± 1.0	10.5 ± 1.0	10.6 ± 0.9
	T[c]-CSF	0.1 ± 0.1	0.2 ± 0.1	0.5 ± 0.4
	T[c]-NPY[b]	5.6 ± 1.3	6.7 ± 1.5[d]	7.5 ± 1.3[d]
Day 25	C-CSF	1.3 ± 0.3	1.5 ± 0.4	2.2 ± 0.7
	C-NPY[b]	9.8 ± 1.1	10.7 ± 1.2	10.8 ± 1.2
	T[b]-CSF	0.7 ± 0.4	0.8 ± 0.4	1.1 ± 0.5
	T[b]-NPY[b]	4.7 ± 1.4[d]	5.8 ± 1.3[d]	6.4 ± 1.1[d]
Day 28	C-CSF	0.9 ± 0.4	1.2 ± 0.3	1.9 ± 0.4
	C-NPY[b]	8.3 ± 1.2	9.8 ± 1.2	10.0 ± 1.3
	T-CSF	0.7 ± 0.2	1.5 ± 0.5	2.2 ± 0.7
	T[c]-NPY[b]	4.6 ± 1.0[d]	6.5 ± 1.1[d]	7.0 ± 0.9[d]
Day 35	C-CSF	1.1 ± 0.3	1.2 ± 0.4	2.2 ± 0.6
	C-NPY[b]	11.0 ± 1.0	11.4 ± 1.1	15.0 ± 1.0
	T[b]-CSF	0.6 ± 0.3	0.8 ± 0.4	1.3 ± 0.5
	T[b]-NPY[b]	3.0 ± 0.9[d]	3.6 ± 1.0[d]	4.1 ± 0.9[d]

[a] The results are grouped according to the number of days after tumor inoculation.
[b] $p < 0.01$.
[c] $p < 0.05$ based on significant main effects and interactions of repeated measures ANOVA.
[d] $p < 0.05$ vs. C-NPY group based on post hoc Tukey's protected T test.

B. Assessment of Blood and Brain NPY Levels in TB Rats

Figure 2 presents food intake by TB and FF rats for the 4 days prior to sacrifice. The TB rats were severely anorectic [1.3 vs. 8.5 g/100 g body weight (BW)] on the day of sacrifice. Mean tumor weight in these rats at sacrifice was 74 ± 3 g, which was 31% of total BW. This degree of cachexia was paralleled by a 32% decrease in the BW of the MCW rats as compared with the FF control rats at sacrifice.

As shown in Figure 3, immunoreactive NPY was reduced significantly ($p < 0.01$) by 47% in the plasma of TB rats and increased ($p < 0.05$) by 44% in the plasma of the MCW control rats. A similar pattern of NPY alterations was observed in the hypothalamus (Figure 4) with levels decreasing ($p < 0.01$) by 80% in TB rats and increasing ($p < 0.01$) by 27% in the MCW rats.

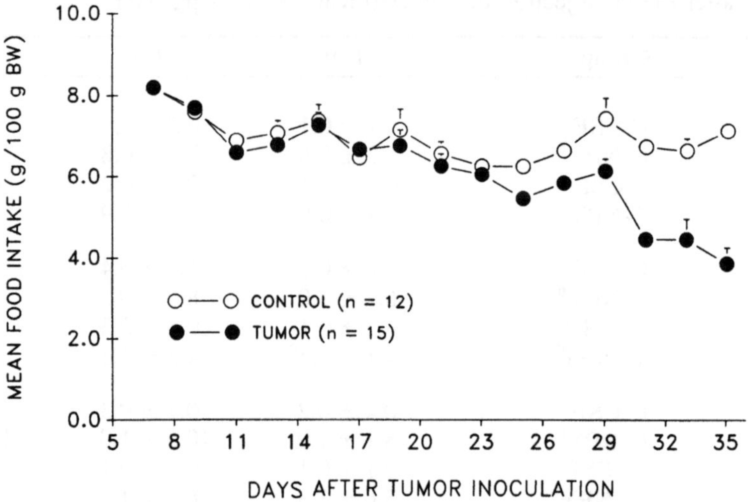

Figure 1. Mean (± SEM) daily intake of rat chow by control and TB rats. Two weeks prior to tumor inoculation cannulae were implanted into the PFH of all rats.

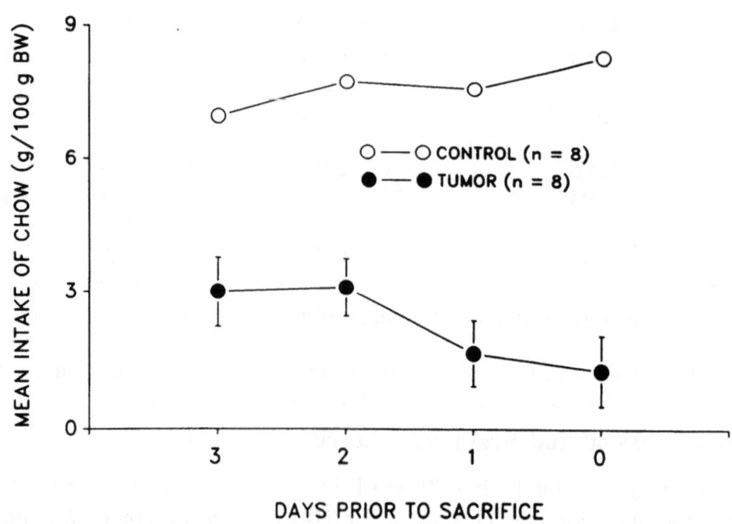

Figure 2. Mean (± SEM) daily intake of rat chow by control and TB rats. The rats were sacrificed after the TB rats exhibited severe anorexia (food intake/100 g BW ≤ 1). Mean day of sacrifice was 32 days after tumor inoculation.

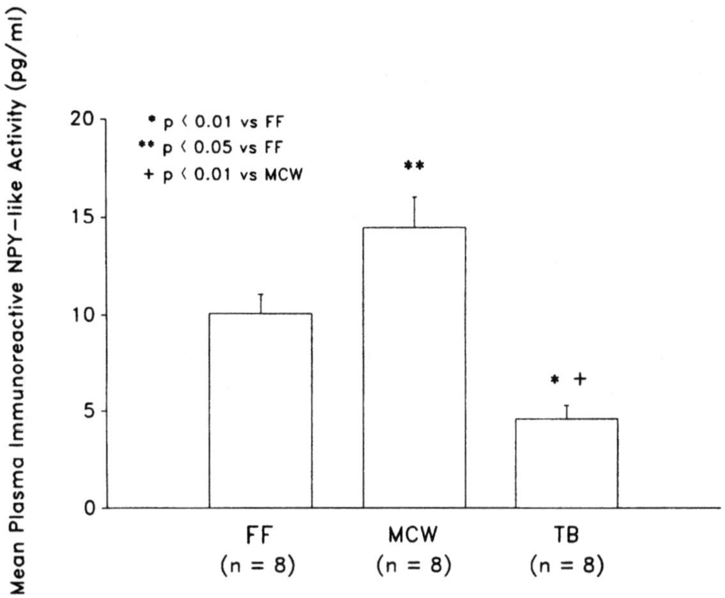

Figure 3. Mean (± SEM) immunoreactive NPY-like activity in plasma taken from FF and MCW control rats and from severely anorectic TB rats.

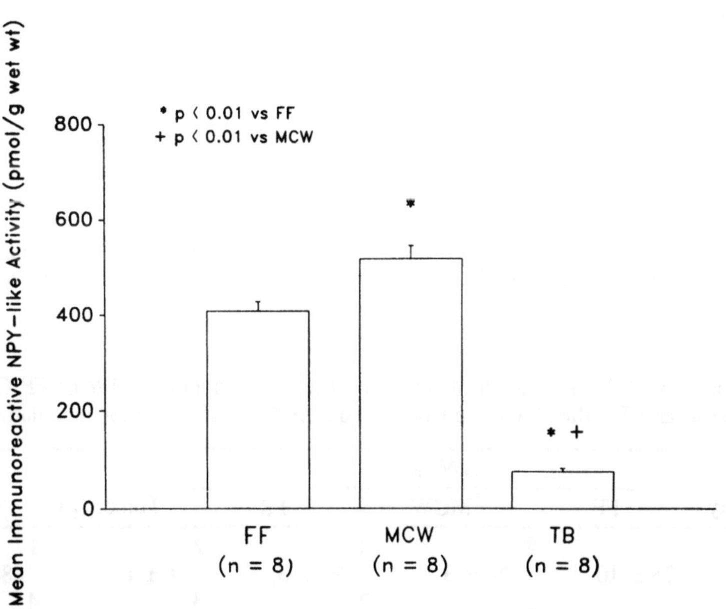

Figure 4. Mean (± SEM) immunoreactive NPY-like activity in hypothalamus taken from FF and MCW control rats and from severely anorectic TB rats.

C. Developmental Analysis of NPY in TB Rats

Mean weekly food intake by TB and control rats is presented in Figure 5. The TB rats were not significantly anorectic at the day 21 sacrifice time point. Significant anorexia was observed in the remaining 9 TB rats beginning on day 24. Again, the degree of cachexia produced by this tumor is reflected in the decreased food allowed the MCW in order to maintain their BWs equal to the nontumor BWs of the TB rats. Tumor weights at sacrifice and nontumor BWs for all groups are presented in Table 2. Since the tumor weights were estimated using planar measurements,[10] as the experiment was being conducted, exact matches were not possible for predicted tumor weight and actual tumor weight as presented in Table 2. Thus, the nontumor BWs were somewhat overestimated for the TB rats euthanized on days 14 and 21. The agreement between the nontumor BW and the MCW rats for day 28, however, was very good.

Plasma immunoreactive NPY concentrations for this developmental study are presented in Figure 6. Although plasma NPY concentrations tended to be decreased in TB rats on days 14 and 21, these differences were not statistically significant. Following the development

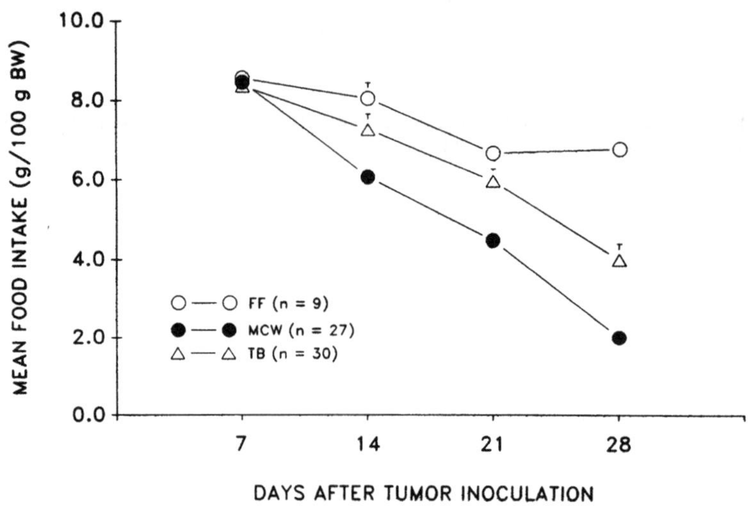

Figure 5. Mean (± SEM) weekly intake of rat chow FF, MCW, and TB rats. Groups of MCW and TB rats were sacrificed on days 14, 21, and 28, while the FF rats were euthanized on day 28.

Table 2. Mean (± SEM) BW, Tumor Weight, and Carcass (Nontumor) BW of FF Control, MCW control, and TB Rats Sacrificed 14, 21, and 28 Days after Tumor Inoculation[a]

	BW (g)				
Sacrifice day	FF	MCW	TB	Tumor (g)	Carcass (g)
14	275 ± 10	268 ± 3	291 ± 3	11 ± 1	280 ± 3
21	284 ± 10	250 ± 3	308 ± 4	40 ± 3	268 ± 6
28	292 ± 10	243 ± 6	316 ± 10	76 ± 5	240 ± 10

[a] The FF group was not sacrificed until day 28. For each data point, $n = 9$.

of anorexia, however, plasma immunoreactive NPY level was decreased significantly ($p < 0.01$) in TB rats sacrificed on day 28. The pattern of immunoreactive NPY change in hypothalamic tissue (Figure 7) was similar to that observed in plasma. The decrease in hypothalamic NPY concentration, however, was statistically significant ($p < 0.01$) at each point as compared with the respective control groups.

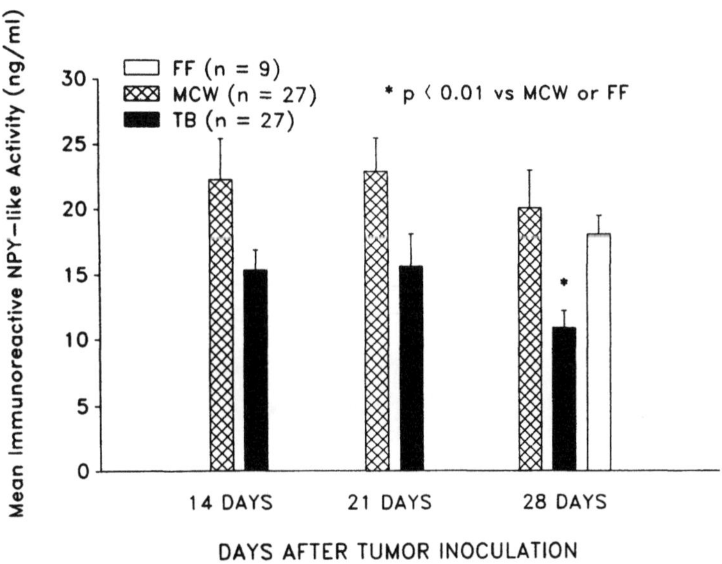

Figure 6. Mean (± SEM) immunoreactive NPY-like activity in plasma taken from FF, MCW, and TB rats sacrificed at increasing times of tumor growth.

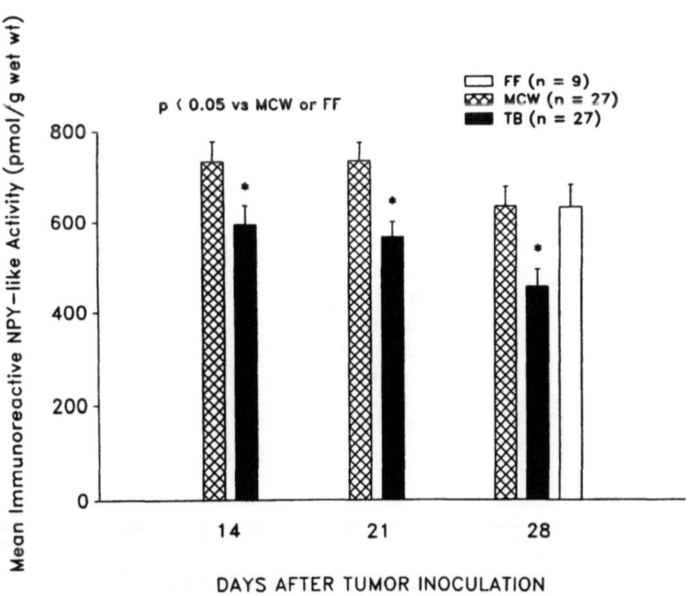

Figure 7. Mean (± SEM) immunoreactive NPY-like activity in hypothalamus taken from FF, MCW, and TB rats sacrificed 14, 21, and 28 days after tumor inoculation.

D. NPY Receptor-Binding Changes in TB Rats

Mean daily intake of rat chow is presented in Figure 8. In this experiment statistically significant anorexia was first observed in TB rats 18 days after tumor inoculation. The apparent increase in feeding in Figure 8 on day 22 is due to the sacrifice of 8 TB rats on day 21 for the binding studies. Again, food allowed to the MCW rats was restricted in order to maintain their BWs close to the nontumor BWs of the TB rats that were sacrificed on day 28. Figure 9 presents the mean BWs for the FF and MCW groups, the nontumor BWs of

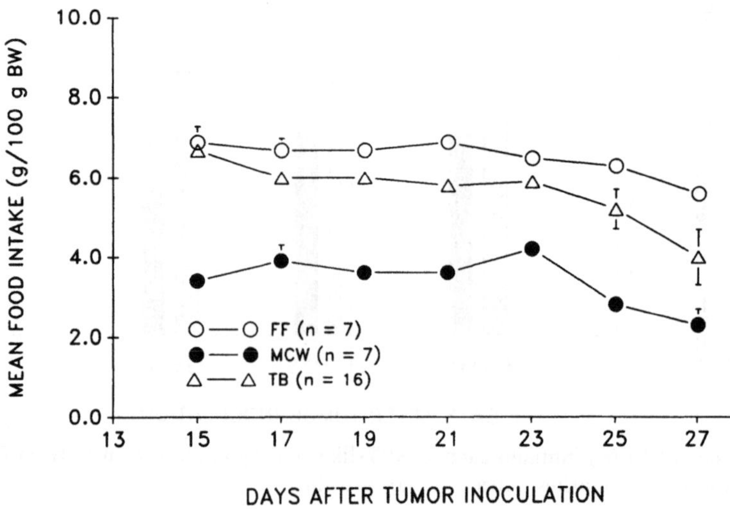

Figure 8. Mean (± SEM) daily intake by FF, MCW, and TB rats. Control rats were euthanized on day 28, while TB rats were sacrificed 21 and 28 days after tumor inoculation.

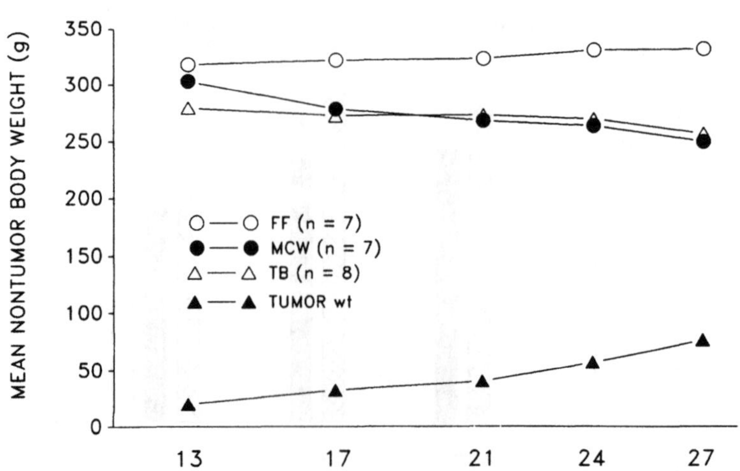

Figure 9. Mean (± SEM) nontumor BWs of control (FF and MCW) and TB rats sacrificed 28 days after tumor inoculation.

day 28 TB rats, as well as the tumor weights of these rats. The day 28 data point for tumor weight is actual weights, while the previous data points are estimated from planar measurements of tumor size.[10] The nontumor BW corresponded very well with the BW of the MCW group, suggesting similar degrees of malnutrition in both groups. Mean plasma NPY concentrations are presented in Figure 10. As in the previous experiment, plasma NPY level was reduced significantly in the TB rats sacrificed on day 28. The results of the NPY binding study are shown in Table 3. There was no change in the number of NPY receptors across these groups, as indicated by the B*max* values. Binding affinity, however, was altered dramatically, with both groups of TB rats exhibiting nearly ten-fold decreases in IC_{50} values. These changes in binding affinity suggest a decrease in the sensitivity of hypothalamic NPY receptors early in the development of anorexia. The lack of change in the IC_{50} values in the MCW group suggests that the alteration in TB rats is not secondary to malnutrition. It should be stated that these observations are preliminary. They do represent, however, the results of three binding replications conducted on hypothalamic membranes taken from this experiment.

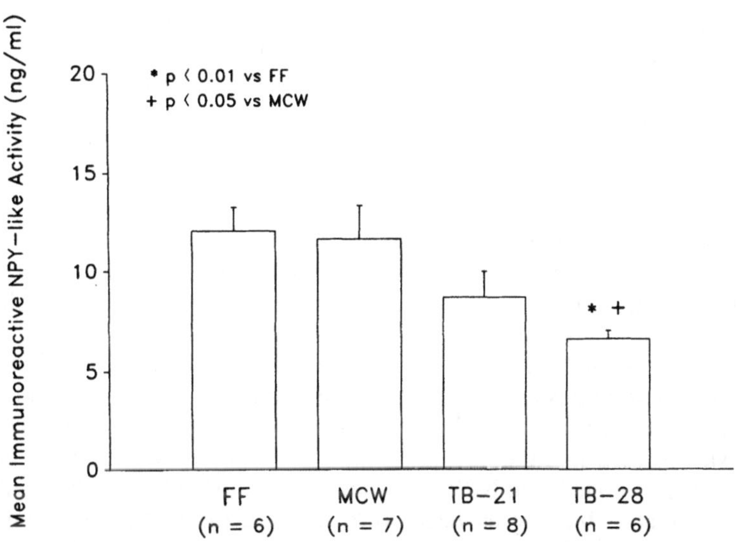

Figure 10. Mean (± SEM) immunoreactive NPY-like activity in plasma of control (FF and MCW) and TB rats sacrificed 21 (TB-21) and 28 (TB-28) days after tumor inoculation.

Table 3. Binding of ^{125}I-NPY to Hypothalamic Membranes Taken from FF, MCW, and Day 21 and Day 28 TB Rats

Group	B*max* (fmol/mg protein)	IC_{50} nMolar
FF	134	0.21
MCW	132	0.35
TB day 21	116	2.27
TB day 28	120	2.18

E. Effect of Ammonia Infusion on NPY-Induced Feeding

If ammonia is a mediator of experimental cancer anorexia, it may also reduce NPY-induced feeding to the degree observed in TB rats. Therefore, ammonium salts were infused at a concentration and rate to cause blood levels to rise to values observed in TB rats (200 to 300 nmol/ml). Mean plasma ammonia concentration in ammonia-infused rats were 257 + 82 nmol/ml, while saline-infused rats had 119 + 17 nmol/ml ammonia in plasma samples. The effect of infusing ammonium salts on food intake is presented in Figure 11. Thus, intake of rat chow began to decrease 2 days after the initiation of ammonia infusion at 0.2 M and continued to decrease to about a 50% reduction as the concentration was increased to 0.3 M. Testing of NPY in these rats also yielded results that were similar to those observed in TB rats in the first experiment (Figure 12). Thus, 1-, 2-, and 4-hr feeding was reduced significantly in ammonia-infused rats as compared with the saline-infused group. Food intake by the ammonia-infused rats following the injection of NPY was not significantly different from the intake of the CSF-treated group in this test.

IV. Discussion

The development of anorexia and loss of lean body mass continue to be primary obstacles to aggressive treatment of neoplastic diseases. Although many hypotheses of cancer anorexia have been advanced[2] few theories have integrated tumor-induced peripheral biochemical aberrations with central mechanisms that control hunger and satiety. In the present series of experiments we have presented several observations suggesting that hypothalamic NPY mechanisms of feeding are altered in TB rats. We have also suggested that these alterations may be secondary to the metabolism of amino acids and release of large quantities of ammonia by these transplantable sarcomas.

Neuropeptide Y has been demonstrated to be a powerful stimulant of feeding rats, with submicrogram intrahypothalamic injections eliciting intake approaching physiological limits.[19] Repeated injections of NPY also elicited sustained overeating and obesity.[20,29] Concentrations of NPY in hypothalamic nuclei also respond to physiological conditions, with food-deprived rats exhibiting elevated NPY levels in the paraventricular (PVN) and arcuate (ARC) regions.[30] In addition, when food-deprived rats were fed, the NPY levels were normalized and NPY was released into the interstitial fluid.[31] Pathological conditions have also been reported to affect hypothalamic NPY concentration, with streptozocin-induced diabetic rats, which also overeat, exhibiting elevated NPY levels in the PVN, ARC, and ventromedial hypothalamic nuclei.[32] Elevated NPY has also been observed in the hypothalamus of genetically obese Zucker rats[33] as has ARC preproneuropeptide Y messenger RNA.[34] Therefore dysfunction of NPY feeding systems appears to be a logical candidate as a mediator of alterations in hunger and satiety.

The present experiments demonstrated significant dysfunction of NPY feeding systems in TB rats. Thus, NPY was less effective as a feeding stimulus in TB rats prior to the onset of overt anorexia. Furthermore, this decrease in feeding worsened as *ad libitum* intake by TB rats decreased into more severe anorexia. These changes suggest that hypothalamic NPY receptor mechanisms, which influence food intake, may be refractory to exogenous peptide.

A large decrease in hypothalamic NPY concentration in severely anorectic TB rats was observed in the second experiment, while the food-restricted MCW group exhibited increased NPY level. This increase in NPY is similar to reported changes in the PVN and ARC of food-deprived rats.[30] Since the primary source of hypothalamic NPY appears to be neuronal,[34] it is unlikely that altered plasma NPY concentrations caused the changes observed in the brain. Plasma NPY level may be decreased in TB rats because of their larger blood

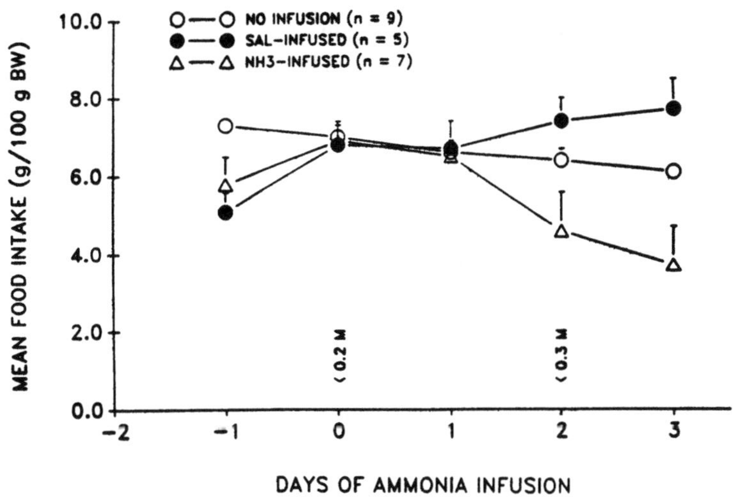

Figure 11. Mean (± SEM) daily intake of rat chow by noninfused, saline-infused, and ammonia-infused rats. Ammonia infusion was initiated on day 0 at a concentration of 0.2 M and a rate of 2 ml/hr. The concentration was increased to 0.3 M on day 2.

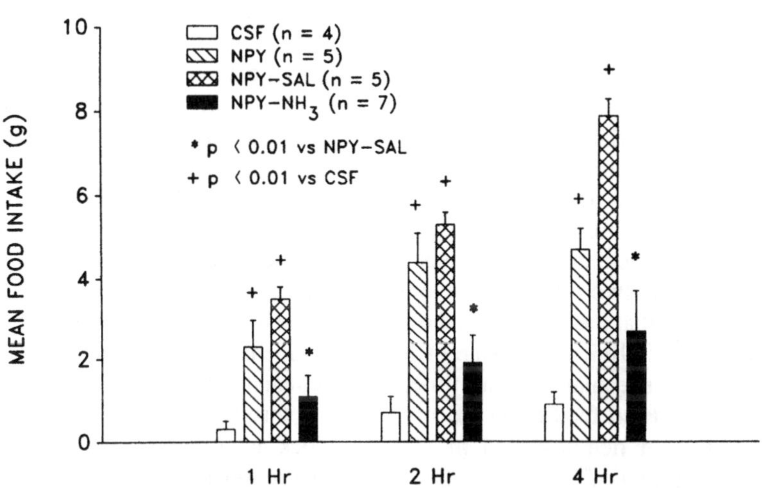

Figure 12. Mean (± SEM) intake of rat chow 1, 2, and 4 hr after the iht injection of 1 μl artificial CSF or 2 μg of NPY in noninfused (CSF and NPY), saline-infused (NPY-SAL), and ammonia-infused (NPY-NH_3) rats.

volume. A more likely cause of the lowered hypothalamic NPY level in TB rats is a decrease in the message for its synthesis due to an unidentified factor.

The third experiment demonstrated significantly decreased NPY content of the hypothalamus prior to the onset of anorexia. Although NPY concentrations did not decrease in this experiment to the same degrees observed previously, it should be pointed out that the anorexia exhibited by these rats was not nearly as severe as in the first experiment. The failure of the MCW rats to exhibit increased NPY concentrations may also be due to less severe cachexia of the TB rats. Alternatively, significant alterations in NPY in specific nuclei could easily have been diluted out, since our assays involved whole hypothalami.

The fourth experiment suggested an alteration in NPY receptor mechanisms shortly after the onset of significant anorexia. In this experiment the rats that were euthanized on day 21 exhibited significantly reduced food intake beginning on day 18. In these rats and the TB rats sacrificed on day 28 the IC_{50} was increased by nearly ten-fold. This alteration in the IC_{50} suggests that the affinity of hypothalamic NPY receptors to NPY was much less in TB rats. The absence of alteration in receptor density (B*max*) is somewhat surprising, since one would predict an increase in receptor number in the presence of decreased NPY level. Therefore, these preliminary results point to some basic dysfunction in NPY receptor-binding mechanisms and a failure of the system to compensate for this change in affinity by increasing receptor numbers and peptide concentration.

The fifth experiment investigated whether hyperammonemia could affect NPY-induced feeding. The effect of ammonia infusion on *ad libitum* intake of rat chow are in agreement with our previous results,[13,21,22] indicating that infusing more than 0.2 M ammonium salts at a rate of 2 ml/hr elicits significant anorexia. We have also demonstrated that infusing ammonium salts elicits neurochemical alterations similar to those observed in TB rats.[13] These neurochemical changes include increased metabolism of dopamine and serotonin in several brain regions and increased concentrations of glutamine and large neutral amino acids (tyrosine, tryptophan, phenylalanine, methionine, histidine, threonine) in the brain. In this experiment ammonia infusion was also shown to reduce NPY-induced feeding similar to that observed in TB rats. Thus, this effect is suggestive that ammonia might be an important mediator of the NPY feeding dysfunction observed in TB rats. Experiments are currently underway to determine if ammonia infusion alters NPY levels and receptors.

In addition to the reports mentioned previously, several experiments suggest hyperammonemia might be involved in cancer anorexia. Surgical resection of MCA sarcomas resulted in the normalization of plasma ammonia and most neurochemical alterations within 24 hr.[35] Associated with this biochemical normalization was an increase in food intake.[6] Determination of arterial-ventricular (A-V) differences in ammonia across MCA sarcomas revealed a five-fold increase in blood ammonia as the tumor tissue was perfused.[36] Blockade of ammonia detoxification with methionine sulfoximine caused an earlier onset of anorexia in MCA sarcoma-bearing rats.[37] Hyperammonemia has also been noted in the Ehrlich acites tumor[38] and in sc inoculated Morris hepatoma.[22] In addition, a recent clinical report noted hyperammonemia in two multiple myeloma patients.[39] *In vitro* culturing of these myeloma cells demonstrated that the tumor was the source of the hyperammonemia.[39] Therefore, there is experimental and clinical evidence of hyperammonenia in TB organisms.

Although only speculative at this time, one mechanism whereby hyperammonemia might influence NPY-induced feeding in TB organisms is through cyclic AMP (cAMP) systems. Ammonia has been demonstrated *in vitro* to block isoproterenol stimulation of cAMP production in astrocytes.[40] Neuropeptide Y also decreases stimulated cAMP production in neural tissue,[41] perhaps through complexing with a pertussis toxin-sensitive G inhibitory protein.[42] We have demonstrated that the iht injection of pertussis toxin antagonizes NPY-induced feeding at a time when *ad libitum* feeding has returned to normal.[43] These observations suggest that hyperammonemia may render this feeding system refractory by depleting the cAMP signal prior to the administration of NPY. Thus, the efficacy of NPY as a feeding stimulus might be proportional to the degree that the cAMP system was deactivated already by ammonia. In future studies we will test this hypothesis in TB rats.

V. Summary

The efficacy of NPY to elicit feeding in TB rats was reduced prior to the onset of overt anorexia, with the feeding response decreasing further as anorexia developed. Hypothalamic concentration of NPY was reduced in TB rats, with the magnitude of the decrease paralleling the degree of anorexia. Binding affinity of NPY to hypothalamic membranes taken from TB rats suggested decreased binding affinity with no change in receptor number. Infusing ammonium salts at a concentration and rate necessary to increase blood ammonia levels to the degree observed in TB rats, produced anorexia and decreased NPY feeding. These results suggest that NPY feeding systems are abnormal in TB rats and that hyperammonemia may be of primary importance in this dysfunction.

Acknowledgment

This research was supported in part by Grant 91B54 from the American Institute for Cancer Research, Grant GM47122 from the National Institutes of Health, and by a Merit Review Grant from the Department of Veterans Affairs.

References

1. G. Costa, Cachexia, the metabolic component of neoplastic diseases, *Cancer Res* 37:2327 (1977).
2. S.D. Morrison, Control of food intake in cancer cachexia: A challenge and a tool, *Physiol Behav* 17:705 (1976).
3. S. Warren, The immediate causes of death in cancer, *Am J Med Sci* 184:610 (1932).
4. W.D. DeWys, C. Begg, P.T. Lavin, Prognostic effect of weight loss prior to chemotherapy in cancer patients, *Am J Med Sci* 69:491 (1980).
5. V.J. Lanzotti, D.R. Thomas, L.E. Boyle, *et al.*, Survival with inoperable lung cancer: An integration of prognostic variables based on simple clinical criteria, *Cancer* 39:303 (1977).
6. W.T. Chance, L. Cao, J.L. Nelson, T. Foley-Nelson, J.E. Fischer, Reversal of neurochemical aberrations after tumor resection in rats, *Am J Surg* 155:124 (1988).
7. J.A. Norton, J.F. Moley, M.V. Green, R.E. Carson, S.D. Morrison, Parabiotic transfer of cancer anorexia/cachexia in male rats, *Cancer Res* 45:5547 (1985).
8. M.C. Stovroff, D.L. Fraker, J.A. Norton, Cachectin activity in the serum of cachectic, tumor-bearing rats, *Arch Surg* 124:94 (1989).
9. F.M. Van Lammeren, W.T. Chance, J.E. Fischer, Cholecystokinin-induced hypophagia is not potentiated by cancer anorexia, *Peptides* 5:97 (1984).
10. W.T. Chancc, F.M. van Lammeren, M.H. Chen, W.J. Chen, R.F. Murphy, S.N. Joffe, J.E. Fischer, Plasma and brain cholecystokinin levels in cancer anorexia, *J Surg Res* 36:490 (1984).
11. W.T. Chance, F.M. van Lammeren, J.E. Fischer, Feeding elicited by cholinergic and adrenergic hypothalamic stimulation of anorectic tumor-bearing rats, *Pharmacol Biochem Behav* 31:209 (1988).
12. W.T. Chance, M. Muggia-Sullam, M.H. Chen, R.F. Murphy, J.E. Fischer, Reversal of tumor-induced biochemical abnormalities by insulin treatment in rats, *J Natl Cancer Inst* 77:497 (1986).
13. W.T. Chance, L. Cao, T. Foley-Nelson, J.L. Nelson, J.E. Fischer, Possible role of ammonia in experimental cancer anorexia, *Brain Res* 486: 316 (1989).
14. W.T. Chance, M. von Meyenfeldt, J.E. Fischer, Serotonin depletion by 5,7-dihydroxytryptamine or para-chloroamphetamine does not affect cancer anorexia, *Pharmacol Biochem Behav* 18:115 (1983).
15. T. Foley-Nelson, W.T. Chance, J.L. Nelson, L. Cao, J.E. Fischer, Delay of cancer anorexia following injection of 6-hydroxydopamine into the corpus striatum, *Proc Soc Neurosci* 12:1450 (1986).
16. K. Tatemoto, M. Carlquist, V. Mutt, Neuropeptide Y-a novel brain peptide with structural similarities to peptide YY and pancreatic polypeptide, *Nature* 296:659 (1982).

17. T. Hokfeldt, J.M. Lundberg, K. Tatemoto, *et al.*, Neuropeptide Y (NPY)- and FMRF-amide neuropeptide-like immunoreactivities in catecholamine neurons of the medulla oblongata, *Acta Physiol Scand* 117:315 (1983).
18. J.M. Lundberg, L. Terenius, T. Hokfeldt, *et al.*, Neuropeptide Y (NPY)-like immunoreactivity in peripheral noradrenergic neurons and effects of NPY on sympathetic function, *Acta Physiol Scand* 116:393 (1982).
19. B.G. Stanley, S.F. Leibowitz, Neuropeptide Y injected in the paraventricular hypothalamus: A powerful stimulant of feeding behavior, *Proc Natl Acad Sci USA* 82:3940 (1985).
20. S.P. Kalra, M.G. Dube, P.S. Kalra, Continuous intraventricular infusion of neuropeptide Y evokes episodic food intake in satiated female rats: Effects of adrenalectomy and cholecystokinin, *Peptides* 9:723 (1988).
21. W.T. Chance, L. Cao, J.L. Nelson, T. Foley-Nelson, J.E. Fischer, Hyperammonemia in anorectic tumor-bearing rats, *Life Sci* 43:67 (1988).
22. W.T. Chance, F.S. Zhang, T. Foley-Nelson, J.E. Fischer, Hyperammonemia and anorexia in Morris hepatoma-bearing rats, *Physiol Behav* 50:397 (1991).
23. W.T. Chance, M.F. von Meyenfeldt, J.E. Fischer, Changes in brain amines associated with cancer anorexia, *Neurosci Biobehav Rev* 50:397 (1991).
24. A. Balasubramaniam, D.W. McFadden, M. Rudnicki, *et at.*, Radioimmunoassay to determine postprandial changes in plasma neuropeptide Y levels in awake dogs, *Neuropeptides* 14:209 (1989).
25. A. Unden, K. Tatemoto, V. Mutt, *et al.* Neuropeptide Y receptor in the rat brain, *Eur J Biochem* 145:526 (1984).
26. G.L. Peterson, A simplification of the protein assay method of Lowry *et al.*, which is generally more applicable, *Anal Biochem* 83:346 (1977).
27. A. Balasubramaniam, S. Sheriff, D.F. Rigel, *et al.*, Characterization of neuropeptide Y binding sites in rat cardiac membranes, *Peptides* 11:545 (1990).
28. W.T. Chance, L. Cao, J.L. Nelson, T. Foley-Nelson, J.E. Fischer, Acivicin reduces tumor growth during total parenteral nutrition (TPN), *Surgery* 102:386 (1987).
29. B.G. Stanley, S.E. Kyrkouli, S. Lampert, S.F. Leibowitz, Neuropeptide Y chronically injected in to the hypothalamus: A powerful neurochemical inducer of hyperphagia and obesity, *Peptides* 7:1189 (1986).
30. A. Sahu, P.S. Kalra, S.P. Kalra, Food deprivation and ingestion induce reciprocal changes in neuropeptide Y concentrations in the paraventricular nucleus, *Peptides* 9:83 (1988).
31. S.P. Kalra, M.G. Dube, A. Sahu, C.P. Phelps, P.S. Kalra, Neuropeptide Y secretion increases in the paraventricular nucleus in association with increased appetite for food, *Proc Natl Acad Sci USA* 88:10931 (1991).
32. G. Williams, J.S. Gill, Y.C. Lee, H.M. Cardoso, B.E. Okpere, S.R. Bloom, Increased neuropeptide Y concentrations in specific hypothalamic regions of streptozocin-induced diabetic rats, *Diabetes* 38:321 (1989).
33. G. Truett, R.J. Martin, Neuropeptide Y in brain areas and tissues of Zucker rats, *J Cell Biochem* 13:243 (1989).
34. G. Sanacora, M. Kershaw, J.A. Finkelstein, J.D. White, Increased hypothalamic content of preproneuropeptide Y messenger ribonucleic acid in genetically obese Zucker rats and its regulation by food deprivation, *Endocrinology* 127:730 (1990).
35. W.T. Chance, L. Cao, J.E. Fischer, Tumor-induced alterations in brain neurotransmitter and plasma ammonia concentrations are normalized 24 hours after tumor resection, *Life Sci* 48:425 (1991).
36. W.T. Chance, L. Cao, J.L. Nelson, T. Foley-Nelson, J.E. Fischer, Hyperammonemia in anorectic tumor-bearing rats, *Life Sci* 43:67 (1988).
37. W.T. Chance, F.S. Zhang, J.E. Fischer, Methionine sulfoximine intensifies cancer anorexia, *Pharmacol Biochem Behav* 39:115 (1991).
38. J.M. Carrascosa, P. Martinez, I.N. de Castro, Nitrogen movement between host and tumor in mice inoculated with Ehrlich ascitic tumor cells, *Cancer Res* 44:3831 (1984).

39. H. Matsuzaki, M. Uchiba, K. Yoshimura, M. Yoshida, Y. Akahoshi, K. Okazaki, K. Takatsuki, Hyperammonemia in multiple myeloma, *Acta Haematol* 84:130 (1990).
40. D.R. Liskowky, L.O.B. Norenberg, M.D. Norenberg, Effect of ammonia on cyclic AMP production in primary astrocyte cultures, *Brain Res* 386:386 (1986).
41. A. Westlind-Danielsson, A. Unden, J. Abens, S. Andell, T. Bartfai, Neuropeptide Y receptors and the inhibition of adenylate cyclase in the human frontal and temporal cortex, *Neurosci Lett* 74:237 (1987).
42. S. Kassis, O. Marjut, L. Terenius, P.H. Fishman, Neuropeptide Y inhibits cardiac adenylate cyclase through a pertussis toxin-sensitive G protein, *J Biol Chem* 262:3429 (1987).
43. W.T. Chance, S. Sheriff, T. Foley-Nelson, J.E. Fischer, A. Balasubramaniam, Pertussis toxin inhibits neuropeptide Y-induced feeding in rats, *Peptides* 10:1283 (1989).

Chapter 15

Metabolic Alterations and Lactate Overproduction in Insulin-Resistant States

MARIO DIGIROLAMO

I. Metabolic Derangements of Cancer

Progressive wasting, weakness, anorexia, and anemia are frequent complications of neoplastic disease.[1-4] Decreases in adipose mass constitute the major body losses and exceed losses seen in simple starvation. Decreases in protein constituents of the body, however, are also observed.[4-5] The mechanisms for depletion of body energy stores are complex and not adequately explained by anorexia, reduced food consumption, or simply by increased energy utilization by tumor tissue.[1,2,6]

Table 1 lists possible causes of cancer cachexia. The first group of three in the table lists depression that frequently accompanies the early stages of cancer and may contribute to the patient's anorexia. Anorexia, however, may also be secondary to metabolic changes such as increased fat mobilization and fatty acid oxidation. Adverse effects of treatment, such as surgery or chemotherapy, can also frequently contribute to reduced food intake by a variety of mechanisms.

Increased caloric expenditure by metabolically hyperactive tumors, or by the host, may accelerate the weight loss observed in cancer patients.[4,6,7] The loss of body weight and fat mass seen in cancer patients, even with apparently adequate nutrition, may reflect alterations in energy balance, with caloric expenditure exceeding caloric intake. Cancer cachexia, like starvation, is accompanied by the decrease in lean body mass and circulating triiodothyronine.[7] Cancer, however, is frequently associated with an enlarged liver[3] and with a relative increase in fasting energy expenditure,[2,4] whereas, in starvation, the liver is reduced in size and a reduced fasting metabolic rate is observed.[8,9] Attempts to replete body tissues by enhanced nutrition, including parenteral hyperalimentation, have not normalized the lean body mass,[2,5,10] even though some repletion of the adipose tissue mass has been observed.

Mario DiGirolamo • Department of Medicine, Emory University School of Medicine, Atlanta, GA 30303

Diet and Cancer: Markers, Prevention, and Treatment
Edited by M.M Jacobs, Plenum Press, New York, 1994

Table 1. Possible Causes of Cancer Cachexia

Depression
Anorexia
Adverse effects of treatment
Decreased caloric intake
Increased caloric expenditure
Tumor-host competition for nutrients
Excessive depletion of stored nutrients (lipolysis, proteolysis, glycogenolysis)
Reduced synthesis and storage of nutrients (LPL, insulin resistance, substrate competition)

The "metabolic mystery" of cancer may find its explanation in a unique imbalance of hormonal and other stimulants (both normal and cancer-specific substances), leading to excessive depletion of stored nutrients by promoting the metabolic processes of lipolysis, proteolysis, and glycogen breakdown, while, at the same time, negatively affecting the synthesis and storage of nutrients.

Several investigations have been directed toward elucidation of abnormal levels of hormones,[6,7,11,12] and additional lipid-mobilizing substances have been found in tumor extracts or bodily fluids of patients and in animals with tumors.[13-16] A recently discovered substance, tumor necrosis factor (TNF), has been found to have multiple metabolic roles.[16,17] Its effect in the liver in stimulating fatty acid and triglyceride synthesis, as well as very low density lipoprotein (VLDL) production, may explain both the hepatomegaly of cancer and the circulating hyperlipidemia (both hypertriglyceridemia and hypercholesterolemia).[18] Tumor necrosis factor has a marked effect on the fat cell; by reducing the activity of the lipogenic enzymes and of the lipoprotein lipase activity, it reduces synthesis and storage of lipids in adipose tissue. A mild lipolytic effect has also been reported,[17] but this finding is not universal. Finally, TNF promotes glucose uptake in the muscle and increases lactate production. The role of TNF in the energy balance of neoplastic diseases is still under investigation.

II. Lactate Elevation in Cancer

Our laboratory has been interested in the metabolic derangements of cancer, particularly lactate overproduction, in view of our work linking lactate overproduction in obesity to insulin resistance.[19]

Evidence for lactate overproduction by tumors[20,21] and for increased circulating levels of lactate in cancer patients is present in the literature.[22,23] With regard to circulating levels of lactate, Holroyde *et al.*[22] have found elevated fasting plasma lactate levels, a mean value of 1.76 mM in 8 patients with metastatic solid tumors and progressive weight loss, whereas 6 patients with tumors but no weight loss showed lactate levels of 0.96 mM (a mean difference of 83%). Russell, *et al.*[23] found mean blood lactate levels of 1.12 mM in 31 untreated patients with small cell lung cancer and 0.62 mM in 20 healthy volunteers (a mean difference of 45%).

We have studied two groups of subjects:

1. In a study carried out in the general clinical Research Center of Emory University Hospital, 10 patients with non-small cell tumors of the lung showed mean fasting blood

lactate levels of 1.06 mM, whereas 5 normal controls had levels of 0.77 mM (a mean difference of 38%).

2. In a second group of patients studies, in the overnight fasted state, at Grady Memorial Hospital, 19 patients with cancer had mean blood lactate values of 1.47 mM, whereas 28 hospitalized (non-tumor-bearing patients) had mean lactate values of 1.16 mM (a mean difference of 27%). (See Table 2).

Table 2. Metabolic Comparison in Patients with Lung Cancer and Hospitalized Controls

	Control (n = 28)	Cancer (n = 19)
Basal glucose (mg/dl)	106 ± 11	110 ± 8
Basal insulin (μU/ml)	13 ± 1.7	21 ± 4.5
Basal lactate (mM)	1.16 ± 0.11	1.47 ± 0.23

Thus, in all four studies listed above, lactate levels were 27-83% higher in cancer patients than in related controls.

One of the possible explanations for elevated circulating levels of lactate may be found in tumor metabolism itself. It has been reported that, partly due to increased blood supply and partly due to the metabolic activity of the tumor, venous effluent blood from tumors has significantly lower glucose concentration and significantly higher lactate concentration than either arterial blood or peripheral venous blood.

For example, Chance *et al.*[20] have reported that blood draining from certain tumors (methylcholanthrene sarcomas) of Fisher 344 rats contained 77% less glucose and 144% more lactate than peripheral venous blood from the same rats. Similar findings have been reported by Grant and Wells.[21]

Glucose intolerance and insulin resistance have been reported in cancer.[7,11,12,22-25] Resistance to the effect of insulin may promote enhancement of catabolic processes (lipolysis, proteolysis, glycogen breakdown) and reduce anabolic processes (lipogenesis, protein synthesis, glycogenesis).

It is of interest that conditions of insulin resistance may be associated with fasting hyperglycemia (resulting from either increased hepatic glucose production or reduced peripheral glucose utilization, or both) and elevated lactate levels.[19] The explanation for elevated lactate levels in conditions associated with insulin resistance is not clear. A possible explanation has been found in obesity.[19]

III. Lactate Overproduction and Insulin Resistance in Obesity

Insulin resistance is defined as a reduced ability of the organism to respond to normal amounts of insulin (see Table 3). As a consequence of insulin resistance, more insulin is needed initially to produce an effect on muscle, liver, and adipose tissue. Later, as the demands on the pancreatic insulin secretion become greater, a "pancreatic exhaustion" leads to reduced insulin secretion and metabolic consequences of insulin deficiency.[26]

Table 3. Insulin Resistance

Definition:	Impaired ability of insulin to stimulate glucose uptake and metabolism, lipid synthesis, and most other anabolic processes.
Consequences:	In the intact organism, insulin resistance leads to slightly elevated glucose levels and marked elevations in insulin levels.

For the past decade, our laboratory has carried out both animal and human studies showing that progressively larger amounts of lactate are produced in adipocytes that become enlarged with the development of obesity.[19,27-30] In the rat, glucose conversion to lactate increases from 5-15% of glucose metabolized in 150 picoliters (pl) fat cells to 50% in 600 pl cells.[19] In the rabbit, even more lactate is produced by large adipocytes (about 80% of the glucose metabolized in 500-600 pl cells).[19] Similar observations have been made in human subjects.[19,31,32]

It is of interest that elevated lactate production in the basal state is frequently associated with insulin resistance.[19,30] For example, in the rat, acute food restriction, high-fat diet, or experimental diabetes leads to increased lactate production by the fat cells and reduced responsiveness to insulin.[19] In humans, lactate overproduction by enlarged adipocytes from obese patients is of quantitative significance[19,32] and may be responsible for reportedly elevated levels of lactate in the overnight fasted state.[19]

A second aspect of the relationship between lactate levels and insulin sensitivity is the magnitude of the rise in lactate levels following a meal or glucose administration. In insulin-sensitive subjects, glucose and insulin administration are followed by a marked rise in lactate production, whereas this production and the insulin response are blunted in the obese insulin-resistant subjects.[30]

Tables 4 and 5 show metabolic manifestations of insulin resistance and possible explanations for the association between elevated basal lactate levels and insulin resistance. A recent review summarizes the evidence available in the literature and proposes a link between lactate overproduction and insulin resistance.[19]

Insulin resistance has recently received wide attention because of its link to the development of diabetes mellitus and other risk factors, such as hyperlipidemia and hypertension.[33-35]

Table 4. Manifestations of Insulin Resistance

Glucose	Progressive elevation in basal levels
	Progressive carbohydrate intolerance
Insulin	Progressive elevation in basal and post-challenge levels until "pancreatic exhaustion"
Lactate	Reduced generation of lactate after glucose administration
	Progressive elevation of basal levels

Table 5. Possible Explanations for the Association between Basal Lactate Elevation and Insulin Resistance

1. Hyperinsulinemia can be secondary to basal glucose elevation and peripheral insulin resistance
2. Insulin resistance in the liver may reduce the insulin-restraining effect on glycogenolysis and increased lactate production
3. Elevated lactate levels may lead to increased gluconeogenesis and hepatic glucose production
4. Lactate may compete with glucose as a substrate for peripheral glucose utilization (glucose-sparing effect of lactate) and provide the basis for peripheral insulin resistance.

In our studies, it has become apparent that the inverse relationship between basal lactate levels and insulin sensitivity is stronger than that between lactate and the degree of obesity.[30] We have therefore searched the literature for additional evidence that elevated lactate levels may be present in other forms of insulin resistance.

IV. Mild, Moderate, and Severe Forms of Insulin Resistance in the Human

Tables 6 and 7 show various conditions in the human that are known to be associated with insulin resistance. They are divided in mild-moderate forms (Table 6) and severe-extreme forms (Table 7).

Table 6. Mild to Moderate Conditions Associated with Insulin Resistance in the Human

Lactate elevation[a]	Mild-moderate forms
ND	Inactivity
+	Pregnancy
+	Obesity
+	High-fat diet
+	Hypertension
+	Non-insulin dependent diabetes
+	Acidosis (elevated free fatty acids and ketone bodies)
+	Infection
+	Cancer
	Excessive production of counter-insulin hormones
+	Glucocorticoids: Cushing's syndrome
ND	Growth hormone: Acromegaly
ND	Thyroid hormone: Thyrotoxicosis
ND	Catecholamines: Pheocromocytoma
ND	Glucagon, etc.: Glucagonoma

[a] ND = Not done

Table 7. Severe Conditions Associated with Insulin Resistance in the Human

Severe-extreme forms (associated with acanthosis nigricans)
Genetic syndromes
Leprechaunism
Lipoatrophic diabetes mellitus
Type A extreme insulin resistance
Automimmune (anti-receptor antibodies)
Type B extreme insulin resistance
High titer of anti-insulin antibodies
Subcutaneous degradation of insulin

Our search for elevation of basal lactate levels in the mild-moderate forms of insulin resistance has shown a positive association not only for obesity and cancer,[11,12,19,22-24,30] but also for pregnancy,[36] diabetes,[37] hypertension,[38] high fat diets, and acidosis. Furthermore, in a recent study of patients with Cushing's disease, evidence of both insulin resistance and basal lactate elevation was found.[39] It is possible that hyperlactatemia also will be found in the other mild-moderate forms of insulin resistance and in the severe-extreme ones.

As mentioned earlier, insulin resistance should lead to lesser stimulation of glucose transport and metabolism and lesser production of lactate by glucose-metabolizing tissues. It is therefore paradoxical that lactate is elevated in conditions associated with insulin resistance. In obesity, this may be explained by overproduction of lactate in enlarged adipocytes. In cancer, this may be explained by excessive conversion of glucose to lactate by the metabolically active tumor tissue. It is also possible that a diminished restraining effect of insulin on glycogenolysis may lead to increased hepatic generation of glucose and lactate. Other, possibly different explanations may be forthcoming for the site and mechanism of lactate overproduction in the lesser studied syndromes of insulin resistance listed in Tables 6 and 7.

V. Concluding Remarks

Our work with obese animals and humans has focused on the relationship between lactate overproduction and insulin resistance. It appears that lactate availability leads to a glucose-sparing effect by glucose-metabolizing tissues. This can be produced by preferential utilization of lactate or by lactate interference with glucose utilization and metabolism. Resistance to the effect of insulin may favor catabolic events over the anabolic processes.

It is possible, but not yet proven, that lactate overproduction in cancer may operate by similar mechanisms. Furthermore, lactate elevation has been associated with reduction in food intake in animals. More studies need to be done to investigate the interaction of the most common metabolic fuels (glucose, free fatty acids, and lactate) on peripheral metabolic functions and hormonal secretions and actions.

Our work, showing certain relationships between lactate production (and circulating levels) and insulin resistance in obesity may provide a framework that may explain other forms of insulin resistance, such as cancer and many of the syndromes listed in Tables 6 and 7.

This small review is by necessity incomplete and fragmented. It is meant only as a provocative stimulus to a broader search for common denominators in the now widely known syndromes of insulin resistance.[34-35]

References

1. D.W. Nixon, S.B. Heymsfield, A. Cohen, M. Kutner, J.O. Ansley, D.H. Lawson, D. Rudman, Protein-calorie undernutrition in hospitalized cancer patients, *Am J Med* 68:683 (1980).
2. C. Waterhouse, L.D. Fenninger, E.H. Keutmann, Nitrogen exchange and calorie expenditure in patients with malignant neoplasms, *Cancer* 4:499 (1951).
3. A. Theologides, Cancer cachexia, *Cancer* 43:2004 (1979).
4. I. Warnold, K. Lundholm, T. Schersten, Energy balance and body composition in cancer patients, *Cancer* 43:2004 (1979).
5. M. Shike, D. McR. Russel, A.S. Detsky, J.E. Harrison, K.G. McNeill, F.A. Shepherd, R. Feld, W.K. Evans, K.N. Jeejeebhoy, Changes in body composition in patients with small-cell lung cancer. The effect of total parenteral nutrition as an adjunct to chemotherapy, *Ann Int Med* 101:303 (1984).
6. R.W. Begg, Tumor-host relations, *Adv Cancer Res* 5:1 (1958).
7. L. Axelrod, J.B. Halter, D.S. Cooper, T.T. Aoki, A.M. Roussell, S.L. Bagshaw, Hormone levels and fuel flow in patients with weight loss and lung cancer. Evidence for excessive metabolic expenditure and for an adaptive response mediated by reduced level of 3,5,3′-triiodothyronine, *Metabolism* 32:924 (1983).
8. M.-U. Yang, T.B. Van Itallie, Composition of weight lost during short-term weight reduction. Metabolic responses of obese subjects to starvation and low-calorie ketogenic and nonketogenic diets, *J Clin Invest* 58:722 (1976).
9. J.O. Hill, S.K. Fried, M. DiGirolamo, Effects of fasting and restricted refeeding on utilizing of ingested energy in rats, *Am J Physiol* 247:R318 (1984).
10. D.W. Nixon, D.H. Lawson, M.H. Kutner, J.O. Ansley, M. Schwartz, S.B. Heymsfield, R. Chawla, T.H. Cartwright, D. Rudman, Hyperalimentation of the cancer patient with protein-calorie undernutrition, *Cancer Res* 41:2038 (1981).
11. W.T. Chance, F.M. van Lammeren, M.H. Chen, R.F. Murphy, S.N. Joffe, J.E. Fischer, Alteration in plasma levels of insulin and glucagon associated with cancer anorexia, *Surg Forum* 34:441 (1983).
12. G. Costa, Cachexia, the metabolic component of neoplastic diseases, *Progr Exp Tumor Res* 3:321 (1963).
13. R.A. Liebelt, A.G. Liebelt, H.M. Johnston, Lipid mobilization and food intake in experimentally obese mice bearing transplanted tumors (35924), *Proc Soc Exp Biol Med* 138:482 (1971).
14. H. Masuno, N. Yamasaki, H. Okuda, Purification and characterization of a lipolytic factor (toxohormone-L) from cell-free fluid of ascites sarcoma 180, *Cancer Res* 41:284 (1981).
15. S. Kitada, E.F. Hays, J.F. Mead, Characterization of a lipid mobilizing factor from tumors, *Progr Lipid Res* 20:823 (1981).
16. B. Beutler, D. Greenwald, J.D. Hulmes, M. Chang, Y.C. Pan, J. Mathison, R. Ulevitch, A. Cerami, Identity of tumor necrosis factor and the macrophage-secreted factor cachectin, *Nature* 316:552 (1985).
17. C. Grunfeld, M. Palladino, Jr., Tumor necrosis factor: Immunologic antitumor, metabolic, and cardiovascular activities, *Adv Int Med* 35:45 (1990).
18. R.J. Spiegel, E.J. Schaefer, I.T. Magrath, B.K. Edwards, Plasma lipid alterations in leukemia and lymphoma, *Am J Med* 72:775 (1982).
19. M. DiGirolamo, F.D. Newby, J. Lovejoy, Lactate production in adipose tissue: A regulated function with extra-adipose implications, *FASEB J* 6:2405 (1992).
20. W.T. Chance, L. Cao, M.W. Kim, J.L. Nelson, J.E. Fisher, Reduction of tumor growth following treatment with a glutamine antimetabolite, *Life Sci* 42:87 (1988).
21. J.P. Grant, S.A. Wells, Tumor resistance in rats immunized to fetal tissues, *J Surg Res* 16:533 (1974).
22. C.P. Holroyde, T.G. Gabuzda, R.C. Putnam, P. Paul, G.A. Reichard, Altered glucose metabolism in metastatic carcinoma, *Cancer Res* 35:3710 (1975).

23. D. McR. Russell, M. Shike, E.B. Marliss, A.S. Detsky, F.A. Shepherd, R. Feld, W.E. Evans, K.N. Jeepeebhoy, Effects of total parenteral nutrition and chemotherapy on the metabolic derangements in small cell lung cancer, *Cancer Res* 44:1706 (1984).
24. J.H. Edmonson, Fatty acid mobilization and glucose metabolism in patients with cancer, *Cancer* 19:277 (1966).
25. G. Costa, W.W. Lane, R.G. Vincent, J.A. Siebold, M. Aragon, P.T. Baxley, Weight loss and cachexia in lung cancer, *Nutr Can* 2:98 (1980/81).
26. R.A. DeFronzo, R.C. Bonadonna, E. Ferrannini, Pathogenesis of NIDDM: A balanced overview, *Diabetes Care* 15:318 (1992).
27. D.L. Crandall, S.K. Fried, A.A. Francendese, M. Nickel, M. DiGirolamo, Lactate release from isolated rat adipocytes: Influence of cell size, glucose concentration, insulin and epinephrine, *Horm Metabol Res* 15:326 (1983).
28. F.D. Newby M. Sykes, M. DiGirolamo, Regional differences in adipocyte lactate production from glucose, *Am J Physiol* 255:E716 (1988).
29. J. Lovejoy, B. Mellon, M. DiGirolamo, Lactate generation following glucose ingestion. Relation to obesity, carbohydrate tolerance and insulin sensitivity, *Int J Obes* 14:843 (1990).
30. J. Lovejoy, F.D. Newby, S.S.P. Gebhart, M. DiGirolamo, Insulin resistance in obesity is associated with elevated lactate levels and diminished lactate appearance following intravenous glucose and insulin, *Metabolism* 41:22 (1992).
31. A. Kashiwagi, M.A. Verso, J. Andrews, B. Vasquez, G. Reaven, J.E. Foley, In vitro insulin resistance of human adipocytes isolated from subjects with noninsulin-dependent diabetes mellitus, *J Clin Invest* 72:1246 (1983).
32. P. Marin, M. Rebuffe-Scrive, U. Smith, P. Bjorntorp, Glucose uptake in human adipose tissue, *Metabolism* 36:1154 (1987).
33. P.-A. Jansson, U. Smith, P. Lonnroth, Evidence for lactate production by human adipose tissue in vivo, *Diabetologia* 33:253 (1990).
34. G.R. Reaven, Role of insulin resistance in human disease, *Diabetes* 37:1595 (1988).
35. R.A. DeFronzo, E. Ferrannini, Insulin resistance: A multifaceted syndrome responsible for NIDDM, obesity, hypertension, dyslipidemia, and atherosclerotic cardiovascular disease, *Diabetes Care* 14:173 (1991).
36. R.B. Fraser, F.A. Ford, G.F. Lawrence, Insulin sensitivity in third trimester pregnancy. A randomized study of dietary effects, *Brit J Obst Gynecol* 95:223 (1988).
37. G.M. Reaven, C. Hollenbeck, C.-Y. Jeng, M.S. Wu, Y.-D.I. Chen, Measurement of plasma glucose, free fatty acid, lactate and insulin for 24 h in patients with NIDDM, *Diabetes* 37:1020 (1988).
38. E. Ferrannini, G. Buzzigoli, R. Bonadonna, M.A. Giorico, M. Oleggini, L. Graziadei, R. Pedrinelli, L. Brandi, S. Bevilacqua, Insulin resistance in essential hypertension, *N Engl J Med* 317:350 (1987).
39. S. Inchiostro, G. Biolo, E. Duner, E. Lori, L. Sabadin, G. Merola, M.G. Fratton, A. Tiengo, F. Mantero, P. Tessari, Metabolic profiles in Cushing's syndrome in the post-absortive state and during euglycemic, hyperinsulinemic clamps, *Diab Nutr Metab* 5:47 (1992).

Poster Abstracts

Altered Expression of Intermediate Biomarkers for Mammary Preneoplasia: Relevance to Cancer Chemoprevention

N. TELANG, H.L. BRADLOW and M.P. OSBORNE

Strang-Cornell Cancer Research Laboratory, Cornell University Medical College, New York, NY

Intermediate biomarkers are molecular, metabolic, endocrine, and cellular end points whose expression precedes the detection of cancer. These end points may predict the chemopreventive efficacy of tumor inhibitors. This study was conducted to validate the relevance of selected molecular, endocrine, and cellular biomarkers for efficacy of prototypic tumor suppressing agents. Mouse mammary epithelial cells were initiated for transformation by chemical carcinogen, *Ras* oncogene and murine mammary tumor virus. The three initiators exhibited at least a two– to five–fold increase in *Ras* p21-GTP binding (molecular marker), a two– to six–fold increase in C16α/C2 hydroxylation of estradiol (endocrine marker), and a 30– to 50–fold increase in anchorage-independent growth (cellular marker). Treatment of initiated cells with the highest noncytotoxic doses of polyunsaturated n-3 fatty acids, retinoid, anti-estrogens, and indole derivatives resulted in downregulation of all the three perturbed biomarkers. Enhanced expression of molecular, endocrine, or cellular markers in initiated, tumorigenic target cells, and their *in vitro* modulation by prototypic tumor suppressing agents demonstrates that the biomarkers provide specific and sensitive end points for evaluating the efficacy of chemopreventive intervention.

Supported by the Wanda Jablonski Fund.

Ascorbic Acid (AA) and Glutathione (GSH) Protect Rat Lung from Chrysotile Asbestos-Induced Oxidative Stress

S.G. KHAN,[1] J.M. ARIF and Q. RAHMAN

Industrial Toxicology Research Centre, Lucknow, India; and
[1]Department of Dermatology, Case Western Reserve University, Cleveland, OH

We reported in earlier studies that the exposure of chrysotile to rat lungs depleted the levels of AA and GSH besides causing fibrosis and hyperplasia around the bronchioles. Based on these observations, we studied the protective role of AA (0.85 mM/day, orally) and GSH (75 µM/day, intraperitoncally) against the asbestos-induced oxidative stress. Rats were treated with AA, GSH, or AA + GSH daily after exposure to a single dose of chrysotile (5 mg, intratracheally) and were sacrificed after 1, 4, 8, and 16 days. Animals either treated with chrysotile or normal saline served as controls. The activities of glutathione peroxidase (GSH-Px), glutathione reductase (GSH-r), catalase, and glucose-6-phosphate dehydrogenase (G-6-PD) were increased significantly in the lung cytosols of chrysotile treated animals, while the levels of GSH and AA were reduced. The levels of H_2O_2 and lipid peroxidation (LPO) significantly increased in the lung microsomes. Ascorbic acid and GSH alone and in combination did not significantly alter the asbestos-mediated increase in the activities of GSH-Px and GSH-r, but activities of catalase and G-6-PD, and levels of H_2O_2 and LPO in the lung significantly decreased. In the animals treated with AA or GSH alone or in combination, the levels of these antioxidants were increased. Ascorbic acid and GSH also ameliorated the asbestos-mediated diminution of the activities of phase I and phase II drug metabolizing enzymes. Our data indicate that AA and GSH, being part of a physiologically significant antioxidant system, can prevent some asbestos-induced biochemical events which subsequently may protect lungs against the development of pathological conditions.

Estimated Breast Cancer Risk Reduction with Weight Loss: Implications in Dietetic Practice

N.B. KUMAR, D.V. SCHAPIRA and G.H. LYMAN

H. Lee Moffitt Cancer Center & Research Institute and University of South Florida College of Medicine, Tampa, FL

Obesity in women can involve predominantly abdominal fat deposition (android) or deposition on the buttocks and thighs (gynoid). Upper body fat localization has been associated with an increased risk of breast and endometrial cancers. The aim of this study was to determine whether calorie reduction and weight loss altered body fat distribution, preferentially reducing abdominal obesity and thus theoretically reducing cancer risk. Anthropometric parameters were determined in 189 apparently healthy, overweight women between ages 21–75. The study demonstrated that 64.2% with at least 4.5 kg weight loss decreased their upper body fat localization as measured by a reduction in suprailliac:thigh skinfold ratios ($p \leq 0.001$) and other skinfold thickness indicative of upper body fat reduction such as suprailliac ($p \leq 0.001$), subscapular ($p \leq 0.002$) and abdomen ($p \leq 0.001$) skinfolds. The reduction in estimated cancer risk based on a previously generated multiple logistic regression model was 45% after a 4.5 kg weight loss or more. Greater weight loss further reduced upper body fat localization and estimated breast cancer risk based on this model.

In conclusion, weight loss with a resultant decrease in upper body fat localization resulted in a significant theoretical reduction in breast cancer risk. Dietitians, who play a critical role in the treatment of cancer, may see their new and significant role in the prevention of this disease by identifying women with predominantly upper body fat distribution and by promoting weight loss, especially in high risk populations with a family history of breast or endometrial cancers.

New Functions for Niacin Imply a Role in Cancer Prevention

E.L. JACOBSON

Markey Cancer Center & Department of Clinical Nutrition, University of Kentucky, Lexington, KY

A new focus for studies of the relationship between niacin and cancer has evolved from the discovery that the principal form of this vitamin, NAD, is consumed as a substrate in ADP-ribose transfer reactions. Four unique classes of ADP-ribosyltransferases that are responsible for the turnover of NAD have been identified, although their functions are yet poorly understood. One of these, poly(ADP-ribose) polymerase, generates polymers of ADP-ribose by the successive transfer of ADP-ribose groups from NAD. This enzyme is activated by DNA strand breaks and functions in repair of DNA. The biochemical characterization of poly(ADP-ribose) polymerase has shown that the K_m of the enzyme for NAD is in the same range as the intracellular concentration of NAD in tissues. The significance of this finding is increased by the observations that: nutritional deprivation of nicotinamide in model *in vitro* systems readily results in decreased intracellular NAD, and limiting niacin intake in humans to levels found in the lowest quartile in the US population also results in a decrease of up to 70% of intracellular NAD [Fu, *et al.*, *J Nutr* 119: 1949 (1989)].

The above studies suggested that maintenance of intracellular NAD is a critical factor in ADP-ribose polymer metabolism and in responses to DNA damage. Thus, the biochemical and biological consequences of DNA damage have been studied in cultured mouse and human cells as a function of progressive nicotinamide depletion. Decreased NAD levels, a selective inhibition of DNA repair replication, and impaired ADP-ribose polymer metabolism were observed. These biochemical abnormalities were accompanied by increased cytotoxicity and malignant transformation along with prolonged cell cycle blocks. These findings suggest that niacin may be a protective factor in carcinogenic events.

A method to assess niacin status that is suitable for screening large populations of humans has been developed and is being tested. The potential for intervention has been demonstrated in patients treated with niacin for hypercholesterolemia.

Changes in Tumor Membrane Structure and Immunologic Activity Induced by Dietary Omega-3 Fatty Acids

L.J. JENSKI, L.K. STURDEVANT, W.D. EHRINGER, A.C. DUMAUAL and W. STILLWELL

Department of Biology, Indiana University-Purdue University at Indianapolis, IN

Omega-3 fatty acids, abundant in fish oil, are associated with health benefits for diseases including cancer. We are exploring how the incorporation of dietary docosahexaenoic acid (DHA, 22:6 ω-3), the most unsaturated fatty acid found in biological systems, into tumor cell plasma membranes alters the tumor cell's susceptibility to immune attack. Mice were fed diets containing 10.5% fat in various combinations of corn oil, hydrogenated coconut oil, and menhaden (fish) oil, and were inoculated intraperitoneally with a murine leukemia. The tumors that grew were tested for susceptibility to T-cell-mediated lysis and lysis with monoclonal anti-H-2 antibody plus complement, expression of H-2 proteins by radioimmunoassay, and changes in membrane structure. The tumor plasma membranes were analysed for phospholipid class and fatty acid content. Tumor cell sensitivity to cell-mediated lysis correlated positively with DHA content in tumor membranes. In contrast, complement-mediated cytolysis directed by a monoclonal antibody against an epitope on D^d was reduced as a function of dietary fish oil, suggesting, as did radioimmunoassay, a DHA-associated decrease in this epitope's expression. The expression of other epitopes, however, was increased or unchanged. DHA-rich membranes showed an increased proportion of liquid crystalline domains, but little change in global fluidity. Although a benefit of fish oil diets has not been linked through epidemiological evidence to this type of cancer, our results showed decreased morbidity and reduced tumor growth when the tumor-bearing hosts were fed diets rich in fish oil. We propose that the structural changes in tumor cell membranes induced by dietary DHA alter the expression of membrane proteins targeted by the immune system, and thus may in part be responsible for decreased tumor growth.

Supported in part by the American Institute for Cancer Research grant 90B06.

Carotenoids Upregulate Cell-to-Cell Communication and Connexin43 Gene Expression in Human Dermal Fibroblast Cells

L-X. ZHANG and J.S. BERTRAM

Molecular Oncology Unit, Cancer Research Center of Hawaii, University of Hawaii, Honolulu, HI

Carotenoids are plant pigments found in vegetables and fruits as common components of the human diet. Evidence from epidemiological and experimental studies has shown that carotenoids have anticancer action. However, their mode of action is unknown. We have previously reported that carotenoids, which inhibit carcinogen-induced neoplastic transformation of mouse fibroblast cells, increase gap junctional communication (GJC) [*Carcinogenesis*, 12:2109 (1991)], induce the gene expression of connexin43 (Cx43), a gene coding for a major gap junctional protein in various cells and tissues, and this activity is highly correlated with their ability to inhibit transformation. Here, we extend these carotenoid effects on murine cells to human dermal fibroblasts. Using a dye microinjection method, we found that carotenoids, β-carotene and canthaxanthin, strongly enhance GJC in a dose- and time-dependent manner. Northern and Western blotting revealed that this enhanced communication was accompanied by an increase in the levels of Cx43 mRNA and protein. Immunofluorescence microscopy using a connexin43 antibody demonstrated that carotenoids induced many junctional plaques in regions of cell-to-cell contact. Since GJC plays an important role in cell growth control, we propose that much of the chemopreventive action of carotenoids in humans could be explained by the enhanced intercellular transfer of growth regulatory signals. Since canthaxanthin, a nonprovitamin A carotenoid, is as active as the provitamin A carotenoid, β-carotene, the induction of GJC and Cx43 gene expression by carotenoids is unlikely to be due to provitamin A properties.

Supported in part by the American Institute of Cancer Research grant 89B60.

Dietary Fat-Fiber Interactions: Effect on Colonic Mediators of Cell Cytokinetics

R.S. CHAPKIN and J.R. LUPTON

Department of Animal Science and the Graduate Faculty of Nutrition, Texas A&M University, College Station, TX

In order to elucidate the biochemical mechanisms by which dietary fat and fiber modulate colonic cell proliferation and thus colon carcinogenesis, groups of 10 rats were fed one of 9 diets for 3 weeks: 3 types of fat at 15% by weight (beef tallow, corn oil, and fish oil) x 2 types of fiber (pectin and cellulose) plus fiber-free as a control group. The overall colonic fatty acid composition in mucosal total phospholipids reflected the incorporation of dietary fatty acids. An increase in the level of mucosal phospholipid arachidonic acid (20:4n-6) mass was associated with increased cell proliferation in the distal colon ($r = 0.70$, $p < 0.05$), measured by *in vivo* incorporation of bromodeoxyuridine (BRDU), whereas, an increase in the total phospholipid levels of fish oil derived n-3 fatty acids (20:5n-3 and 22:6n-3) was associated with decreased cell proliferation ($r = -0.73$, $p < 0.05$). The synthesis of prostaglandin E_2 (PGE_2) and PGI_2 derived from 20:4n-6, was significantly reduced by fish oil feeding ($p < 0.01$). In distal mucosa, pectin supplementation significantly increased PGE_2 and PGI_2 biosynthesis *ex vivo* ($p < 0.05$). Interestingly, there was a site-specific relationship between PGI_2 synthesis and cell proliferation. PGI_2 levels were negatively correlated in the proximal colon ($r = -0.67$, $p = 0.048$) and positively correlated in the distal colon ($r = 0.65$, $p = 0.059$), implying the effects of PG in the colon may be site specific. The main effect of fiber on cell proliferation was noted in the proximal colon, where pectin stimulated proliferation compared to cellulose and fiber-free diets ($p < 0.05$). Comparatively, the main effect of fat was confined to the distal colon, where beef tallow was more promotive with respect to cell proliferation compared to fish and corn oil fed animals ($p < 0.05$). However, in the proximal colon, the effect of fiber was highly dependent on the source of fat in the diet. Therefore, dietary fiber and fat modulate colonic cell proliferation in an interactive-site-specific manner. Since the effect of fiber on mucosal cell proliferation was confined to the proximal colon, protein kinase C (PKC) activity in membrane and cytosolic fractions was determined in proximal colonic mucosa. Protein kinase C activity was modulated by both the type of dietary fat and the source of fiber ($p < 0.05$). Increased membrane PKC activity ($r = 0.76$, $p = 0.02$) and increased PKC membrane/cytosol ratio ($r = 0.64$, $p = 0.06$) were positively associated with increased colonic crypt proliferative zone. These data suggest that the effects of dietary fibers and fats on colonic cell proliferation may be mediated in part through the modulation of mucosal PKC activity.

Supported in part by the American Institute for Cancer Research grant 91A19.

Comparison of the Effect of Selenite on Drug-Resistant and -Sensitive Tumor Cells

P.B. CAFFREY and G.D. FRENKEL

Biological Sciences Department, Rutgers University, Newark, NJ

The development of tumor cell drug resistance is considered a major cause for the failure of cancer therapy. In many cases drug resistant tumor cells have been shown to contain high levels of glutathione (GSH). Selenium compounds are known to act as anti-cancer agents and compounds such as selenite have been shown to induce cytotoxic effects in tumor cells. The available evidence suggests that for selenite, in contrast to most xenobiotics, a high level of intracellular GSH is correlated with a high degree of cytotoxicity. Based upon these observations, we hypothesized that tumor cells with high GSH levels, which are resistant to chemotherapeutic agents, will prove to be extremely sensitive to selenite. We examined this hypothesis *in vitro* by comparing the effects of selenite on drug resistant human ovarian tumor (NIH:OVCAR-3) cells, which have high GSH levels, and non-resistant human ovarian tumor (A2780) cells with low GSH levels. We employed three assays which measure different aspects of cytotoxicity *in vitro*: cell viability (ability of cells to exclude trypan blue), proliferation of viable cells, and attachment of cells to a solid matrix. In each of these assays the drug resistant cells proved to be significantly more sensitive to selenite than the drug sensitive cells. This study represents the first step in the evaluation of selenite as a treatment for drug-resistant tumors.

Supported by the American Institute for Cancer Research grant 91A40 and a grant from the National Institutes of Health.

Chemoprevention of DMBA-Induced Mammary Carcinoma by Inositol Compounds

I. VUCENIK,[1] M. BANSAL[1] and A.M. SHAMSUDDIN

Department of Medical and Research Technology,[1] Department of Pathology, University of Maryland School of Medicine, Baltimore, MD

Inositol (Ins) and inositol hexaphosphate ($InsP_6$ or phytic acid), are components of cereals. $InsP_6$ is particularly abundant in rice, and wheat bran. Both $InsP_6$ and Ins have been shown to inhibit experimentally induced colon cancers, in both rats and mice; Ins potentiating the action of $InsP_6$. The effect of $InsP_6$ and Ins on 7,12-dimethylbenz(a)anthracene (DMBA)-induced mammary carcinoma has been examined in female Sprague-Dawley rats. The rats were started on a treatment regimen of either Na-$InsP_6$, Ins or Na-$InSP_6$6+Ins (15 mM each) in drinking water one week prior to intragastric administration of a single dose of 20 mg DMBA. After 17 weeks of treatment our results were as follows:

Treatment	Number of rats	Number of rats with tumor	Number of tumors	Number of tumors per rat	Number of tumors per tumor bearing rat	Tumor diameter (mm)	Rats with ≥5 tumors (%)
DMBA	10	7	28	2.8 ± 3.9	4.0 ± 4.2	12.6 ± 5.5	20.0
DMBA + 15 mM $InsP_6$	21	17	36	1.7 ± 1.5	2.1 ± 1.4	12.4 ± 7.6	4.8
DMBA + 15 mM Ins	25	20	43	1.7 ± 1.5	2.2 ± 1.4	10.6 ± 6.1	8.0
DMBA + 15mM $InsP_6$ 15 mM Ins	21	12	37	1.8 ± 1.8	3.1 ± 1.2	10.8 ± 5.8	0.0

We show a moderate reduction of mammary tumorigenesis; particularly interesting was the apparent beneficial effect of $InsP_6$+Ins. The number of tumors per tumor bearing rats was reduced by half, the tumor diameter was reduced by 25%, and the number of rats that developed five or more tumors was reduced in all treated groups. Since in this study a very high dose of DMBA (20 mg) was used, and the number of animals was small, not surprisingly the results were not statistically significant. Judging from body weight gains, food intake, general appearance of the control animals, and serum levels of various divalent cations, the compounds are not toxic. Additional investigations with better study design are needed to confirm these preliminary observations. Because inositol compounds are –effective against various cancers and carcinogens, found naturally (cereals and legumes), and nontoxic– they have great potential as antineoplastic agents and are recommended for clinical trials.

Supported in part by the American Institute for Cancer Research grant 91SG04.

The Influence of Folic Acid Content of Media on TPA-Induced Transformation of JB6 Cells

Y.W. CHIEN, C.W. PRINCE and R.J. HINE

Department of Nutrition Sciences, University of Alabama at Birmingham, AL; and Arkansas Cancer Research Center; and Department of Dietetics and Nutrition, University of Arkansas for Medical Sciences at Little Rock, AR

The influence of folates on carcinogenesis has been studied for many years, but there are still few published data about the effects of folates on carcinogenic events in cells. The JB6 murine epidermal cell line is well developed and characterized as a model of late stage tumor development. Transformation of these cells in the presence of a promoting agent gives rise to new phenotypes that overcome normal growth controls. Anchorage-independent growth is a phenotypic characteristic of transformed cells. The objectives of these studies were: 1) to examine the growth of JB6 mouse epidermal cells cultured in media with variable folate content and 2) to assess the effects of folate content of media on 12-o-tetra-decanoylphorbol-13-acetate (TPA)-induced anchorage-independent growth of JB6 cells. During the growth studies, the cells were cultured for up to twelve days in several media with folate content ranging from 1.8 nM to 3.0 μM. Growth studies showed a positive, dose response relationship between the folate level of the media and cell growth. Before the anchorage-independent assays were performed, to test cell growth in soft agar with and without 5 ng/ml TPA, the cells were cultured in media containing different levels of folates for four passages. The number of colonies at the end of the fourteen–day incubation period was significantly higher ($p<0.05$) with each ten-fold increase in media folate content. The increase in colony number was especially marked between the cells grown in media containing 230 nM versus 2.3 μM folic acid. Higher folate content of cell culture media facilitated TPA-induced transformation of JB6 mouse epidermal cells. The mechanism responsible for this finding remains to be identified.

Supported in part by the American Institute for Cancer Research grant 90SG12.

Diets Containing High Levels of Corn Oil Suppress Immunoresponsiveness and Exacerbate UV-Carcinogenesis

H.S. BLACK and J. GERGUIS

Baylor College of Medicine
Veterans Affairs Medical Center, Houston, TX

Although ultraviolet radiation (UV) is considered a "complete" carcinogen, an irradiation protocol has been devised that approximates the classic two-stage chemical carcinogenesis protocol, allowing inferences to be made as to the segment of the carcinogenic continuum at which dietary lipid exerts its principal effect upon UV-induced skin cancer. Employing this protocol, we have determined that enhancement of UV-carcinogenesis by high levels of polyunsaturated fat (12% corn oil) occurs during the promotion/progression stage; that replacement with a low fat (0.75% corn oil) diet after initiation will negate the exacerbating effect of high fat; and that an omega-3 fatty acid source (12% menhaden oil) inhibits UV-carcinogenesis even at high dietary levels, although apparently exerting its inhibitory effect during UV-initiation. In addition, both dietary lipid sources (corn oil vs. menhaden oil) and level of corn oil markedly influence delayed-type hypersensitivity (DTH) to the hapten, dinitrochlorobenzene. As level of corn oil also influences tumor expression and prostaglandin E_2 levels in a near linear fashion and as PGE_2 has been reported to modulate immune hypersensitivity responses, we have examined the influence of low/high corn oil levels on DTH during various time intervals within the UV-carcinogenesis protocol. DTH response was significantly ($p < 0.01$) suppressed in animals receiving high fat diet prior to irradiation. Although both dietary groups exhibited UV-induced immunosuppression, there remained a significant ($p = 0.01$) difference in response between the dietary groups through nine weeks of irradiation. These data indicate that level of dietary lipid has a pronounced effect upon immunoresponsiveness during carcinogenesis and may partially explain the exacerbation of carcinogenic expression by high corn oil diets.

Supported by the American Institute for Cancer Research grant 90BW03.

Nutritional Problems in Patients with Advanced Cancer

P.J. DEETER,[1] P.S. GOLDSTEIN[2] and T.D. WALSH[3]

[1]*Department of Nutrition Services,* [2]*Hospice of The Cleveland Clinic,* [3]*Palliative Care Program, Cleveland Clinic Foundation, Cleveland, OH*

In order to determine the incidence of nutrition related signs and symptoms, a prospective study of 100 patients with advanced cancer was conducted using a standardized nutrition assessment tool. The patients were admitted to The Hospice of the Cleveland Clinic. The group was comprised of 42 males and 58 females, with a median age of 70 and a range of 24 to 92 years. Common primary sites were lung (20%), breast (13%), pancreas (8%), colorectal (7%), and prostate (7%). Common gastrointestinal symptoms were weight loss (79%), early satiety (70%), anorexia (67%), constipation (46%), xerostomia (41%), and taste changes (40%). Data were analyzed for relationships between occurrence of nutritional problems and primary site, gender, age, and weight loss (including percentage of usual body weight lost and acuity of weight loss, or weight lost over time), as well as relationships among nutrition related signs and symptoms.

Primary head and neck cancers were more common among males ($p = 0.01$) and they suffered weight loss of greater acuity ($p = 0.03$), as well as more frequent dysphagia ($p = 0.01$). They experienced, however, fewer complaints of anorexia ($p = 0.00$) and early satiety ($p = 0.03$), while those patients with pancreatic cancer experienced a greater incidence of early satiety ($p = 0.05$). Weight loss was significantly less common ($p = 0.01$) among patients with breast cancer.

Nutrition related signs and symptoms showed no statistically significant correlation with age or gender, with one exception. Our data shows a greater incidence of dyspepsia among women ($p = 0.05$).

Weight loss was assigned an acuity factor, based on Blackburn's criteria for mild, moderate, and severe weight loss. The number of signs and symptoms was tallied and each patient given a symptom score. A mild correlation was made between acuity of weight loss and symptom score. Weight loss $< 10\%$ and $\geq 10\%$ showed no significant relationship to symptom score. There was, however, a highly significant ($p = 0.001$) increase in symptom score for patients with weight loss, in general.

Gastrointestinal signs and symptoms occurring in $\geq 20\%$ of the patients were analyzed for concurrent problems. Though several interrelationships exist, it is important to note that taste changes, early satiety, and nausea contribute to anorexia and, either directly or indirectly, weight loss and cachexia. Therefore, future research in the management of weight loss and cachexia should emphasize the alleviation of these symptoms.

Chemopreventative Activity of Chlorophyllin: Inhibition of Mutagenicity and Covalent DNA Binding of Benzo[a]pyrene-7,8-dihydrodiol-9,10-epoxide

Y.J. SURH

Yale University School of Medicine, Department of Epidemiology and Public Health, New Haven, CT

A variety of plant and vegetable constituents have been found to possess substantial inhibitory effects against some chemical carcinogens and mutagens. One of the major antimutagenic factors in plant and vegetable extracts is chlorophyll. Because of the insolubility of chlorophyll, its water soluble sodium/copper salt, chlorophyllin (CHL), has been used as an alternative in studying the biological effects of chlorophyll. CHL strongly inhibits the genotoxicity of certain environmental mutagens. However, mechanisms by which CHL exerts its antimutagenic effects remain unclear.

In the present study, protective properties of CHL against a direct-acting mutagen, benzo[a]pyrene-7,8-dihydrodiol-9,10-epoxide (BPDE), was investigated. When [^{3}H]BPDE (0.25 μM) was incubated with calf thymus DNA in the presence of CHL (100 μM), covalent BPDE-DNA binding was significantly reduced (66% inhibition). Hemin, which is structurally related to CHL, also inhibited the covalent binding of BPDE to DNA. In a parallel experiment, the effect of CHL on BPDE-induced bacterial mutagenicity was investigated using *Salmonella typhimurium* TM677 as a tester strain. BPDE was a direct mutagen in a dose-dependent manner at concentrations from 0.5 μM to 5 μM. CHL (0.1 mM) completely inhibited the mutagenic activity of BPDE at all concentrations tested. CHL also protected bacterial cells against cytotoxicity induced by BPDE; cell survival was maintained at the control level in the presence of CHL. The inhibition by CHL of mutagenicity and covalent DNA binding of BPDE appears to be associated with formation of an inactive complex between CHL and BPDE. A possible protective role of CHL in carcinogenesis by benzo[a]pyrene and BPDE is under investigation.

Antimutagenic Activity of Natural Carotenoid Extracts

G. LOZOVIK,[1] M. SHLYANKEVICH[2] and A. SERGEEV[2]

[1]*Innovative Foods, Inc., San Francisco, CA*
[2]*Cancer Research Center, Russian Academy of Medical Science, Moscow, Russia*

The carotenoids, contained in natural oil extracts, are considered by us as the possible resource for cancer prevention in unfavorable areas in Russia, especially for victims of the Chernobyl disaster. Recently, we showed the preventive effect of natural carotenoids in spontaneous and chemically induced carcinogenesis models. In this work, we investigated the antimutagenic activity of natural oil extracts containing carotenoids.

We prepared oil extracts of rosehips (*Rosa cinnamomea L.*), rowan (*Sorbus aucuparia L.*), and buckthorn (*Hippophal rhamnoides L.*) and compared their activity with Russian synthetic β-carotene.

We used the chromosome aberration activity in marrow cells of mice for evaluation of antimutagenic properties. Male CBA mice received chow diet with natural extract or oil solution of synthetic β-carotene at the same dose (10 mg/kg). Induction of chromosome aberrations was made by intraperitoneal injection of 15, 30, or 60 mg/kg cyclophosphamid (CP) at 7, 12 and 18 days. The marrow cells were taken away after 24 hours of CP injections, and metaphases were prepared and stained by standard procedure.

Natural carotenoid extracts decreased the level of chromosome aberrations by 30% to 50% at all doses of CP. The solution of synthetic β-carotene in oil with antioxidants showed approximately the same activity. Soybean oil decreased aberrations, alone and in a mixture with antioxidants. Our results showed that natural oil carotenoid extracts are effective and better antimutagens than oil solution of synthetic β-carotene. This can be explained partly by synergistic activity of β-carotene with other soluble components (e.g., carotenoids and tocopherols), and partly by the better bioavailability in natural extracts.

Effects of RRR-α-Tocopheryl Succinate on IL-1 and PGE_2 Production by Avian Macrophages

E.H. ROMACH, B.G. SANDERS and K. KLINE

Department of Zoology and Division of Nutritional Sciences
University of Texas, Austin, TX

Vitamin E is an immunoregulatory agent that enhances both cellular and humoral immunity. Vitamin E has been postulated to function as a potent immune enhancer by increasing stimulatory cytokine production and by down-regulating suppressive prostaglandin synthesis. In an effort to better understand the mechanism whereby vitamin E ameliorates retrovirus-induced immune suppression, studies of vitamin E's effects on macrophages and their soluble products were conducted. PGE_2 levels produced by avian peritoneal exudate cells (PEC) cultured with avian erythroblastosis virus (AEV) in the presence or absence of vitamin E succinate were quantitated by a radioimmunometric assay. Interleukin-1 (IL-1) levels were quantified by a thymocyte co-mitogenesis assay using the avian macrophage cell line, HD11, cultured with and without vitamin E succinate supplementation. Supernatant from PEC obtained from normal chickens, exposed to AEV for 45 minutes, exhibited a 256% increase in the levels of PGE_2 when compared to replica PEC cultures not exposed to AEV. Pretreatment of PEC with different amounts of vitamin E succinate prior to exposure to AEV decreased PGE_2 levels in a dose dependent manner. PEC cultured with AEV and with 0.1 and 0.01 μg/ml of vitamin E succinate exhibited 55% and 77% reductions in PGE_2 levels, respectively. Vitamin E succinate treated HD11 cells exhibited IL-1 activity 196% greater than that observed with untreated and vehicle controls. Vitamin E succinate and lipopolysaccharide (LPS) acted synergistically to increase IL-1 activity. LPS treatment alone gave a stimulation index 135% higher than controls, while a combination of LPS and vitamin E succinate gave a stimulation index 220% higher than controls. Down regulation of negative immune function regulator, PGE_2, and upregulation of positive immune function regulator, IL-1, by vitamin E succinate could account for the ability of vitamin E succinate to ameliorate AEV-induced immune suppression.

Supported by PHS Grant Number CA45422, Awarded by the National Cancer Institute, DHHS.

Modulation of Phorbol Ester Induced Events by Dietary Corn Oil

M.F. LOCNISKAR, S.M. FISCHER and M.A. BELURY

Nutritional Sciences, University of Texas, Austin, TX;
University of Texas, M.D. Anderson Cancer Center,
Science Park-Research Division, Smithville, TX

Increased consumption of dietary corn oil, rich in linoleic acid, has been shown to have a protective effect on 12-O-tetradecanoylphorbol-13-acetate (TPA)-induced tumor promotion in mouse skin [Leyton *et al., Cancer Res.* 51:907-15 (1991)]. To investigate the molecular basis of this observation, it was hypothesized that the dietary lipids incorporated into the epidermal cell membrane modulate events associated with tumor promotion, specifically ornithine decarboxylase activity and hyperplasia, and subcellular distribution of protein kinase C activity. To test this hypothesis, weanling female SENCAR mice were fed modified AIN-76 diets containing 15% total fat for approximately four weeks. Corn oil and coconut oil were added to achieve three levels of linoleic acid: 0.8%, 4.5%, and 8.4% (1%, 7.9%, and 15% corn oil, respectively). Fatty acid analysis of each epidermal phospholipid fraction by gas chromatography revealed that increasing dietary linoleic acid was associated with an increase in phospholipid linoleate. Although TPA induction of ornithine decarboxylase was not significantly modified by dietary linoleate, TPA induced hyperplasia was decreased in the epidermis of mice fed the 8.4% linoleate diets compared to the other groups. Subcellular distribution of protein kinase C in the epidermis was altered by the different diets. In mice fed 0.8% linoleate, 69% of the protein kinase C activity was found in the cytosol compared to 78% and 74% for the mice fed 4.5% and 8.4% linoleate diets, respectively. It was also found that partially purified protein kinase C isolated from mouse epidermis was activated by linoleic acid *in vitro*, however, the extent of activation was less than that for arachidonic acid. These data suggest that the effect of dietary lipid on TPA tumor promotion may be mediated through modulation of protein kinase C.

Supported by National Institutes of Health grant CA 46886; M.A. Belur supported by National Institute of Environmental Health Sciences grant ES07247.

Declining Serum Cholesterol Levels Indicate Cancer Progression: Efforts for Increasing These Values up to 250mg per Treatment

E.K.M. BOSKAMP

Mary's House, Oldsum, Germany

In a check-up of 298 tumor-patient histories with gastrointestinal tract cancer at the UKE in Hamburg, a statistical correlation between tumor stage and cholesterol level was found: T1 = 250 mg/dliter, T2 = 214 mg/dliter, T3 = 201 mg/dliter, and T4 = 170 mg/dliter. Metastasis was observed at smaller levels than 185 mg/dliter and no progression was observed at levels higher than 245 mg/dliter. Similar results have been obtained from other authors for different kinds of cancer.

Cancer, cells in both solid tumors and leukemia, are characterized as undifferentiated, remaining in a juvenile state. Potentially, a lack of certain essential nutritional factors block complete differentiation of these cells. Considering the fact that only completely differentiated cells have the ability to emit signals to stop further cell production, the key-factor behind the uncontrolled growth of cancer cells, may be due to these cells having never reached their stage of complete differentiation.

Certain nutritional deficiencies in the diet of cancer-patients result from an insufficient intake of lipids, including the sterols and vitamins A, D, and E in natural fats. These nutritional deficiencies may cause cancer and its metastasis. A decline in cholesterol, vitamin A, and vitamin E level has been reported by others, both before and during the progression of cancer.

Unfortunately, many nutritional factors are eliminated during the processing of oilseeds to oil, during the milling of wheat to meal, or they are avoided by nutritional habits.

Dietetic therapy is proposed with the aim to increase cholesterol-levels to stop further progression and the development of metastasis.

In Germany, "1 2 3 V Capsules" (Ph.Z.No. 3226503) contain natural oil sources, in which the above mentioned lipids are in high concentration. These capsules and a change in nutritional habits are used in the treatment of Germans in a study arranged by the "Central Research Institute for Roentgenology and Radiology of the Ministry of Public Health" in St. Petersburg, Russia.

Anticancer Effect of *Ganoderma lucidum* on Human Cancer Cell Survival

E-C. JHEE

Chonbuk National University, College of Dentistry, Chonju, South Korea

Ganoderma lucidem, a widely distributed mushroom in Europe, America, Africa, and Asia has been greatly valued in China, Korea, and Japan as a traditional medicine. It has been reported that *Ganoderma* has both a hypotensive effect through the inhibition of an angiotensin converting enzyme, and an antiallergic effect through the inhibition of mast cells. It also inhibits lipid peroxidation and promotes the immune function.

Attempts were made to detect the anticancer effect of *Ganoderma* extracted in boiled water. A commonly used natural application in oriental medicine is a boiled water extract of *Ganoderma*. Our studies employed human cancer cell lines from hematopoietic origin K562, Raji and Molt-4.

The anticancer effect of *Ganoderma* was also investigated using other human cancer cells of anchorage-dependent cultures derived from liver, lung, kidney, and stomach cancers.

The anticancer effect was compared in three different preparations. These were a naturally grown *Ganoderma*, the most favored form used in herb medicine, a cultivated one and a molded one from malstorage.

The survival rate of cancer cells was measured by the ^{3}H-thymidine uptake and by a colorimetric assay.

A Meta-Analysis of Studies of Dietary Fat and Breast Cancer Risk

N.F. BOYD, L.J. MARTIN, M. NOFFEL,
G. LOCKWOOD and D.L. TRITCHLER

Division of Epidemiology and Statistics
Ontario Cancer Institute, Toronto, Ontario, Canada

We have carried out a meta-analysis to summarize quantitatively the large published literature on dietary fat in the etiology of breast cancer. The summary relative risk for the highest compared to the lowest level of intake for the 20 studies that examined fat as a nutrient was 1.15 (95% CI 1.05-1.25). Cohort studies had a summary relative risk of 0.96 (95% CI 0.80-1.15) and case control studies a relative risk of 1.21 (95% CI 1.10-1.34). The difference between cohort and case control results was entirely attributable to two of the four published cohort studies, the other two giving results similar to those of case control studies. Summary estimates of risk excluded unity for monounsaturated fat and saturated fat (case control studies only), but not for polyunsaturated fat. For the 17 studies that examined food intake, the summary relative risks were 1.22 (95% CI 1.09-1.37) for meat, 1.35 (95% CI 1.19-1.54) for milk, and 1.17 (95% CI 1.02-1.36) for cheese. Summary relative risks for total fat intake were generally higher in studies of higher quality, that adjusted for energy intake and risk factors, that employed population based comparison groups, and that were done in Europe.

Efficacy of Ethylvanillin in the Prevention of Mammary Tumors Induced by Methylnitrosourea (MNU) in the Rats

C.J. GRUBBS, T.L. CASEBOLT, I. ETO, M.M. JULIANA, L.M. WHITAKER, G.J. CANFIELD, V.E. STEELE[1] and G.J. KELLOFF[1]

Department of Nutrition Sciences, University of Alabama at Birmingham, AL;
[1]Chemoprevention Branch, National Cancer Institute, Bethesda, MD

Ethylvanillin is a synthetic flavoring agent that has been shown to have an inhibitory effect against mutagenic agents such as 4-nitroquinoline 1-oxide [Ohta, *et al. Fd Chem Toxic 24*:51 (1986)]. Ethylvanillin was evaluated in the present study in the MNU-induced mammary cancer model to determine the chemopreventive efficacy of this agent. Prior to this study, a six-week dose selection study confirmed that a dose of 20 grams of ethylvanillin/kg of diet (Teklad) was not toxic. Two dose levels (20 and 10 g/kg of diet) of ethylvanillin were administered in the chemoprevention study to female Sprague-Dawley rats beginning at 43 days of age and continuing until the end of the study. Carcinogen treated rats received one injection of MNU (50 mg/kg BW) via the jugular vein at 50 days of age. Rats were palpated for mammary tumors two times per week and checked daily for signs of toxicity. Diets were analyzed twice during the study to ensure accurate preparation and compound stability. At termination of the study (100 days after MNU), all grossly observed mammary tumors were weighed.

In rats that received only MNU, the incidence of mammary tumors was 95% (average number per rat equaled 3.4). The high dose of ethylvanillin caused a 20% decrease in tumor incidence and a 22% decrease in tumor number. Similar decreases in mammary tumor incidence and number were observed in rats receiving the low dose of this agent. The time of appearance of the mammary tumors was delayed in rats receiving ethylvanillin and the weight of the mammary tumors was decreased (61% less in rats receiving the high dose). Rats receiving ethylvanillin had normal estrus cycles, and a detailed histological evaluation revealed no compound-related changes in the animals. The mechanism responsible for the observed inhibition of mammary carcinogenesis is not known; the reduced growth rate of the tumors, however, suggests an effect during the promotional stage.

Supported by National Cancer Institute grants NO1-CN-95156-02 and CA28103-14.

Efficacy of Pyridoxal Therapy in Controlling the Growth of Melanomas in Cell Culture and an Animal Pilot Study

A.B. MAKSYMOWYCH, N.M. ROBERTSON and G. LITWACK

Department of Pharmacology, Jefferson Cancer Institute, Thomas Jefferson University, Philadelphia, PA

We have demonstrated, using confocal laser scanning microscopy, that pyridoxal treatment of B16C3 murine melanoma cells inhibits triamcinolone acetonide induced translocation of the glucocorticoid receptor (GR) to the nucleus. Further investigations extended our work with murine cell cultures, demonstrating that pyridoxal also kills WM9-83A human melanoma cells in culture. In studies with glucocorticoid antagonists, cortexolone functions to inhibit cell proliferation. This mechanism, however, appears to initiate in the GR signal transducing cascade at a point prior to the impact of pyridoxal treatment alone. RU486 alone has no detrimental effect on melanoma cell viability. However, in combination with pyridoxal, RU486 extends cell viability.

Based an our data that pyridoxal can kill melanoma cells in culture, we completed a pilot study examining the efficacy of using topical application of a pyridoxal cream to inhibit the growth and/or cause regression of (B16C3) xenograft melanoma tumors in an immunocompetent (Hairless Rhino-J^3) and an immunocompromised (Crl: nu/nu (CD1®) BR) murine animal model. The results of the study with immunocompetent animals are encouraging. Tumors are brought under control by pyridoxal treatment, however, further work needs to be done in order to determine the most efficacious treatment regimen and to establish formal concentrations for pyridoxal in our topical ointments. Data from the trials using immunocompromised animals indicated that although some qualitative differences may be detected between the control and experimental animals, tumor growth in these animals is so aggressive that multiple applications or higher concentrations of pyridoxal may be needed to obtain useful data from this system.

Supported by the American Institute for Cancer Research grant 87B73R90B.

Effect of Selenium on the Expression and Function of the Interleukin-2 Receptor

M. ROY, L. KIREMIDJIAN-SCHUMACHER, H.I. WISHE, M.W. COHEN and G. STOTZKY[1]

New York University Dental Center and Department of Biology,[1] New York, NY

This study examined the effect of dietary (2ppm for 8 weeks) or *in vitro* (1×10^{-7}M) supplementation with Se (as sodium selenite) on the expression of the p55 and p75 subunits of the interleukin-2 receptor (I12R) and the function of p55/p75 on the surface of lymphoblasts from 12 week-old C57B1/6J male mice. Scatchard analyses of ^{125}I-I12 binding (2nM-6pM) on the surface of Con A induced (5μg/ml; 48h) pooled spleen lymphoblasts indicated that cells from Se-supplemented animals expressed significantly higher numbers per cell of low (53,687 ± 6160; Kd 3.31×10^{-9}M) and high affinity (3,235 ± 216; Kd = 6.91×10^{-11}M) I12R as compared to cells from Se-normal animals (34,648 ± 2923, Kd = 1.78×10^{-9}; 2,169 ± 95, Kd = 4.36×10^{11}M, respectively). The numbers of both p55 (56,908 ± 5980) and p75 (12,736 ± 839) expressed per cell were also significantly higher than on cells from Se-normal animals (37,203 ± 2988; 9663 ± 1009, respectively), but their ratios were comparable. Supplementation with Se *in vitro* resulted in similar changes. The high affinity receptors on both cell types internalized I12 at the same rate (15 ± 0.9 and 14 ± 1.2 min 50% I12 internalization) but a significantly higher number of ^{125}I-I12 molecules were internalized by cells from Se-supplemented animals (1305 ± 107 vs. 835 ± 96) at 60 min after exposure. The results indicated that Se regulates the expression of both subunits of the I12R and thus can modulate I12-dependent responses.

Supported by the American Institute for Cancer Research grant 91A01.

Effect of Selenium Supplementation on Human Lymphocyte Tumor Cytotoxicity

L. KIREMIDJIAN-SCHUMACHER, M. ROY, H.I. WISHE, M.W. COHEN and G. STOTZKY[1]

New York University Dental Center and Department of Biology,[1] New York, NY

This study examined the effect of dietary (200 µg/day for 8 weeks) supplementation with Se (as sodium selenite) on the ability of human peripheral blood lymphocytes to respond to stimulation with antigen, develop into cytotoxic lymphocytes, and to destroy tumor cells. Sixteen male and six female volunteers were matched by age (mean age = 26.2 years; range = 23-37), sex, weight, and nutritional habits, and given selenite or placebo tablets. Lymphocytes (2.5×10^6), isolated prior to and 8 weeks following supplementation, were activated by allogeneic stimulation with 8.3×10^5 mitomycin C-treated Ragi cells for 5 days. Lymphocyte mediated tumor cytotoxicity was evaluated using a 4h–^{51}Cr release assay and effector: target cell ratios of 1.25-20:1. The results indicated that supplementation with selenite in humans results in a significant (118%) increase in the ability of a given lymphocyte population to destroy a fixed number of tumor cells, i.e., 45.57 ± 5.96 vs. 20.87 ± 2.62% cytotoxicity at 20:1 ratio. While the cytotoxic efficiency of both lymphocyte populations remained the same, the number of lymphocytes required to destroy 2×10^5 tumor cells decreased significantly after supplementation ($1.53 \times 10^6 \pm 1.98 \times 10^5$ vs. $7.10 \times 10^5 \pm 1.49 \times 10^5$). The results indicate that supplementation with Se in humans results in a significant increase in the number of cytotoxic lymphocytes within a cell population.

Supported by the American Institute for Cancer Research grant 91A01.

Reversibility of Changes in Rat Liver DNA Methylation and Gene Expression Induced by a Methyl-Deficient Diet

J.K. CHRISTMAN, G. SHEIKHNEJAD, M. DIZIK, S. ABILEAH and E. WAINFAN

Michigan Cancer Foundation and the Meyer L. Prentis Comprehensive Cancer Center of Metropolitan Detroit, Detroit, MI

As we have previously reported, both t-RNA and DNA become rapidly hypomethylated in livers of rats fed a cancer promoting, methyl-deficient diet [MDD-an amino acid defined, semisynthetic diet (AAD) devoid of choline, methionine, folate and vitamin B12 and supplemented with 9 g/kg D, L-homocystine]. Hypomethylation is detectable within one week and increases over a four week period of continued feeding of MDD.[1] Hypomethylation of specific CCGG sites is observed within several genes for which mRNA levels are increased (c-*myc*, c-*fos* and c-Ha-*ras*).[2] Interestingly, even though methylation of both DNA and tRNA is inhibited in the absence of a dietary source of methyl groups, the activity of DNA and tRNA MTases increases 2-3 fold.[3]

To determine the effects of restoring sources of methyl groups to the diet, rats fed MDD for 4 weeks were placed on a complete semi-synthetic diet (CSD-AAD supplemented with 2 g choline chloride, 5.2 g DL-methionine, 5 mg folic acid and 100 µg vitamin B12/kg). Within 1–3 weeks of feeding CSD, DNA and tRNA methyltransferase activities returned to the levels found in livers of age matched animals fed CSD continuously. Overall levels of DNA and t-RNA methylation, measured as ability of the isolated nucleic acids to serve as substrates for methylation *in vitro*, also returned to control values. However, examination of methylation at specific CCGG sites in c-*fos*, c-*myc* and c-Ha-*ras* genes in liver DNA indicated that remethylation of these sites occurred more slowly than remethylation of total DNA. At some sites, hypomethylation was still readily detectable after 3 weeks feeding of CSD. If persistent alterations in methylation of growth regulatory genes allow hepatocytes to more readily escape normal constraints on cell division, our finding could explain why prolonged feeding of less severely methyl deficient diets than those used in our studies can be sufficient to induce hepatocarcinogenesis in rats.

Supported by the American Institute for Cancer Research grants 86A48 to E. Wainfan and 92A35 to J.K. Christman, and the Lloyd and Marilyn Smith fund.

---

[1]E. Wainfan, M. Dizik, M. Stender, J.K. Christman, Rapid appearance of hypomethylated DNA in livers of rats fed cancer-promoting, methyl-deficient diets, *Cancer Res* 49:4094 (1989).

[2]M. Dizik, J.K. Christman, E. Wainfan, Alterations in expression and methylation of specific genes in livers of rats fed a cancer promoting methyl-deficient diet, *Carcinogenesis* 7:1307 (1991).

[3]J.K. Christman, L. Chen, M. Dizik, E. Wainfan, Regulation of gene expression by DNA methylation: a possible link between dietary methyl deficiency and hepatocarcinogenesis, *in*: *The role of nutrients in cancer treatment*, Report of the 9th Ross Conference on Medical Research, pp 60-65 (1991).

Cancer Inhibiting Effects of Bean Dietary Fiber and Soy Protein Isolate in Rats

J.S. HUGHES and C.H. GANTHAVORN

Food and Nutrition Science
Northern Arizona University, Flagstaff, AZ

Both epidemiological and experimental evidence indicate that dietary factors in general and dietary fiber and fat in particular are important in the etiology of colon cancer. Adequate dietary fiber consumption has been correlated with reduced risk of several cancers of the gastrointestinal tract, particularly colon cancer. Total dietary fat and saturated fat have been implicated as enhancing factors in colon cancer development.

Research to date has focused primarily on dietary fibers from cereals (i.e., wheat bran), or dietary fiber extracts rich in insoluble (i.e., cellulose) or soluble (i.e., pectin and guar gum) dietary fiber. Little is known about the potential protective effect of bean dietary fiber despite the fact that beans are widely consumed in unindustrialized countries (especially Latin America) where colon cancer incidence is low. Beans (*Phaseolus vulgaris*, also known as common beans or dry beans) are the most commonly consumed legume in the world and are a good source of both soluble and insoluble dietary fiber. The purpose of this research is to examine the role of dietary fiber in colon cancer prevention by using bean dietary fiber while controlling for the influence of dietary fat.

Sixty-five 7-week-old weanling female F344 rats were randomly assigned to five experimental groups and one control group. Experimental groups were treated with AOM, and fed diets containing varying levels of bean dietary fiber. Experimental diets contained varying amount of precooked pinto beans calculated to provide 0, 5, 10, 15, and 20% bean dietary fiber. The protein content of the experimental diets was adjusted to 18% with soy protein isolate. Fat content was adjusted to 5% with corn oil, and corn starch was added to make up the remainder of the diet.

Preliminary data indicate that both bean dietary fiber and the soy protein isolate appear to inhibit carcinogenesis. The lowest incidence of total malignant tumors was observed in the experimental animals fed 20% bean dietary fiber or 15% to 20% soy protein isolate. However, no apparent relationship was observed between the dietary fiber content of the diets and the number of animals bearing colon tumors. Additional analysis of the data is needed to determine if any relationship exists between the soluble and insoluble dietary fiber content of the diets and tumor incidence.

Supported by the American Institute for Cancer Research grant 91SG05.

Association Between Dietary Vitamin A Intake and Bronchial Metaplasia in Asbestos Exposed Workers

C.A. REDLICH, S.N. MOHR, S.T. MAYNE, W.W. MERRILL, D. CARTER, and M.R. CULLEN

Departments of Medicine, Epidemiology and Public Health, and Pathology, Yale University School of Medicine, New Haven, CT

Asbestos-exposed workers and cigarette smokers have an increased risk for lung cancer. Experimental and epidemiologic data suggest that vitamin A (retinol and/or β-carotene) has anticancer effects. It is unclear, however, whether intervention with supplemental vitamin A can reduce the risk of lung cancer.

Fifty volunteers with asbestosis, mostly former or current smokers, underwent detailed clinical, physiologic and exposure evaluation, followed by bronchoscopy with bronchoalveolar lavage (BAL) and multiple airway biopsies. Exposure to β-carotene and retinol was assessed by a previously validated questionnaire. Bronchial biopsies were read blindly by the study pathologist using a predesigned scoring system.

About one-third of subjects (n = 15) had histologic evidence of squamous metaplasia with keratinization, a potential early marker for lung cancer. This correlated strongly with diet and with the number of neutrophils on BAL, and weakly with history of smoking, or extent of asbestos exposure or disease. Vitamin A intake was markedly lower among those with metaplasia (53,000 IU total vitamin A equivalent weekly) compared to those without metaplasia (over 87,000 IU, $p < 0.003$), with approximately equal contributions of β-carotene and retinol to the effect.

These data suggest that reduced vitamin A intake may result in an increased risk of squamous metaplasia in asbestos exposed subjects. Currently in progress is a double blind placebo controlled trial of vitamin A in these subjects to determine whether this dietary intervention will reduce airway metaplasia and/or neutrophil alveolitis. Bronchoscopy with bronchial biopsies and BAL will be performed before and after 6 months of supplemental retinol (25,000 IU/day) and β-carotene (30 mg/day). Whether altering airway metaplasia modifies the risk for lung cancer remains unknown.

Supported by the American Institute for Cancer Research grant 92A69.

Cellular Events Involved in Dietary Sodium Deprivation Induced Inhibition of B16 Melanoma Proliferation

B.P. FINE, K.A. HANSEN, T.N. DENNY and T.R. WALTERS

Department of Pediatrics, UMDNJ-New Jersey Medical School, Newark, NJ

Dietary sodium deprivation inhibits both normal growth and solid tumor proliferation. This study proposed to ascertain the presence of circulating or cellular factors induced by dietary Na^+-deprivation that might alter *in vitro* growth of B16 melanoma cells. Serum and melanoma cells were harvested from 24 mice who were on either a Na^+-containing diet (control) or Na^+-restricted diet. The cells were isolated and then propagated in tissue culture. Cytoplasmic calcium ($[Ca^{2+}]_i$) was measured in cultured cells by dual wavelength image fluoroscopy using the indicator fura-2. Progression through the cell cycle was measured by the fluorescence activated cell sorter using propidium iodide. Cells in culture were serum-free for 24 hours; an alteration in cytoplasmic calcium $[Ca^{2+}]_i$ after a serum challenge was the early membrane event studied, and thymidine uptake, change in cell number, and progression through the cell cycle were the later nuclear events evaluated. A 2-way factorial analysis (diet vs. serum/cells) showed the main effect of serum from the Na^+-restricted mice was a lesser rise in $[Ca^{2+}]_i$ than from control serum (53 ± 22 vs. 111 ± 29 nM/liter (mean ± SEM)). There were also differences in the cellular response from mice on different diets.

Melanoma cells from mice on:	Na^+-restricted diet	Control diet
Baseline $[Ca^{2+}]_i$ (nM)	83 ± 10.4	43 ± 8.5
After serum stimulation: $[Ca^{2+}]_i$ increase to (nM)	237 ± 29	103 ± 16
Thymidine uptake % serum free	37 ± 6.1	53 ± 9.7
Rate of exit from G_0/G_1 during first 3 hours after serum stimulation (%)	20 ± 4.3	27 ± 5.5

In the cells from the controls there was a rapid exit from G_0/G_1 during the first 3 hours and then a second rapid egress at 16 hours. In contrast, fewer cells from the Na^+-deprived group showed the early exit from G_0/G_1; most cells exited at 14 to 16 hours. These data show that dietary Na^+-deprivation induces both circulating and cellular factors which alter cellular proliferation.

Supported in part by the American Institute for Cancer Research grant 89B19.

Effect of Specific Fatty Acids on p53 Expression in Rat Mammary Tumor Cells

J.K. TILLOTSON, L.A. COHEN and Z. RONAI

American Health Foundation, Valhalla, NY

In previous studies, we have demonstrated that, while dietary fats were not found to affect the frequency of *ras* mutation in rat mammary tumors induced by N-nitrosomethylurea (NMU), they did affect final incidence of tumors, supporting the idea that the diet mediates promotion-related changes. As altered p53 expression has been described in 45–55% of human breast cancers, we have used a rat mammary tumor cell line, NMU (ATCC CRL 1743) as a model to determine whether specific fatty acids might affect expression and, in turn, activity of the p53 tumor suppressor gene. Under standard growth conditions (10% FBS), NMU cells over-express p53. Metabolic pulse-chase labeling revealed p53 protein with a prolonged half-life of about 2 hours. To study the effects of specific fatty acids on p53 protein expression, NMU cells were exposed to medium containing high serum (10% FBS), low serum (1% FBS), or low serum plus 1 µg/ml linoleic acid (LA), oleic acid (OA), or docosahexaenoic acid (DHA)). Growth rates for NMU cells were reduced by 30% to 60% in
1% FBS compared with 10% FBS, while addition of LA resulted in an intermediate growth rate and no changes were observed after treatment with OA or DHA. After 3 days in the experimental medium, p53 expression was found to be higher in NMU cells maintained in low serum medium than in those grown in high serum. In the presence of low serum, LA was found to decrease p53 levels, an effect which was not seen with either OA or DHA. Increased p53 expression in cells grown in low serum correlated with their slower growth rate and, similarly, LA was able to decrease p53 expression as well as increasing the growth rate.

Supported in part by the American Institute for Cancer Research grant 92A24.

Colon Carcinogenesis is Inhibited More Effectively by Phytate than by Selenium in F344 Rats Given 30 mg/kg Azoxymethane

T.O. PRETLOW, M.A. O'RIORDAN and T.G. PRETLOW

Institute of Pathology, Case Western Reserve University School of Medicine, Cleveland, OH

Diets supplemented with either 2.5 ppm selenium (Se) or 2% phytate (P) in the drinking water have been reported to inhibit colon carcinogenesis in F344 rats treated with azoxymethane. Aberrant crypt foci (ACF) are putative preneoplastic lesions that can be quantified readily in unembedded segments of colon a short time after treatment with carcinogen. We wanted to test the validity of the ACF assay to assess chemopreventive agents that are given after carcinogen. Male F344 rats (age 6 weeks) were injected sc with 30 mg/kg azoxymethane (AOM) and one week later put on one of three diets until sacrifice: AIN-76A formulation of semipurified rodent chow and plain drinking water (AOM alone), AIN-76A diet supplemented with 2.5 ppm selenium and plain drinking water (AOM+Se), or AIN-76A diet and drinking water with 2% sodium phytate (AOM+P). Control rats on the three diets were injected with saline. Rats were killed at 4, 12, and 36 weeks after injection. No tumors were seen in any control group; two ACF were seen in 36-week control rats on Se; no ACF were seen in any other control rats. The tumor incidence in the AOM alone and the AOM+SE groups was 83% (10 of 12 rats) in both cases compared to 25% (3 of 12 rats) in the AOM+P group. The percentage and number of large ACF (those with 4 or more crypts) were greater ($p = 0.0008$ and $p = 0.014$, respectively, Mann-Whitney) in rats with tumors than in those without tumors at 36 weeks after injection. At 12 weeks after injection, the number of large ACF in the AOM alone or the AOM+SE group was greater ($p = 0.005$, Mann-Whitney test) than in the AOM+P group. Our data suggest that the number of large ACF is predictive of tumor incidence.

Supported by American Institute for Cancer Research grant 89B48, the American Cancer Society grant PDT-407, and the National Cancer Institute grant CA48032.

Studies on the Metabolism of Retinol and Retinol-Binding Protein (RBP) in Transthyretin-Lacking Mice Produced by Homologous Recombination

S. WEI, V. EPISKOPOU, R. PIANTEDOSI, M.E. GOTTESMAN, E.J. ROBERTSON and W.S. BLANER

College of Physicians and Surgeons
Columbia University, New York, NY

In the circulation, retinol-binding protein (RBP) is present as a one-to-one protein complex with transthyretin (TTR). It has been hypothesized that the binding of RBP to TTR prevents filtration of RBP in the kidney and may play a role in mediating the secretion of RBP from hepatocytes. To examine these possibilities, we have studied RBP metabolism and retinol metabolism in a strain of mice (aTTR) produced by targeted integration to lack TTR. In the plasma of aTTR and control mice, mean RBP levels ($N = 6$) were measured to be respectively, 0.11 ± 0.09 and 3.41 ± 1.21 mg RBP/dliter. Mean hepatic levels of RBP for the aTTR and control mice were respectively, 39.8 ± 4.5 and 25.0 ± 2.3 µg RBP/g tissue. Plasma retinol could not be detected (< 2.0 µg/dliter) in the aTTR mice; whereas, plasma retinol levels in control animals averaged 30.0 ± 1.2 µg/dliter. The level of retinol present in the livers of control mice was found to be 229.8 ± 29.6 µg retinol eq./g tissue. For the livers of aTTR mice, this level was found to be significantly elevated, at 505.5 ± 144.2 µg/g tissue. Urine levels of RBP appeared to be elevated in the aTTR mice. RBP mRNA levels in both the liver and adipose tissue of the aTTR and control mice were identical. In addition, tissue CRBP mRNA levels in the liver and testis, which are known to be regulated by retinoid availability, were the same for the aTTR and control animals. Our studies suggest that TTR prevents the filtration of RBP in the kidney and the elevated levels of retinol and RBP in the livers of the aTTR mice may suggest that TTR plays a role in aiding the secretion of RBP from the liver. However, since these animals have a normal phenotype, it would seem that even in the absence of TTR and with essentially no retinol or RBP, tissues are able to obtain sufficient retinoid to maintain normal function.

Supported in part by the American Institute for Cancer Research grant 92A34.

Calcium Regulates the Differentiation of Vitamin A-Responsive Human Papillomavirus-Immortalized Cervical Keratinocytes, But Not the Expression of the Papillomavirus E6 and E7 Oncogenes

C.K. CHOO, E.A. RORKE and R.L. ECKERT

Departments of Physiology and Biophysics, Dermatology, Reproductive Biology, Environmental Health Sciences and Biochemistry, Case Western Reserve University School of Medicine, Cleveland, OH

Transcription of the papillomavirus virus open reading frame encoding oncogenes E6 and E7 is necessary for cell immortalization and is known to be regulated as a function of cell differentiation in cervical tumors *in vivo*. Thus, it is important to understand how exogenous factors regulate transcription from the HPV P_{97} promoter. Vitamin A and other retinoids are important dietary agents that regulate normal ectocervical epithelial cell differentiation. Moreover, these agents have also been shown to be clinically useful in reducing the mass of cervical tumors. In a previous study supported by the AICR, we examined the effects of vitamin A and other retinoids on the differentiation of an HPV16-immortalized cervical cell line, ECE16-1. We showed that HPV16-immortalization sensitizes HPV16 transformed cells to the effects of retinoids [Agarwal *et al., Cancer Res* 51: 3982-3989 (1991)].

In vivo, vitamin A must act to regulate cervical tumor cell differentiation in the context of other agents that regulate cervical cell differentiation. We therefore examined the effects calcium, another dietary agent that regulates cervical cell function, on ECE16-1 cell differentiation. We show that keratin K13, transglutaminase (TG) and involucrin levels are increased when calcium is shifted from 0.06 to 1.4 mM. These changes can be accounted for by changes in the level of each corresponding mRNA. Maximal induction of each gene is observed at calcium levels ≥ 0.4 mM. Increased mRNA levels can be detected by 24h and are maximal by 72hr. Nuclear runoff experiments reveal an absence of transcriptionally regulation, suggesting that the concentration of each mRNA is controlled by changes in mRNA stability, nuclear RNA processing or transport of RNA from nucleus to cytoplasm.

Parallel experiments indicate that the transcription rate and mRNA level of the reading frame encoding the HPV16 E6/E7 oncogenes is not regulated by calcium. Thus, in ECE16-1 cells, expression of the reading frame encoding E6/E7 is not coupled to changes in cell differentiation. Detection of the presence of cytokeratins K6 and K7, the presence of lowered levels of involucrin, the presence of integrated HPV16 DNA and the absence of regulation of E6/E7 levels by calcium suggests that ECE16-1 cells represent an *in vitro* correlate of moderate to advanced cervical-intraepithelial neoplasia (CIN II/III).

Supported in part by the American Institute for Cancer Research grant 91AO3.

Inhibitory Effects of Organosulfur Compounds in Garlic on Canine Mammary Tumor Cells

S.G. SUNDARAM and J.A. MILNER

Department of Nutrition, The Pennsylvania State University, University Park, PA

Six organosulfur compounds found in garlic were examined for their ability to alter the growth of canine mammary tumor cells (CMT-13) in culture. S-allyl-cysteine, S-ethyl-cysteine, or S-propyl-cysteine (water soluble organosulfur compounds) did not significantly alter the growth of CMT-13 cells when added at 1.0 mM or less. However, diallyl sulfide, diallyl disulfide, and diallyl trisulfide (oil soluble organosulfur compounds) addition markedly inhibited growth. Diallyl disulfide induced growth depression was accompanied by an increased proportion of cells in the G_0/G_1 phase compared to untreated cultures. Exposure to diallyl disulfide (1 mM) increased the intracellular glutathione by 57% and the activity of glutathione-S-transferase by 300%. Addition of glutathione prior to diallyl disulfide markedly decreased the severity of the diallyl disulfide-mediated growth inhibition. Simultaneous treatment with sodium selenite (9.6 µM) and diallyl disulfide (1 or 5 µM) enhanced the growth depression compared to that caused by either compound alone. Studies with supplemental glutathione and selenium suggest an involvement of membrane sulfhydryls in the growth inhibition caused by the tested oil soluble sulfhydryls in garlic. These studies document that some sulfhydryls found in garlic are effective inhibitors of the growth of neoplastic cells. Additional studies are warranted to determine the physiological significance of these organosulfur compounds as agents against human tumors.

Supported in part by the American Institute for Cancer Research grant 91A75 and a grant from Wakunaga of America Co., Ltd.

Impact of Dietary Lipid and Garlic on the Binding of 7,12-dimethylbenz[a]anthracene (DMBA) to Mammary DNA

H. AMAGASE and J.A. MILNER

Department of Nutrition, The Pennsylvania State University
University Park, PA

The present studies determined the impact of dietary lipids in the presence or absence of supplemental garlic on the binding of DMBA to mammary cell DNA in female Sprague-Dawley rats. In all studies rats were fed for 2 weeks prior to DMBA treatment (25 mg/kg body weight). DMBA-DNA adducts were quantitated using a ^{32}P-postlabeling technique. In experiment 1, rats were feed diets with or without 2% garlic and varying amounts of corn oil (5, 10, or 20%). Increasing dietary corn oil resulted in a proportional increase in DMBA-DNA binding. The depression in total and individual DMBA-DNA adducts resulting from garlic consumption was greatest in rats fed 20% corn oil. Experiment 2 examined the influence of source of lipid (15% corn, olive, coconut, or fish oil) added to basal diets containing 5% corn oil. Regardless of source, consumption of diets containing 20% lipid increased DMBA-DNA binding by approximately 100% compared to diets containing 5% dietary corn oil in experiment 2. Garlic depressed DMBA-DNA binding by 50, 29, and 70% when diets contained 20% corn, olive, or fish oil, respectively. Although coconut oil increased the formation of DMBA-DNA adducts, garlic failed to reduce the binding when this source of lipid was provided. These studies demonstrate an effect of several sources of lipids on the initiation phase of carcinogenesis. Furthermore, these studies suggest the effect of garlic is dependent upon the degree of unsaturation of the dietary lipid fed.

Supported in part by American Institute for Cancer Research grant 91A75 and a grant from Wakunaga of America Co., Ltd.

Contributors

Rajesh Agarwal, Department of Dermatology, University Hospitals of Cleveland, Case Western Reserve University, Cleveland, Ohio 44106

Cheryl J. Aslakson, Department of Microbiology, Montana State University, Bozeman, Montana 59717

Ambikaipakan Balasubramaniam, Department of Surgery, University of Cincinnati Medical Center, Cincinnati, Ohio 45267

Stephen Barnes, Departments of Biochemistry and Pharmacology, University of Alabama at Birmingham, Birmingham, Alabama 35294

Ayse Batova, Department of Chemistry and Biochemistry, University of South Carolina, Columbia, South Carolina 29208

C. Anita H. Bigger, Microbiological Associates, 9900 Blackwell Road, Rockville, Maryland 20850

William T. Chance, Department of Surgery, University of Cincinnati Medical Center, and Veterans Affairs Medical Center, Cincinnati, Ohio 45267

S. Chau Cheng, Clinical Research Associates, 50 Madison Avenue, New York, New York 10010

Kim E. Creek, Department of Chemistry and Biochemistry, University of South Carolina, Columbia, South Carolina 29208

Mario DiGirolamo, Department of Medicine, Emory University School of Medicine, Atlanta, Georgia 30303

Anthony Dipple, Chemistry of Carcinogenesis Laboratory, ABL-Basic Research Program, NCI-Frederick Cancer Research and Development Center, Frederick, Maryland 21702

Zora Djuric, Department of Internal Medicine, Wayne State University, Detroit, Michigan 48201

Catherine A. Elstad, Department of Pharmaceutical Sciences, College of Pharmacy, and the Pharmacology/Toxicology Graduate Program, Washington State University, Pullman, Washington 99164

Josef E. Fischer, Department of Surgery, University of Cincinnati Medical Center, Cincinnati, Ohio 45267

Mary Forbes, Cancer Biology Laboratories, Department of Pathology, College of Veterinary Medicine, Cornell University, Ithaca, New York 14853

Tracy French, Cancer Biology Laboratories, Department of Pathology, College of Veterinary Medicine, Cornell University, Ithaca, New York 14853

John D. Groopman, Department of Environmental Health Sciences, The Johns Hopkins University School of Hygiene and Public Health, Baltimore, Maryland 21205

Clinton Grubbs, Department of Nutrition Sciences, University of Alabama at Birmingham, Birmingham, Alabama 35294

Ronald W. Hart, National Center for Toxicological Research, Jefferson, Arkansas 72079

Lance K. Heilbrun, Department of Internal Medicine, Wayne State University, Detroit, Michigan 48201

John R. Hodam, Department of Chemistry and Biochemistry, University of South Carolina, Columbia, South Carolina 29208

Jun-Yan Hong, Laboratory for Cancer Research, College of Pharmacy, Rutgers University, Piscataway, New Jersey 08855

G. Ronald Jenkins, Department of Chemistry and Biochemistry, University of South Carolina, Columbia, South Carolina 29208

Santosh K. Katiyar, Department of Dermatology, University Hospitals of Cleveland, Case Western Reserve University, Cleveland, Ohio 44106

Thomas W. Kensler, Department of Environmental Health Sciences, The Johns Hopkins University School of Hygiene and Public Health, Baltimore, Maryland 21205

Mohammad A. Khan, Department of Pathology, University of South Carolina School of Medicine, Columbia, South Carolina 29208

Shiriki Kumanyika, Center for Biostatistics and Epidemiology, College of Medicine, Pennsylvania State University, Hershey, Pennsylvania 17033

Silvana Martino, Department of Internal Medicine, Wayne State University, Detroit, Michigan 48201

Gary G. Meadows, Department of Pharmaceutical Sciences, College of Pharmacy, and the Pharmacology/Toxicology Graduate Program, Washington State University, Pullman, Washington 99164

Hasan Mukhtar, Department of Dermatology, University Hospitals of Cleveland, Case Western Reserve University, Cleveland, Ohio 44106

Richard M. Niles, Department of Biochemistry and Molecular Biology, Marshall University School of Medicine, Huntington, West Virginia 25755

Kimmo Peltonen, Institute of Occupational Health, Department of Industrial Hygiene and Toxicology, Topeliuksenkatu 41 aA, SF-00250, Helsinki, Finland

Greg Peterson, Department of Biochemistry, University of Alabama at Birmingham, Birmingham, Alabama 35294

Lucia Pirisi, Department of Pathology, University of South Carolina School of Medicine, Columbia, South Carolina 29208

Bill D. Roebuck, Department of Pharmacology and Toxicology, Dartmouth Medical School, Hanover, New Hampshire 03755

Helen Ross, Chemistry of Carcinogenesis Laboratory, ABL-Basic Research Program, NCI-Frederick Cancer Research and Development Center, Frederick, Maryland 21702

Karen Russell, Cancer Biology Laboratories, Department of Pathology, College of Veterinary Medicine, Cornell University, Ithaca, New York 14853

Kosaku Sakamoto, First Department of Surgery, Gunma University School of Medicine, Maebashi, Gunma 371, Japan

Ken Setchell, Clinical Mass Spectrometry Center, Children's Hospital, Cincinnati, Ohio 45229

Abulkalam M. Shamsuddin, Department of Pathology, University of Maryland School of Medicine, Baltimore, Maryland 21201

Sulaiman Sheriff, Department of Surgery, University of Cincinnati Medical Center, Cincinnati, Ohio 45267

Jean R. Starkey, Department of Microbiology, Montana State University, Bozeman, Montana 59717

William H. Tolleson, Department of Pathology, University of South Carolina School of Medicine, Columbia, South Carolina 29208

Frank M. Torti, Department of Medicine, Stanford University and Palo Alto Veterans Affairs Medical Center, Palo Alto, California 94304

Suzy V. Torti, Department of Medicine, Stanford University and Palo Alto Veterans Affairs Medical Center, Palo Alto, California 94304

Susi Varvayanis, Cancer Biology Laboratories, Department of Pathology, College of Veterinary Medicine, Cornell University, Ithaca, New York 14853

Zhi-Yuan Wang, Laboratory for Cancer Research, College of Pharmacy, Rutgers University, Piscataway, New Jersey 08855

Chung S. Yang, Laboratory for Cancer Research, College of Pharmacy, Rutgers University, Piscataway, New Jersey 08855

Andrew Yen, Cancer Biology Laboratories, Department of Pathology, College of Veterinary Medicine, Cornell University, Ithaca, New York 14853

Index

MIX
Papier aus verantwortungsvollen Quellen
Paper from responsible sources
FSC® C105338

If you have any concerns about our products,
you can contact us on
ProductSafety@springernature.com

In case Publisher is established outside the EU,
the EU authorized representative is:
Springer Nature Customer Service Center GmbH
Europaplatz 3, 69115 Heidelberg, Germany

Printed by Libri Plureos GmbH
in Hamburg, Germany